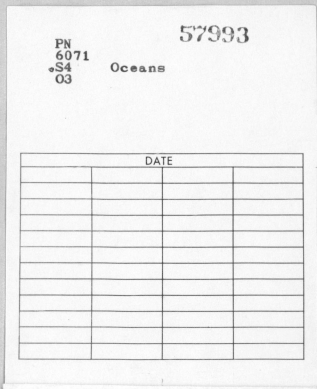

57993

PN
6071
.S4
O3   Oceans

DATE			

© THE BAKER & TAYLOR CO.

OUR CONTINUING FRONTIER

OCEANS

OUR CONTINUING FRONTIER

A Courses by Newspaper Reader

Edited by H. William Menard and Jane L. Scheiber

Courses by Newspaper is a project of University Extension, University of California, San Diego.
Funded by The National Endowment for the Humanities

Publisher's Inc.  *Del Mar, California*

Library of Congress Cataloging in Publication Data
Main entry under title:

Oceans: our continuing frontier.

1. Ocean—Literary collections. 2. Oceanography —Addresses, essays, lectures. 3. Marine biology— Addresses, essays, lectures. I. Menard, Henry William, 1920– II. Scheiber, Jane L.
PN6071.S403   551.4'6'008   76-7108
ISBN 0-89163-015-5
ISBN 0-89163-014-7 pbk.

iv

Acknowledgments

Cover photograph: Warren Bolster/Photophile

1: 2: From Samuel Eliot Morison, *The Maritime History of Massachusetts, 1783–1860,* Houghton Mifflin Company, used with permission of the publisher; 10: From pp. 5–8 in *The Silent World* by Captain Jacques Y. Cousteau with Frederic Dumas. Copyright 1953 by Harper & Row, Publishers, Inc. By permission of the publisher; 13: From Jacques Cousteau, "The Perils and Potentials of a Watery Planet," *Saturday Review/World,* August 24, 1974. Used by permission of The Cousteau Society.

2: 27 (top) and 29: Reprinted with permission of Macmillan Publishing Co., Inc. from *Poems* by John Masefield. Copyright 1912 by Macmillan Publishing Co., Inc., renewed 1940 by John Masefield; 27 (bottom): Copyright 1937 and renewed 1965 by W.H. Auden. Reprinted from *Collected Shorter Poems 1929–1957* by W.H. Auden. By permission of Random House, Inc. and Faber and Faber Ltd.; 28: From *Margins* by Philip Booth. Copyright © 1966 by Philip Booth. All rights reserved. Reprinted by permission of Viking Penguin, Inc.; 58: From *Kingdom of the Octopus* by Frank W. Lane, used by permission of Sheridan House, Inc.; 63: From Eugenie Clark, "The Red Sea's Sharkproof Fish," *National Geographic,* November 1974; 71: From *Seascape and the American Imagination* by Roger B. Stein. © 1975 by the Whitney Museum of American Art. Used by permission of Clarkson N. Potter, Inc.; 73: From *Art and the Seafarer,* edited by Hans Jurgen Hansen. Copyright © 1966 by Gerhard Stalling Verlag. Copyright © 1968 by Faber and Faber Ltd. Reprinted by permission of Viking Penguin, Inc.

3: 97: From H. William Menard, *The Anatomy of an Expedition.* Copyright © 1969 by McGraw-Hill, Inc. Used by permission of the publisher; 105: Reprinted by permission of Collins-Knowlton-Wing. Copyright © 1961 by Jacques Piccard and Robert S. Dietz; 110: From William Wertenbaker, *The Floor of the Sea.* Copyright © 1974 by William Wertenbaker. Used by permission of Little, Brown and Company. This material appeared originally in somewhat different form in *The New Yorker;* 125: From Allen L. Hammond, "Minerals and Plate Tectonics: A Conceptual Revolution," *Science,* September 5, 1975. Copyright 1975 by the American Association for the Advancement of Science; 129: From J.V. Luce, *The End of Atlantis,* Thames and Hudson Ltd., 1973. Used by permission of the publisher; 137: Excerpted from *Deep Water, Ancient Ships,* © 1976 by Willard Bascom. Reprinted by permission of Doubleday & Company, Inc.; 143: From *Rock Strata and the Bible Record* by Paul A. Zimmerman. © 1970 by Concordia Publishing House. Used by permission.

4: 150: Copyright 1974 by Newsweek, Inc. All rights reserved. Reprinted by permission; 153: © 1975 by The New York Times Company. Reprinted by permission; 162: From Luther J. Carter, "Outer Continental Shelf: Congress Weighs Oil Needs and Environment," *Science,* September 12, 1975. Copyright 1975 by the American Association for the Advancement of Science; 164: From Allen L. Hammond, "Manganese Nodules," *Science,* February 8 and February 15, 1974. Copyright 1974 by the American Association for the Advancement of Science; 170: Reprinted with permission from *Pacem in Maribus,* Dodd, Mead & Company, 1972, a publication of the Center for the Study of Democratic Institutions, Santa Barbara, California; 172: From *The Sea Around Us* by Rachel L. Carson. Copyright © 1950, 1951, 1961 by Rachel L. Carson. Reprinted by permission of Oxford University Press, Inc.; 174: Copyright © 1969 by Thomas Y. Crowell Co., Inc. Excerpted from *Exploring the Ocean World* by C.P. Idyll and Hiroshi Kasahara with permission of Thomas Y. Crowell Co., Inc., Publisher; 179: Excerpted from Elisabeth Mann Borgese, *The Drama of the Oceans.* Copyright © 1976 by Elisabeth Mann Borgese. Used with permission of Harry N. Abrams, Inc.; 190: From Thor Heyerdahl, "How to Kill an Ocean," *Saturday Review,* November 29, 1975. Copyright © 1975 by Saturday Review; 191: From M. Blumer, "Oil Pollution of the Ocean," *Oceanus,* October 1969; 192: From Bostwick Ketchum, "A Realistic Look at Ocean Pollution," *Marine Technology Society Journal,* October-November 1973; 200: From *Man's Impact on Terrestrial and Oceanic Systems,* edited by W.H. Matthews, *et al.* By permission of MIT Press.

5: 215: Reprinted with permission from *Foreign Affairs,* January 1974. Copyright 1973 by Council on Foreign Relations, Inc.; 226: Reprinted by permission of Dr. Arvid Pardo and Dr. Robert E. Osgood; 244: From *Sea Power in the 1970s,* edited by George H. Quester. Copyright © 1972 by Dunellen Publishing Co.; 252: From *Supership* by Noël Mostert. Copyright © 1974 by Noël Mostert. Reprinted by permission of Alfred A. Knopf, Inc. Portions of this material originally appeared in *The New Yorker.*

6: 268: From *The European Reconnaissance* by John H. Parry, published by Walker & Co., Inc., New York, © 1968 by John H. Parry; 273: From J.H. Parry, *Trade and Dominion,* copyright © 1971 by J.H. Parry. Used by permission of Praeger Publishers, Inc.; 284: Reprinted with permission of Macmillan Publishing Co., Inc. from *The Bird of Dawning* by John Masefield. Copyright 1933 by John Masefield, renewed 1961 by John Masefield; 286: Reprinted by permission of Charles Scribner's Sons from *The Way of a Ship* by Alan Villiers. Copyright 1953 by Alan Villiers; 290: From Jeremy Tunstall, *The Fisherman,* MacGibbon & Kee, 1962; 294: Excerpted from Luis Marden, "Ama, Sea Nymphs of Japan," *National Geographic,* July 1971; 297: From Francis Chichester, *Along the Clipper Way,* Coward, McCann & Geohegan, 1966; Ann Davison excerpts from *My Ship Is So Small,* A.M. Heath Ltd.; 301: © 1972 by The New York Times Company. Reprinted by permission; 310: From *The Cruise of the Snark* by Jack London. Used by permission of the Irving Shepard Estate; 316: © 1930 by The New York Times Company. Reprinted by permission; 318: From the book *Zane Grey: Outdoorsman* by George Reiger (ed.). © 1972 by Zane Grey Estate. Published by Prentice-Hall, Inc., Englewood Cliffs, New Jersey; 323: From *Kon-Tiki* by Thor Heyerdahl. Copyright 1950 by Thor Heyerdahl. Published in the United States by Rand McNally & Co.; 325: From André Maurois, *Byron,* D. Appleton, 1930.

Contents

Unit Five: Policy and Sea Power   *209*

Unit Six: Men and Women at Sea   *265*

Epilogue. *H. William Menard*   *328*

The Authors   *334*

Preface

This is the fifth in a series of books prepared for Courses by Newspaper. A national program originated and administered by University Extension, University of California, San Diego, and funded by the National Endowment for the Humanities, Courses by Newspaper develops materials for college-level courses that are presented to the general public through the nationwide cooperation of newspapers and participating colleges and universities. The program offers three levels of participation: interested readers can follow a series of newspaper articles that comprise the course "lectures"; they can purchase supplementary books—a Reader and Study Guide; and they can enroll for credit.

This Reader supplements the sixteen newspaper articles especially written for the fifth course, *Oceans: Our Continuing Frontier;* the articles will be published weekly in newspapers across the nation in the fall of 1976.

Courses by Newspaper has been made possible through the efforts of many people and organizations. We should like to acknowledge them here. Hundreds of newspaper editors and publishers across the country have contributed precious newspaper space to bring the specially prepared articles to their readers. The faculties and administrations of the many colleges and universities participating in the program have cooperated to make credit available to students throughout the nation.

Deserving special mention at the University of California, San Diego, are Caleb A. Lewis, project consultant and director of Media Programs at University Extension, who originated the idea; and C. David Hellyer, university editor, now retired, who helped develop the program in its earliest phases.

Paul D. Saltman, vice-chancellor for academic affairs and professor of biology at the University of California, San Diego, has enthusiastically supported Courses by Newspaper from its inception. He served as the first academic coordinator and is chairman of the University of California, San Diego, Faculty Committee. Dr. Kenneth Nealson of the Scripps Institution of Oceanography helped in conceptualizing the course, and Gilven Slonim, executive vice-president of the Oceanic Educational Foundation in Falls Church, Virginia, consulted with us in the development of the course and made many constructive suggestions regarding content. Vital to the success of this year's program were the dedicated members of the Courses by Newspaper staff: Cecilia Solis, Yvonne Hancher, Stephanie Giel, Susan Rago, Patricia Murphy, and Mary Robinson. Brad Tebo of the University of California, San Diego, provided helpful research assistance.

We also wish to thank the authors of the newspaper articles—George Elliott, Eugenie Clark, John Wilmerding, Sir Edward Bullard, Willard Bascom, Don Kash, C.P. Idyll, Bostwick Ketchum, William Burke, Herman Kahn, John H. Parry, Constantina Safilios-Rothschild, and Heywood Hale Broun—for their suggestions for this volume. Several of the headnotes in Unit Two drew heavily on Professor Elliott's work, and Professor Parry contributed the headnotes for several of the selections in Unit Six of the Reader.

Finally, we wish to express our gratitude to the National Endowment for the Humanities for funding Courses by Newspaper. The Endowment, a federal agency created in 1965 to support education, research, and public activity in the humanities, has generously supported this nationwide program from its beginning. We wish particularly to acknowledge the support and advice of James Kraft, program officer in the Office of Planning and Analysis, who worked closely with us on this course.

Although Courses by Newspaper is a project of University Extension, University of California, San Diego, and is supported by the National Endowment for the Humanities, the views expressed in this course and the accompanying articles are those of the authors only and do not necessarily reflect those of the Endowment or of the University of California.

Introduction

The ocean is a thing, a thing to be viewed or studied with all the enormous diversity of human perception and thought. It is a place, a place to work or play in the innumerable ways of our society.

This Reader contains some of the best expressions of different viewpoints and concerns about the sea. The sea is seen in myth and science, in word and image, as a place to put things and as a place to fish them out. Our human relation to the sea is exposed in all its complexity, the sea as a killer of people, people as killers of the sea. Many issues about the sea are controversial—otherwise there would be little need for the Reader. Within the limited space available we have tried to present many sides of each question. At the very least the existence of different views is identified, and reference is made in the Study Guide to additional reading for those who wish to pursue a subject.

Inevitably anyone involved with the sea—and who is not—brings to it a wide range of attitudes and objectives. In this book we attempt not only to illustrate different viewpoints, problems, and discoveries, but how they all tie together. Thus we have poems by men who went to sea, paintings of the sea, and a poem, "Dauber," about a man who goes to sea to paint it. He wants to perceive its spirit instead of just its surface.

At the other end of the spectrum of human activities we have readings about mineral resources, commerce, sea power, and the law of the sea. As one would anticipate, it rapidly becomes apparent that these subjects can hardly be discussed separately. Admiral Mahan justifies a navy in terms of the protection or interdiction of commerce. Later authors show the growing conflicts of interest between the various users of the sea. The law and sea power have the potential to permit optimum use of the sea, but, from another aspect, they may have the opposite effect.

Some of the words and concepts in the book may be new to the reader and missing from the dictionary or incorrectly defined. Moreover, authors may use the same words in different and confusing ways. "Continental shelf" is a prime example. This problem is addressed in the epilogue, "New Interests, New Words."

Oceans: Our Continuing

Frontispiece from T. Beale's *Natural History of the Sperm Whale*, 1839

Frontier

Unit One

The first four selections in this book serve as an introduction to those that follow. Among them, they span American history and gaze into the future. Each is written by a sailor, each by an acclaimed author, each by the person best qualified to address the subject.

The most distinguished of nautical historians is Admiral Samuel Eliot Morison. In the selection from *The Maritime History of Massachusetts* he describes the beginnings of the American enterprise on the sea—almost three and a half centuries ago.

Admiral Alfred Thayer Mahan was a theoretician, the Clausewitz of naval warfare. His book *The Influence of Sea Power Upon History* had an enormous impact on political and military thinking at the turn of the present century. He writes of America when its maritime affairs had floundered in the slough of disinterest. Many of his insights, however, are strikingly applicable to the present time.

Captain Jacques Cousteau, who needs no introduction, is the author of two of these selections. In the first, from *The Silent World,* he conveys to the reader his delight as he first experienced freedom underwater, which his inventiveness has since made available to all of us. His second selection is a recent magazine article in which he expresses his deep human-itarian hopes for what the sea can offer to humankind. He also presents his intense concern about what people are doing to the sea. Unlike historians Morison and Mahan, Cousteau is concerned primarily with the present and the years ahead. Thus he must evaluate incomplete data and predict what might happen in a murky future. Other authors whose works appear later in the Reader make different evaluations and perceive calms or storms ahead where Cousteau sees storms or calms. This first section of the Reader, thus, is but the beginning.

They Went to Sea

Samuel Eliot Morison

Throughout its history, the American nation has been intimately involved with the sea. For most of the early colonists, the sea was both a route of escape to greater opportunities—religious, political, and economic—and a formidable barrier to be crossed before those opportunities could be realized. For one in six, however—those of African descent—it was the terrifying and tragic route to slavery. While the Atlantic barrier was eventually to encourage in America a sense of independence and separateness from Europe, the ocean was also the only highway connecting the young colonies to the rest of the world on which they were still dependent, and for more than 150 years the settlers clung to the Atlantic seaboard. The southern colonies depended on ocean trade to export their tobacco and import manufactured goods, while farther north the colonists found it easier to harvest the products of the sea than those of the rock-strewn land. In the following account from The Maritime History of Massachusetts, *first published in 1921, Pulitzer Prize-winning historian Samuel Eliot Morison relates how Massachusetts in particular became a maritime colony.*

*I*n 1630, ten years after its settlement, the Plymouth Colony contained but three hundred white people. At that time the Colony of Massachusetts-Bay, founded only at the end of 1628, had over two thousand inhabitants. Within thirteen years the numbers had reached sixteen thousand, more than the rest of the continental colonies combined; and the characteristic maritime activities of Massachusetts—fishing, shipping, and West India trading—were already well under way.

It was not the intention of the founders of Massachusetts-Bay to establish a predominantly maritime community. The first and foremost object of Winthrop and Dudley and Endecot and Saltonstall was to found a church and commonwealth in which Calvinist Puritans might live and worship according to the Word of God, as they conceived it. They aimed to found a New England, purged of Old England's corruptions, but preserving all her goodly heritage. They intended the economic foundation of New England, as of Old England and Virginia, to be large landed estates, tilled by tenants and hired labor.

In this they failed. The New England town, based on freehold and free labor, sprang up instead of the Old English manor. And for only

a decade was agriculture the mainstay of Massachusetts. The constant inflow of immigrants, requiring food and bringing goods, enabled the first comers to profit by corn-growing and cattle-raising. This could not continue. "For the present, we make a shift to live," wrote a pessimistic pioneer in 1637; "but hereafter, when our numbers increase, and the fertility of the soil doth decrease, if God discover not means to enrich the land, what shall become of us I will not determine."

God performed no miracle on the New England soil. He gave the sea. Stark necessity made seamen of would-be planters. The crisis came in 1641, when civil war in England cut short the flow of immigrants. "All foreign commodities grew scarce," wrote Governor Winthrop, "and our own of no price. Corn would buy nothing; a cow which cost last year £20 might now be bought for 4 or £5 . . . These straits set our people on work to provide fish, clapboards, plank, etc., . . . *and to look out to the West Indies for a trade* . . ."

In these simple sentences, Winthrop explains how maritime Massachusetts came to be. The gravelly, boulder-strewn soil was back-breaking to clear, and afforded small increase to unscientific farmers. No staple of ready sale in England, like Virginia tobacco or Canadian beaver, could be produced or easily obtained. Forest, farms, and sea yielded lumber, beef, and fish. But England was supplied with these from the Baltic, and by her own farmers and fishermen. Unless a new market be found for them, Massachusetts must stew in her own juice. It was found in the West Indies—tropical islands which applied slave labor to exotic staples like sugarcane, but imported every necessity of life. More and more they became dependent on New England for lumber, provisions, and dried fish. More and more the New England ships and merchants who brought these necessities, controlled the distribution of West-India products.

Massachusetts went to sea, then, not of choice, but of necessity. Yet the transition was easy and natural. "Farm us!" laughed the waters of the Bay in Maytime, to a weary yeoman, victim of the "mocking spring's perpetual loss." "Here thou may'st reap without sowing—yet not without God's blessing; 't was the Apostles' calling." And with sharp scorn spake the waters to an axeman, hewing a path from river landing to new allotment: "Hither thy road! And of the oak thou wastest, make means to ride it! Southward, dull clod, and barter the logs thou would'st spend to warm thy silly body, for chinking doubloons, as golden as the sunlight that bathes the Spanish main."

Materials and teachers for a maritime colony were already at hand. The founders had been careful to secure artisans, and tools for all useful trades, that Massachusetts might not have the one-sided development of Virginia. Fishing had not ceased with the failure of the Gloucester experiment. Dorchester, the first community "that set upon the trade of fishing in the bay," was little more than a transference to New England soil of Dorset fishing interests. Scituate was settled by a similar company. The rocky peninsula of Marblehead, with its ample harbor, attracted fisherfolk from Cornwall and the Channel Islands, who cared neither for Lord Bishop nor Lord Brethren. Their descendants retained a distinct dialect, and a jealous exclu-

siveness for over two centuries. Marblehead obeyed or not the laws of the Great and General Court, as suited her good pleasure; but as long as she "made fish," the Puritan magistrates did not interfere. Literally true was the Marblehead fisherman's reproof to an exhorting preacher: "Our ancestors came not here for religion. Their main end was to catch fish!" . . .

Shipping was the other key industry of the colony. Fishing would have brought little wealth, had Massachusetts depended on outside interests for vessels—as she must to-day for freight-cars. Distribution, not production, brought the big returns in 1620 as in 1920. Massachusetts shipbuilding began with the launching in 1631 of Governor Winthrop's *Blessing of the Bay,* on the same Mystic River that later gave birth to the beautiful Medford-built East-Indiamen. By 1660 shipbuilding had become a leading industry in Newbury, Ipswich, Gloucester, Salem, and Boston. The great Puritan emigration brought many shipwrights and master builders, such as William Stephen, who "prepared to go to Spayne, but was persuaded to New England." A four-hundred-ton ship *Seafort* was built at Boston in 1648, but wrecked on the Spanish coast, decoyed by false lights ashore.

Few Massachusetts-built vessels were so large as this; four hundred tons meant a great ship as late as 1815. The colonial fleet for the most part consisted of small single-decked sloops, the usual rig for coasters, and lateen-rigged ketches, the favorite rig for fishermen, of twenty to thirty tons burthen, and thirty-five to fifty feet long. Good oak timber and pine spars were so plentiful that building large ships on order or speculation for the English market soon became a recognized industry. Rope-walks were established, hempen sailcloth was made on hand looms, anchors and coarse ironwork were forged from bog ore, and wooden "trunnels" (tree nails) were used to fastening planking to frame.

The English Navigation Act of 1651, restraining colonial commerce to English and colonial vessels, gave an increased impetus to New

England shipbuilding; for the Dutch, with their base at New Amsterdam, had been serious competitors. In another generation, vessels built and owned in New England were doing the bulk of the carrying trade from Chesapeake Bay to England and southern Europe. "Many a fair ship had her framing and finishing here," wrote Edward Johnson about 1650, "besides lesser vessels, barques and ketches; many a Master, beside common Seamen, had their first learning in this Colony."

The shipmaster's calling has always been of high repute in Massachusetts. Only the clergy, the magistracy, and the shipowning merchants, most of whom were retired master mariners, enjoyed a higher social standing in colonial days. The ship *Trial* of two hundred tons, one of the first vessels built at Boston, was commanded by Mr. Thomas Coytmore, a gentleman of good estate, "a right godly man, and an expert seaman," says Governor Winthrop—who made his fourth matrimonial venture with Captain Coytmore's widow. The foremast hands were recruited in part from English seaports, but mostly from the adventure-loving youth of the colonies. When Captain John Turner came back from the West Indies in a fifteen-ton pinnace, with so many pieces of eight that the neighbors hissed "Piracy!"; when the *Trial* "by the help of a diving tub," recovered gold and silver from a sunken Spanish galleon; what ploughboy did not long for a sea-change from grubbing stumps and splitting staves? When gray November days succeeded the splendor of Indian summer, the clang of wild geese overhead summoned the spirit of youth to wealth and adventure.

The Influence of Sea Power Upon History

Alfred Thayer Mahan

Seaborne commerce continued to be a major element in the development of the American economy during much of the nineteenth century, and the young nation relied on its navy to protect that vital trade. American ships fought pirates off the North African coast from 1801 to 1805, and interference with shipping and impressment of American seamen were major causes of the War of 1812. By the end of the Civil War, America had the largest naval force in the world. But in the latter part of the nineteenth century, while other nations were building colonial empires, the United States was taking advantage of the natural isolation afforded it by the ocean to develop its own continent. The navy, regarded as a defensive force only, suffered a sharp decline in strength.

It was in such an atmosphere that Alfred Thayer Mahan, naval officer and historian, published in 1890 his influential book, The Influence of Sea Power Upon History, 1660–1783. *First delivered as a set of lectures at the Naval War College in 1886, this book and his subsequent volume,* The Interest of America in Sea Power *(1897), advanced a program of mercantile imperialism, including the expansion of overseas markets, of the merchant marine, of the navy, and of overseas bases that would allow the fleet to operate throughout the world. His arguments were cited by advocates of the Spanish-American War of 1898 in which America became a colonial power; and his influence led to the expansion not only of the American fleet, but of the German and Japanese fleets as well, contributing to a naval arms race that culminated during World War I.*

The first and most obvious light in which the sea presents itself from the political and social point of view is that of a great highway; or better, perhaps, of a wide common, over which men may pass in all directions, but on which some well-worn paths show that controlling reasons have led them to choose certain lines of travel rather than others. These lines of travel are called trade routes; and the reasons which have determined them are to be sought in the history of the world.

Notwithstanding all the familiar and unfamiliar dangers of the sea, both travel and traffic by water have always been easier and cheaper than by land. The commercial greatness of Holland was due not only to her shipping at sea, but also to the numerous tranquil water-ways which gave such cheap and easy access to her own interior and to that of Germany. This advantage of carriage by water over that by land was yet more marked in a period when roads were few and very bad, wars frequent and society unsettled, as was the case two hundred years ago. Sea traffic then went in peril of robbers, but was nevertheless safer and quicker than that by land. A Dutch writer of that time, estimating the chances of his country in a war with England, notices among other things that the water-ways of England failed to penetrate the country sufficiently; therefore, the roads being bad, goods from one part of the kingdom to the other must go by sea, and be exposed to capture by the way. As regards purely internal trade, this danger has generally disappeared at the present day. In most civilized countries, now, the destruction or disappearance of the coasting trade would only be an inconvenience, although water transit is still the cheaper. Nevertheless, as late as the wars of the French Republic and the First Empire, those who are familiar with the history of the period, and the light naval literature that has grown up around it, know

Thomas Birch, *Naval Engagement Between the Frigates "United States" and the "Macedonian," 1812*, 1813
"The necessity of a navy . . . springs, therefore, from the existence of a peaceful shipping."—Mahan

how constant is the mention of convoys stealing from point to point along the French coast, although the sea swarmed with English cruisers and there were good inland roads.

Under modern conditions, however, home trade is but a part of the business of a country bordering on the sea. Foreign necessaries or luxuries must be brought to its ports, either in its own or in foreign ships, which will return, bearing in exchange the products of the country, whether they be the fruits of the earth or the works of men's hands; and it is the wish of every nation that this shipping business should be done by its own vessels. The ships that thus sail to and fro must have secure ports to which to return, and must, as far as possible, be followed by the protection of their country throughout the voyage.

This protection in time of war must be extended by armed shipping. The necessity of a navy, in the restricted sense of the word, springs, therefore, from the existence of a peaceful shipping, and disappears with it, except in the case of a nation which has aggressive tendencies, and keeps up a navy merely as a branch of the military establishment. As the United States has at present no aggressive purposes, and as its merchant service has disappeared, the dwindling of the armed fleet and general lack of interest in it are strictly logical consequences.

6

When for any reason sea trade is again found to pay, a large enough shipping interest will reappear to compel the revival of the war fleet. It is possible that when a canal route through the Central-American Isthmus is seen to be a near certainty, the aggressive impulse may be strong enough to lead to the same result. This is doubtful, however, because a peaceful, gain-loving nation is not far-sighted, and far-sightedness is needed for adequate military preparation, especially in these days. . . .

The principal conditions affecting the sea power of nations may be enumerated as follows: I. Geographical Position. II. Physical Conformation, including, as connected therewith, natural productions and climate. III. Extent of Territory. IV. Number of Population. V. Character of the People. VI. Character of the Government, including therein the national institutions. . . .

Number of Population

After the consideration of the natural conditions of a country should follow an examination of the characteristics of its population as affecting the development of sea power; and first among these will be taken, because of its relations to the extent of the territory, . . . the number of the people who live in it. It has been said that in respect of dimensions it is not merely the number of square miles, but the extent and character of the sea-coast that is to be considered with reference to sea power; and so, in point of population, it is not only the grand total, but the number following the sea, or at least readily available for employment on ship-board and for the creation of naval material, that must be counted.

For example, formerly and up to the end of the great wars following the French Revolution, the population of France was much greater than that of England; but in respect of sea power in general, peaceful commerce as well as military efficiency, France was much inferior to England. In the matter of military efficiency this fact is the more remarkable because at times, in

point of military preparation at the outbreak of war, France had the advantage; but she was not able to keep it. Thus in 1778, when war broke out, France, through her maritime inscription, was able to man at once fifty ships-of-the-line. England, on the contrary, by reason of the dispersal over the globe of that very shipping on which her naval strength so securely rested, had much trouble in manning forty at home; but in 1782 she had one hundred and twenty in commission or ready for commission, while France had never been able to exceed seventy-one. . . .

A contrast such as this shows a difference in what is called staying power, or reserve force, which is even greater than appears on the surface; for a great shipping afloat necessarily employs, besides the crews, a large number of people engaged in the various handicrafts which facilitate the making and repairing of naval material, or following other callings more or less closely connected with the water and with craft of all kinds. Such kindred callings give an undoubted aptitude for the sea from the outset. There is an anecdote showing curious insight into this matter on the part of one of England's distinguished seamen, Sir Edward Pellew. When the war broke out in 1793, the usual scarceness of seamen was met. Eager to get to sea and unable to fill his complement otherwise than with landsmen, he instructed his officers to seek for Cornish miners; reasoning from the conditions and dangers of their calling, of which he had personal knowledge, that they would quickly fit into the demands of sea life. The result showed his sagacity, for, thus escaping an otherwise unavoidable delay, he was fortunate enough to capture the first frigate taken in the war in single combat; and what is especially instructive is, that although but a few weeks in commission, while his opponent had been over a year, the losses, heavy on both sides, were nearly equal. . . .

. . . The United States has not that shield of defensive power behind which time can be gained to develop its reserve of strength. As for a seafaring population adequate to her possible needs, where is it? Such a resource, proportionate to her coast-line and population, is to be found only in a national merchant shipping and its related industries, which at present scarcely exist. It will matter little whether the crews of such ships are native or foreign born, provided they are attached to the flag, and her power at sea is sufficient to enable the most of them to get back in case of war. When foreigners by thousands are admitted to the ballot, it is of little moment that they are given fighting-room on board ship.

Though the treatment of the subject has been somewhat discursive, it may be admitted that a great population following callings related to the sea is, now as formerly, a great element of sea power; that the United States is deficient in that element; and that its foundations can be laid only in a large commerce under her own flag.

National Character

The effect of national character and aptitudes upon the development of sea power will next be considered.

If sea power be really based upon a peaceful and extensive commerce, aptitude for commercial pursuits must be a distinguishing feature of the nations that have at one time or another been great upon the sea. History almost without exception affirms that this is true. Save the Romans, there is no marked instance to the contrary.

All men seek gain and, more or less, love money; but the way in which gain is sought will have a marked effect upon the commercial fortunes and the history of the people inhabiting a country.

If history may be believed, the way in which the Spaniards and their kindred nation, the Portuguese, sought wealth, not only brought a blot upon the national character, but was also fatal to the growth of a healthy commerce; and so to the industries upon which commerce lives, and ultimately to that national wealth which was sought by mistaken paths. The desire for gain rose in them to fierce avarice; so they sought in the new-found worlds which gave such an impetus to the commercial and maritime development of the countries of Europe, not new fields of industry, not even the healthy excitement of exploration and adventure, but gold and silver. They had many great qualities; they were bold, enterprising, temperate, patient of suffering, enthusiastic, and gifted with intense national feeling. When to these qualities are added the advantages of Spain's position and well-situated ports, the fact that she was first to occupy large and rich portions of the new worlds and long remained without a competitor, and that for a hundred years after the discovery of America she was the leading State in Europe, she might have been expected to take the foremost place among the sea powers. Exactly the contrary was the result, as all know. Since the battle of Lepanto in 1571, though engaged in many wars, no sea victory of any consequence shines on the pages of Spanish history; and the decay of her commerce sufficiently accounts for the painful and sometimes ludicrous inaptness shown on the decks of her ships of war. Doubtless such a result is not to be attributed to one cause only. Doubtless the government of Spain was in many ways such as to cramp and blight a free and healthy development of private enterprise; but the character of a great people breaks through or shapes the character of its government, and it can hardly be doubted that had the bent of the people been toward trade, the action of government would have been drawn into the same current. . . .

The tendency to trade, involving of necessity the production of something to trade with, is the national characteristic most important to the development of sea power. Granting it and a good seaboard, it is not likely that the dangers of the sea, or any aversion to it, will deter a people from seeking wealth by the paths of ocean commerce. Where wealth is sought by other means, it may be found; but it will not necessarily lead to sea power. . . .

Character of the Government

In discussing the effects upon the development of a nation's sea power exerted by its government and institutions, it will be necessary to avoid a tendency to over-philosophizing, to confine attention to obvious and immediate causes and their plain results, without prying too far beneath the surface for remote and ultimate influences.

Nevertheless, it must be noted that particular forms of government with their accompanying institutions, and the character of rulers at one time or another, have exercised a very marked influence upon the development of sea power. The various traits of a country and its people which have so far been considered constitute the natural characteristics with which a nation, like a man, begins its career; the conduct of the government in turn corresponds to the exercise of the intelligent will-power, which, according as it is wise, energetic and persevering, or the reverse, causes success or failure in a man's life or a nation's history.

It would seem probable that a government in full accord with the natural bias of its people would most successfully advance its growth in every respect; and, in the matter of sea power, the most brilliant successes have followed where there has been intelligent direction by a government fully imbued with the spirit of the people and conscious of its true general bent. Such a government is most certainly secured when the will of the people, or of their best natural exponents, has some large share in making it; but such free governments have sometimes fallen short, while on the other hand despotic power, wielded with judgment and consistency, has created at times a great sea commerce and a brilliant navy with greater directness than can be reached by the slower processes of a free people. The difficulty in the latter case is to insure perseverance after the death of a particular despot. . . .

To turn now from the particular lessons drawn from the history of the past to the general question of the influence of government upon the sea career of its people, it is seen that that influence can work in two distinct but closely related ways.

First, in peace: The government by its policy can favor the natural growth of a people's industries and its tendencies to seek adventure and gain by way of the sea; or it can try to develop such industries and such sea-going bent, when they do not naturally exist; or, on the other hand, the government may by mistaken action check and fetter the progress which the people left to themselves would make. In any one of these ways the influence of the government will be felt, making or marring the sea power of the country in the matter of peaceful commerce; upon which alone, it cannot be too often insisted, a thoroughly strong navy can be based.

Secondly, for war: The influence of the government will be felt in its most legitimate manner in maintaining an armed navy, of a size commensurate with the growth of its shipping and the importance of the interests connected with it. More important even than the size of the navy is the question of its institutions, favoring a healthful spirit and activity, and providing for rapid development in time of war by an adequate reserve of men and of ships and by measures for drawing out that general reserve power which has before been pointed to, when considering the character and pursuits of the people.

Menfish

Jacques Cousteau

Although men and women had learned to use the sea for human needs—for food, for communications, and, as we shall see in the next unit, for artistic inspiration—the sea remained, until 1943, an essentially alien element. We could fly over it, sail on it, submerge beneath it, but only in confining vehicles or apparatuses. The invention of the aqualung by Jacques Cousteau and Emile Gagnan in 1943 revolutionized our relationship to the sea. For the first time an individual diver could swim freely through the underwater world, invading its depths and exploring its secrets.

The aqualung itself was a development of World War II. Cousteau, working for the underground French Naval Intelligence against the occupying German forces, had been experimenting with a self-contained underwater breathing apparatus (SCUBA). Gagnan, an expert on industrial gas equipment, had been working on a valve to feed cooking gas automatically into automobile engines, thus saving precious gasoline. Within a few months, they had adapted the demand valve, or regulator, for human breathing with cylinders of compressed air. In the following selection from The Silent World, *written with his diving companion Frédéric Dumas and published in 1950, Cousteau describes the magical moment of his first dive with the aqualung.*

I looked into the sea with the same sense of trespass that I have felt on every dive. A modest canyon opened below, full of dark green weeds, black sea urchins and small flowerlike white algae. Fingerlings browsed in the scene. The sand sloped down into a clear blue infinity. The sun struck so brightly I had to squint. My arms hanging at my sides, I kicked the fins languidly and traveled down, gaining speed, watching the beach reeling past. I stopped kicking and the momentum carried me on a fabulous glide. When I stopped, I slowly emptied my lungs and held my breath. The diminished volume of my body decreased the lifting force of water, and I sank dreamily down. I inhaled a great chestful and retained it. I rose toward the surface.

My human lungs had a new role to play, that of a sensitive ballasting system. I took normal breaths in a slow rhythm, bowed my head and swam smoothly down to thirty feet. I felt no increasing water pressure, which at that depth is twice that of the surface. The aqualung automatically fed me increased compressed air to meet the new pressure layer. Through the fragile human lung linings this counter-pressure was being transmitted to the blood stream and instantly spread throughout the incompressible body. My brain received no subjective news

of the pressure. I was at ease, except for a pain in the middle ear and sinus cavities. I swallowed as one does in a landing airplane to open my Eustachian tubes and healed the pain. (I did not wear ear plugs, a dangerous practice when under water. Ear plugs would have trapped a pocket of air between them and the eardrums. Pressure building up in the Eustachian tubes would have forced my eardrums outward, eventually to the bursting point.)

I reached the bottom in a state of transport. A school of silvery sars (goat bream), round and flat as saucers, swam in a rocky chaos. I looked up and saw the surface shining like a defective mirror. In the center of the looking glass was the trim silhouette of Simone,★ reduced to a doll. I waved. The doll waved at me.

I became fascinated with my exhalations. The bubbles swelled on the way up through lighter pressure layers, but were peculiarly flattened like mushroom caps by their eager push against the medium. I conceived the importance bubbles were to have for us in the dives to come. As long as air boiled on the surface all was well below. If the bubbles disappeared there would be anxiety, emergency measures, despair. They

---

★ *Ed. note:* Cousteau's wife.

10

MENFISH

roared out of the regulator and kept me company. I felt less alone.

I swam across the rocks and compared myself favorably with the sars. To swim fishlike, horizontally, was the logical method in a medium eight hundred times denser than air. To halt and hang attached to nothing, no lines or air pipe to the surface, was a dream. At night I had often had visions of flying by extending my arms as wings. Now I flew without wings. (Since that first aqualung flight, I have never had a dream of flying.)

I thought of the helmet diver arriving where I was on his ponderous boots and struggling to walk a few yards, obsessed with his umbilici and his head imprisoned in copper. On skin dives I had seen him leaning dangerously forward to make a step, clamped in heavier pressure at the ankles than the head, a cripple in an alien land. From this day forward we would swim across miles of country no man had known, free and level, with our flesh feeling what the fish scales know.

I experimented with all possible maneuvers of the aqualung—loops, somersaults and barrel rolls. I stood upside down on one finger and burst out laughing, a shrill distorted laugh. Nothing I did altered the automatic rhythm of air. Delivered from gravity and buoyancy I flew around in space.

I could attain almost two knots' speed, without using my arms. I soared vertically and passed my own bubbles. I went down to sixty feet. We had been there many times without breathing aids, but we did not know what happened below that boundary. How far could we go with this strange device?

Fifteen minutes had passed since I left the little cove. The regulator lisped in a steady cadence in the ten-fathom layer and I could spend an hour there on my air supply. I determined to stay as long as I could stand the chill. Here were tantalizing crevices we had been obliged to pass fleetingly before. I swam inch-by-inch into a dark narrow tunnel, scraping my chest on the floor and ringing the air tanks on the ceiling.

In such situations a man is of two minds. One urges him on toward mystery and the other reminds him that he is a creature with good sense that can keep him alive, if he will use it. I bounced against the ceiling. I'd used one-third of my air and was getting lighter. My brain complained that this foolishness might sever my air hoses. I turned over and hung on my back.

The roof of the cave was thronged with lobsters. They stood there like great flies on a ceiling. Their heads and antennae were pointed toward the cave entrance. I breathed lesser lungsful to keep my chest from touching them. Above water was occupied, ill-fed France. I thought of the hundreds of calories a diver loses in cold water. I selected a pair of one-pound lobsters and carefully plucked them from the roof, without touching their stinging spines. I carried them toward the surface.

Simone had been floating, watching my bubbles wherever I went. She swam down toward me. I handed her the lobsters and went down again as she surfaced. She came up under a rock which bore a torpid Provençal citizen with a fishing pole. He saw a blonde girl emerge from the combers with lobsters wriggling in her hands. She said, "Could you please watch these for me?" and put them on the rock. The fisherman dropped his pole.

Simone made five more surface dives to take lobsters from me and carry them to the rock. I surfaced in the cove, out of the fisherman's sight. Simone claimed her lobster swarm. She said, "Keep one for yourself, *monsieur*. They are very easy to catch if you do as I did."

Lunching on the treasures of the dive, Tailliez and Dumas questioned me on every detail. We reveled in plans for the aqualung. Tailliez pencilled the tablecloth and announced that each yard of depth we claimed in the sea would open to mankind three hundred thousand cubic kilometers of living space. Tailliez, Dumas and I had come a long way together. We had been eight years in the sea as goggle divers. Our new key to the hidden world promised wonders. We recalled the beginning.

12

The Perils and Potentials of a Watery Planet

Jacques Cousteau

New discoveries and new technologies are changing our perceptions of the oceans. We have long viewed them as commercial highways, defensive barriers, and a source of food and such occasional riches as pearls and sunken treasures. But scientific exploration is gradually removing the mask of mystery that has surrounded the sea, revealing the potentials for mining its floor, farming its waters, and harnessing its energy. As the increasing population of the world threatens to exhaust available land resources, we turn increasingly to the sea to satisfy our human wants. Exploitation of the sea, however, like that of the land, bears an environmental cost that cannot be ignored if we are to preserve the oceans for the use of future generations. Here Jacques Cousteau, who has become the leading publicist of ocean concerns through his books and television films, gives his own assessment of the promises and dangers inherent in our new relationship to the sea.

The ocean cannot be dissociated from any of our problems. Though not always properly "billed," it is nonetheless a vital actor in the "production" of climate, storms, agriculture, health, war and peace, trade, leisure, and creative art. It is not merely a weather-regulating system and a source of food, cattle feed, fuel, and minerals. More generally, it absorbs vast quantities of the carbon dioxide generated by the combustion of fossil fuels, it releases a major part of the oxygen we breathe, and it acts as a powerful buffer to slow down or to avoid such calamities as quick variations in the sea level. Moreover, the human body is made up of much more water than all its components combined. A dehydrated human being would weigh little more than 30 pounds. Our flesh is composed of myriads of cells, each one of which contains a miniature ocean, less salty than today's ocean but comprising all the salts of the sea, probably the built-in heritage of our distant ancestry, when some mutating fish turned into reptiles and invaded the virgin land.

Life was born in the ocean, evolved in the ocean, then occupied land wherever water was available, and produced two top-of-the-line creatures, both as smart as they were physically vulnerable—the ancestors of the dolphin and of the human. Dolphins returned to the sea, where they enjoy almost the status of playboys.

A Painful Hangover

As far as humans are concerned, they have only very recently achieved the domination of nature, and they suddenly realize that the long and difficult conquest of the planet and the enthusiasm of victory have spread worldwide devastation and ruin. As recently as seven or eight generations ago, Western civilization triggered various kinds of explosions: more children, more food, more tonnage of goods, more energy—which created an exhilarating climate of pride, of overconfidence. The rapture of growth. But after such a great wild party, we are just awakening with a painful hangover, and all around us our home is littered with the sad remains of the "morning after." While we are slowly and painfully attempting to clean the place and to return to normal life, we realize that growth, at least in quantity, has limits, that our conventional resources can, and will soon, be exhausted, and that our very life depends on the quality of water.

13

Eating Our Capital

Today, the ocean is sick—very sick. To my best estimate, in the past 25 years overall productivity of sizable creatures (shrimp to whales) has decreased by 30 to 40 percent. Such a tragic situation is substantiated not only by undersea observation and photography but also by the analysis of the world fish catch: the global tonnage reached a ceiling in 1969 and has decreased slightly ever since, although there are more better-equipped fishing vessels, the captains of which have learned how to use the latest scientific findings about migrations. In fact, on their traditional grounds, in three years (1969 to 1972) catches of cod decreased by 22 percent, of haddock by 42 percent, of herring by 64 percent. The only way that man could maintain the world's fish harvest above the asymptotic figure of 60 million tons was by searching restlessly for new grounds and by switching to species that had hitherto never been fished commercially. The whale population is today less than 10 percent of what it was at the turn of the century. Coral reefs are dying rapidly all around the world. On a number of popular beaches, a swimmer may be exposed to skin diseases or even to hepatitis.

The Price to Pay

The sickness of the ocean has its origins in our heredity: ignorance, superstition, thirst for individual power. The diagnosis points to uncontrolled proliferation and growth, a disease of the cancer family. The cancer that plagues the ocean is *our* cancer. At this stage of emergency, the "drugs" used recently (such hallucinogens as "better living by owning more") have proved to be inoperative; major surgery is needed, although we have not yet decided to turn to surgery. The cancer operation will be painful; it will not include the use of nuclear weapons, but it will result in spectacular transfers of wealth and in radical changes in our living habits and our moral standards. Today's option for the coming 50 years is dramatic: either we will let the ocean die and mankind will die soon thereafter, or surgery will be performed successfully and our new relationship with the sea will bring about the dawn of a golden age, as early as the year 2024.

World Weather Control

Within 50 years, the oceans, if managed with imagination and care, could provide the clues to "weather-making," suppression of droughts and famines, and taming of tropical storms and tornadoes. A computerized, world "Weather-Production Center" would continuously receive two sets of data from two different fleets of satellites. The first fleet would give an accurate description of the existing weather, somewhat in the manner of present-day meteorological satellites. The other fleet would interrogate thousands of instrumental buoys scattered over all of the oceans and seas, thus gathering detailed measurements concerning the evaporation and condensation factors: humidity, temperature differences, currents, and winds. If rain was badly needed in a famine-threatened agricultural province, the computers, fed with all the necessary data, would recommend that evaporation of seawater should be accelerated or reduced in a corresponding area of the ocean. Special rockets could then be sent to the selected area of the sea. They would spread a molecular layer of harmless and biodegradable dye or oil on the surface. A few hours or days later, falling rain would avert a famine.

A Bonanza of Food and Drugs

Farming the sea could provide 10 times the amount of food and proteins obtained today from the world's fishing industries. But the traditional entries into the marine biological pyramid offer very limited possibilities and will be abandoned. Most "commercial" fish, for example, are thriving at the third, fourth, and even sixth or seventh level of the life cycle, while on land the cow feeds directly on grass. This means that at least 10,000 tons of algae have to be consumed in order to produce one single ton of tuna meat. This compares very unfavorably with the 10-to-1 efficiency ratio of the cow, or with the 50-to-1 ratio of the whale. The dugong, or sea cow, is the only marine mammal that feeds directly on vegetables. It is likely that 50 years from now, the bulk of human food originating in the sea will be obtained by farming massive quantities of vegetable plankton to be fed to such land mammals as cows, pigs, or, more particularly, sheep.

The basic (primary) production of planktonic algae reads in the trillions of tons. The energy required for producing such masses of vegetables through photosynthesis is delivered by the sun to the upper layers of the sea, but the crop is limited by the very slow turnover of nutrients, such as nitrates and phosphates. Deep, cold waters are rich in nutrients. In certain regions "upwelling" currents bring these rich waters to the surface, automatically increasing the abundance of algae and fish. Artificial upwellings will be caused, and the productivity of the ocean will be improved substantially.

The marine world is physiologically extremely complex, and we are learning that fish and invertebrates produce a great variety of substances ranging from poisons to tranquilizers, antispasmodics, antibiotics, and antiseptic chemicals. These will serve as models for the development of families of synthetic drugs of great therapeutic value.

Inexhaustible Liquid Mines

The continental shelf can provide some sources of heavy metals from "placers," but recent advances in ocean exploration and geophysics have opened new avenues to the exploitation of minerals from the ocean. The famous "manganese nodules" that litter vast areas of the abyssal plains contain not only manganese but also iron, copper, nickel, and many other metals. These nodules are not the result of some mysterious phenomenon that occurred in a distant past; they are still being formed today. The recent theory of "plate tectonics," explaining the mechanism of continental drift, also indicates the most prob-

able areas where mineral deposits are formed on the sea floor. The study of the "hot brines" of the Red Sea is also casting a new light on how the concentration of minerals from seawater naturally occurs. In the future, we will probably know how artificially to trigger such concentrations, so as to directly extract precious ores from inexhaustible liquid mines.

Clean Energy From the Sea

Tides have already been successfully tapped for the production of clean energy in a pilot plant at the Rance River, in France. A total of about 100,000 megawatts could be produced along a similar scheme in the Northern Hemisphere. Even more promising, though, it the project for rationally taming the temperature differences between both surface and deep layers of the ocean in most of the tropical regions. The sea absorbs the heat from the sun and, thanks to the currents, acts as a natural concentrator of solar energy. A small experimental thermal plant actually produced electricity from the sea as early as 1929. Bigger and improved plants could generate, without any pollution, practically all the energy needed by a developing world provided that population and wastefulness are controlled.

An Era of Marine Urbanism

Efficient prevention of all pollutions would make it possible to revitalize the dying coral reefs and to create undersea "acclimatization parks." Artificial shorelines would be developed, as well as huge floating surfaces for cities, airports, industries, and leisure parks. Careful interventions in genetic processes would allow us greatly to enrich the overall vitality of the sea and, consequently, of the planet. On land, deserts would be progressively refertilized. Such goals are not dreams; they are definitely within reach during the next half-century.

Social Drift and Communication Gap

Only science and technology can help us cure what the misuse and abuse of the same science and technology has threatened. But, unfortunately, scientists have great difficulty in communicating with the public. Herman Kahn and other futurologists have described the divorce between the specialists of all kinds, including the executives of government and industry, and the people. The two categories of population are separated by their style of life, their sources of information, and even their language; scientists, as lawyers, are using a jargon that the man in the street does not always understand.

The communication gap has its origin in many causes, among which is the lack of clear, simple education about the ocean and the life cycle to which we belong. Also, because information is often biased under the influence of interest groups in all countries today, the evaluation of the human impacts of such vital activities as nuclear plants, offshore drilling, and so on is always and exclusively made by the agency or by the company involved in the production. Obviously, independent and respected sources of evaluation are needed. Furthermore, research is funded today almost only if it fits the policy or the short-range plans of the government apparatus. This is especially true in the environmental protection field. The result is that people realize how little they know, and they feel frustrated. They no longer trust statements from the United Nations because they suspect that these are dictated by some majority of unconcerned states; they don't believe announcements made by government agencies, such as those comparing the impact of radioactive tests with that of the luminous dials of a wristwatch. They suspect the data from pollution analyses published by oil or power companies. They desperately miss the existence of a completely independent source of such information.

Independent bureaus of evaluation and control exist in such fields as shipbuilding, engineering, and bridge or building construction, but are lacking in fields involving the quality of our life and the future of the coming generations. Whenever such an organization exists, its

duty will be to widely publish, through all existing mass media, the results of both independent evaluation and research.

Who is actually responsible for the accelerated deterioration of our water planet? Is it progress itself, as a few druids claim? Is it industry, as could be suggested by the murky effluents that murder our rivers and our seas? Is it publicity when it promotes artificial needs? Is it we, as individuals, through neglect, ignorance, selfishness? None of these accusations could be seriously substantiated. The responsibility lies almost exclusively at the level of government. Progress offers as much good as evil, and it is the *use* of progress that is to be corrected. The only thing industry refuses is to be penalized in a highly competitive world—industrial firms would welcome international regulations, the same for all in all countries, imposing the same water and smoke standards—even drastic ones! Such regulations would finally create a climate of fair competition, and the additional environmental burden would naturally be included in overhead costs. Publicists have openly stated that they would just as happily publicize recycling as they do "throwaways." And the public is rapidly developing a strong desire to make a personal contribution to a better natural and social environment.

Governments alone have the duty to protect us. To manage the future, they must enter long-range planning and drastic measures today. Governments must unite to determine the acceptable levels of discharge for those toxic products that will slowly decompose in the sea; but they must also unite to impose the absolute-zero level in all effluents for the non-degradable poisons, as these would pile up forever in the sea and would eventually become concentrated in the food chain, ending up on our plates.

O.H.A.—*Ocean High Authority*

It is too early to turn to the United Nations, because it would be unfair to apportion a share of the cost of the big cleanup to the developing countries, which have no responsibility yet in today's spoliations. The solution lies in the hands of the 15 to 20 most polluting countries: they should—they must—very soon delegate a fraction of their national sovereignty to a High Authority fully empowered to decide, implement, and control the safety measures that will protect us and our heirs. The other countries will join as soon as their industry grows. Such a High Authority is not a utopian concept; it has precedents. For example, in Europe, after the war, the "Six" delegated their sovereignty to the Coal and Steel Authority, which became the nucleus of the European Community. A strong, enthusiastic public opinion is the only force capable of influencing our governments in the creation of such a High Authority.

The riches of the sea must belong to *everybody*, not to *nobody* and not to *anybody*—it is high time to establish a world policy and an international control of the high seas, using such modern devices as monitoring satellites, which could become "space sheriffs." It is only within such a framework that we can consider taking advantage of the incredible marine resources without destroying them forever.

We must all work hard toward this goal; this is the price of our liquid future.

SEA CITY

A concrete and glass city at sea, housing 30,000 people in terraced apartments, could be a realistic answer to problems of living space, according to the Pilkington Glass Age Development Committee of Great Britain.

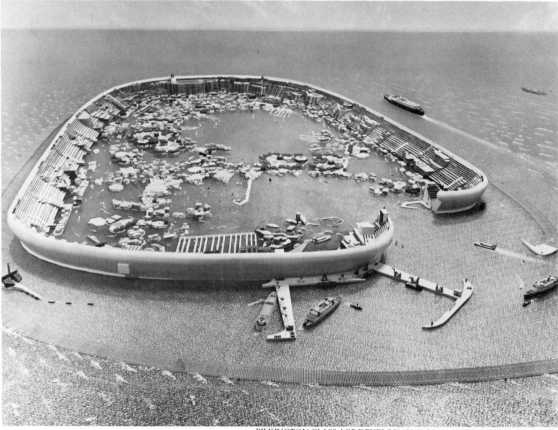

PILKINGTON GLASS AGE DEVELOPMENT COMMITTEE OF GREAT BRITAIN

Anonymous, *Meditation by the Sea*, circa 1855

Unit Two

Restless and changing, awesome and mysterious, the sea has gripped the human imagination throughout the ages. The visions of the sea expressed in art and literature fill whole museums and libraries. Since we cannot hope for a complete sample through space and time in the few pages available in this Reader, we have opted for some examples of the best that might be representative of many cultures or many times. In addition, we have spread examples of the fine arts, largely American painting, throughout the Reader in which they illuminate the subject matter. As a unifying theme in this part of the Reader, we use the varying perceptions of the deep and of its creatures—friendly, threatening, mysterious, familiar—depending on the culture and the individual. Our present visions of the sea, shaped by our knowledge of science and the technology of the underwater motion-picture camera, differ markedly from those of the past.

Fishhooks and other artifacts indicate that humans invaded the margins of the sea very early in their history. One of the earliest known portraits of a ship capable of leaving the shore behind was engraved on a rock by a Bronze Age Dane some 3,000 to 4,000 years ago. The reason for the engraving is obscure, but it confirms the existence of a sea culture long ago

Literature and Art

even in the harsh cold of the north. What then of the warm south? "There," said Homer,

> "There is a land, named Crete, fruitful and fair,
> Set like a jewel in the wine-dark sea,
> Peopled by countless multitudes of men
> And ninety cities . . ."

And Homer wrote not of ancient Greece in which he lived, but of an already legendary, half-mythical land that had vanished by his day. It was the land of King Minos, of Theseus and the Minotaur, of Daedalus who fashioned wings and flew toward the sun, of bull dancers, and ships, always ships. Crete was the first sea power, moated, like Shakespearean England, by the sea—the ninety cities had no walls. The Minoans loved the sea and its creatures and from them they felt no threat. On their walls and furnishings, the quiet octopi undulate harmoniously and the exuberant dolphin sport half in the water and half—like Daedalus—in the air.

This mighty sea power vanished suddenly. As will be shown later in Unit III, it was probably destroyed by a great natural disaster and thereby created the legend of the lost continent of Atlantis.

Greek culture replaced Minoan, and it too was intimately related to the sea that was the highway between its isles. Homer wrote the first great sea epic, the *Odyssey,* in which the sea has a new image, full of peril, and peopled by monsters like Scylla, a terror that is now downgraded to a mere huge octopus.

The ancients also peopled the sea with gods and spirits whose wrath accounted for the fury of the oceans. Counterparts of these deities are still worshiped among some littoral cultures today. In the Judeo-Christian tradition, the power and the mystery of the oceans become the work of a single God.

But even for those who now worship science instead of a deity, the menace of the sea always remains. Men have learned to live with it as a part of daily life, but still it is there. In our Reader it is expressed by Stephen Crane in

"The Open Boat." This concerns a relatively modern episode, experienced by Crane in 1896 after a shipwreck in Florida. Crane was not a seafarer—merely a passenger—but he was a gifted author, and in his short story he captures feelings that most sailors could not. The men in the open boat know that they are almost powerless against the sea—as all sailors ultimately come to know. But the men find within themselves the reserves of strength to keep fighting, and they become united by the struggle—as sailors do in such circumstances.

In our Reader we skip from ancient times to fairly modern ones, using a few poems as stepping stones. In the narrative poem "Dauber," John Masefield, poet laureate of England and of the sea, drew on his life at sea to recount a voyage around Cape Horn—the nightmare of sailing ships. On board is a painter who learns to fight the sea and share the comradeship of shipmates like the men in "The Open Boat."

American paintings touch every aspect of our interaction with the sea. Ships are shown in peaceful commerce and in fierce combat. The lone meditator stares at the placid sea, while elsewhere Watson is attacked by a shark in a furious scene painted by John Singleton Copley. For Winslow Homer—one of America's finest painters—the sea was variably protagonist and playmate, threatening life and buoying spirit.

While the painters ashore in the nineteenth century used the sea as background, something new was happening in the life of sailors. The American whaling fleet was ranging the world in voyages that lasted for years. Sailors before—and mainly since—were merely at sea long enough to reach the next port, but these Yankees worked on great waters, and often they waited there as well. From their waiting time, for no Yankee had idle time, came the folk art called "scrimshaw," which is described in a selection by Gerhard Timmermann and Helen L. Winslow.

From the whaling voyages also came the novel *Moby-Dick* by Herman Melville. Of his own work, Melville wrote, "[It] is of the horrible texture of a fabric that should be woven of ships' cables and hawsers. A Polar wind blows through it, & birds of prey hover over it." It was disdained as folk art of "humble artistic standards," according to the eleventh edition of the *Encyclopaedia Britannica,* published in 1911, sixty years after it was written. Now it is "art" in the highest rank of American literature.

The whale, however, was only one monster of the deep. Other popular menaces at the time were the sea serpent and the "kraken," a legendary beast that could drag a ship down with its great arms and terrible suckers. This beast had become known in medieval times in another type of folk art, the sailor's yarn, which was famed more for embellishment than for truth. The tales of the kraken were collected in natural histories between the sixteenth and eighteenth centuries by clerics who firmly believed in miracles and were not troubled by a few more. Scientists did not take the accounts seriously because they were so wild and miraculous. As it later developed, the scientists became so wedded to doubts that they maintained them despite the presence of facts. How and when the matter was resolved is given in the selection on the kraken by Frank W. Lane. A famous fictional account of an attack by a similar monster has been selected from *Twenty Thousand Leagues Under the Sea* by Jules Verne. The sea serpent is still a mystery, although it is easy to explain away if, like the scientists who doubted the existence of the kraken, you ignore the reported sightings.

The giant squid, a real monster, has often been confused with the octopus, which is not. Indeed anyone who was brought up on Hollywood motion pictures can hardly fail to remember the octopus with horror. Remember the snaky black tentacle slithering toward the unsuspecting diver in his heavy suit? Remember, for that matter, the pearl diver with his foot trapped by a giant clam? Monsters filled the matinee sea.

The octopus is easily distinguished from the squid once you realize that he is not a monster. He is the one the Minoans used for cheerful decorations. He is the shy playmate of divers. He is Cousteau's "soft intelligence" for he has a large brain for his ilk. We eat the octopus, so that soft brain has cause to view us with horror, but not the reverse.

The visual arts of this century portray the sea and the shore as a place for sailing and sport. The sea is increasingly familiar and the menace seems far away. Perhaps that is the reason the book *Jaws* by Peter Benchley and the motion picture based on it have had such an impact.

Why are sharks conceived of as so menacing when they were such unimportant monsters even a century ago? Is it that we have a natural desire to find menace and mystery remaining in the sea as we sterilize the land and poison the air?

Perhaps so, but the selection from *Shark Attacks Against Man* by David Baldrige suggests otherwise. Sharks attack in proportion to the number of people in the water. Until the modern rise of aquatic sports, the shark was not much of a menace. The exception was in the South Seas where people swam—and made statues of shark gods.

Perhaps the terrible shark has reached his peak as a menace. Eugenie Clark, the famous diving biologist, describes for us a remarkable fish in the Red Sea, the Moses sole, which is immune to shark attack because it exudes a poison. Oddly enough, the computerized analysis of shark attacks described by Baldridge reveals a partial immunity that is more mysterious. Although half the people in the water at any time are women, nine-tenths of the shark attacks are on men.

The sea that yielded the coelacanth, long thought to be extinct, still contains many mysteries to intrigue the scientist. It echoes with the whale song. It flashes with color for the underwater photographers. Can we believe that it will not remain the inspiration for great artists? Can we believe the mystery is gone?

Sir Patrick Spens

Stories of the sea abound in folk literature as well as in more formal writings. Seafaring adventure was a favorite theme of the ballads that had become popular in Scotland and England by the fifteenth century. Handed down and changed through oral tradition, these ballads were finally collected and printed in the seventeenth century.

The king sits in Dumferline town,
    Drinking the *blude-reid* wine: blood-red
"O whar will I get a guid sailor
    To sail this ship of mine?"

Up and spak an *eldern knicht*   ancient knight
    Sat at the king's richt knee:
"Sir Patrick Spens is the best sailor
    That sails upon the sea."

The king has written a *braid* letter   broad (plain)
    And signed it wi' his hand,
And sent it to Sir Patrick Spens,
    Was walking on the sand.

The first line that Sir Patrick read,
    A loud *lauch* lauchéd he;   laugh
The next line that Sir Patrick read,
    The tear blinded his *ee*.   eye

"O *wha* is this has done this deed,   who
    This ill deed done to me,
To send me out this time o' the year,
        To sail upon the sea?

"Make haste, make haste, my *mirry men* all,   companions, followers
    Our guid ship sails the morn."
"O say *na sae,* my master dear,   not so
    For I fear a deadly storm.

"Late late *yestre'en* I saw the new moon   last evening
    Wi' the *auld* moon in her arm,   old
And I fear, I fear, my dear master,
    That we will come to harm."

O our Scots nobles were richt *laith*   loath (reluctant)
    To *weet* their cork-heeled *shoon*,   wet / shoes
But *lang owre a'* the play were played   long ere all
    Their hats they swam *aboon*.   above

O lang lang may their ladies sit,
    Wi' their fans into their hand,
Or *e'er* they see Sir Patrick Spens   before
    Come sailing to the land.

O lang lang may the ladies stand,
    Wi' their gold *kembs* in their hair,   combs
Waiting for their *ain* dear lords,   own
    For they'll see *thame no mair*.   them no more

Half o'er, half o'er to Aberdour   halfway over
    It's fifty *fadom* deep,   fathoms
And there lies guid Sir Patrick Spens
    Wi' the Scots lords at his feet.

Psalm 107

From the familiar stories of the Creation, the Flood, and Jonah in the Old Testament to those of the fisherman Peter in the New Testament, there are recurrent images of the sea in the Bible. The Book of Psalms contains one of the most moving descriptions of man's relationship to the sea and, through it, to God.
The following passage is from the King James version of the Bible.

Those who go down to the sea in ships,
    Who do business in great waters;
They see the works of the Lord,
    and His wonders in the deep.
For he commandeth, and raiseth the stormy wind,
    Which lifteth up the waves thereof.
They mount up to the heaven, they go down
    again to the depths: their soul is melted
    because of trouble.
They reel to and fro, and stagger like a drunken
    man, and are at their wit's end.
Then they cry unto the Lord in their trouble,
    and He bringeth them out of their distresses.
He maketh the storm a calm, so that the waves
    thereof are still.
Then they are glad because they be quiet;
    so He bringeth them unto their desired haven.

Washington Allston, *The Rising of a Thunderstorm at Sea,* 1804

24

Ariel's Song

(*The Tempest*, Act I, Scene 2)

William Shakespeare

The poet's rendition of man's relationship to the sea is perhaps nowhere more lyrical and magical than in Ariel's song from The Tempest. *The young Ferdinand, believing his father to have been drowned in the same shipwreck that has cast him ashore on a strange island, hears this song from an invisible fairy spirit. Man and the sea become one in this dream, as the human form is changed into the riches of the sea.*

Full fathom five thy father lies,
    Of his bones are coral made:
Those are pearls that were his eyes:
    Nothing of him that doth fade,
But doth suffer a sea-change
Into something rich and strange.
Sea-nymphs hourly ring his knell.
        (Ding-dong.)
Hark now I hear them—ding-dong bell.

Dover Beach

Matthew Arnold

In "Dover Beach," the poet evokes the sounds of the sea as well as a visual image. A noted literary critic and one of the greatest poets of Victorian England, Matthew Arnold used the image of the sea to express his religious doubts and melancholy view of a joyless, uncertain world.

The sea is calm to-night.
The tide is full, the moon lies fair
Upon the straits;—on the French coast the light
Gleams and is gone; the cliffs of England stand,
Glimmering and vast, out of the tranquil bay.
Come to the window, sweet is the night-air!
Only, from the long line of spray
Where the sea meets the moon-blanch'd sand,
Listen! you hear the grating roar
Of pebbles which the waves draw back, and fling,
At their return, up the high strand,
Begin, and cease, and then again begin,
With tremulous cadence slow, and bring
The eternal note of sadness in.

Sophocles long ago
Heard it on the Aegean, and brought
Into his mind the turbid ebb and flow
Of human misery; we
Find also in the sound a thought,
Hearing it by this distant northern sea.

The sea of faith
Was once, too, at the full, and round earth's shore
Lay like the folds of a bright girdle furl'd.
But now I only hear
Its melancholy, long, withdrawing roar,
Retreating, to the breath
Of the night-wind, down the vast edges drear
And naked shingles of the world.

Ah, love let us be true
To one another! for the world, which seems
To lie before us like a land of dreams,
So various, so beautiful, so new,
Hath really neither joy, nor love, nor light,
Nor certitude, nor peace, nor help for pain;
And we are here as on a darkling plain
Swept with confused alarms of struggle and flight,
Where ignorant armies clash by night.

SCRIMSHAW JAGGING WHEEL OR PASTRY CRIMPER. *Sailors produced their own images of the sea in beautifully wrought handicrafts made from whale teeth and jawbone.*

Sea Fever

John Masefield

The eternal lure of the sea is captured in this poem by John Masefield, poet laureate of England from 1930 until his death in 1967. Masefield himself could not resist that lure—he ran away to sea at the age of thirteen. One of the best-loved of all sea poems, "Sea Fever" is from his first collection, Salt Water Ballads, *published in 1902.*

I must go down to the seas again, to the lonely sea and the sky,
And all I ask is a tall ship and a star to steer her by,
And the wheel's kick and the wind's song and the white sail's shaking,
And a grey mist on the sea's face and a grey dawn breaking.

I must go down to the seas again, for the call of the running tide
Is a wild call and a clear call that may not be denied;
And all I ask is a windy day with the white clouds flying,
And the flung spray and the blown spume, and the sea-gulls crying.

I must go down to the seas again to the vagrant gypsy life,
To the gull's way and the whale's way where the wind's like a whetted knife;
And all I ask is a merry yarn from a laughing fellow-rover,
And quiet sleep and a sweet dream when the long trick's over.

On This Island

W. H. Auden

W. H. Auden, poet of England and of America, was awarded both King George's Gold Medal for poetry (1937) and the Pulitzer Price for poetry (1948). In "On This Island," published in 1936, he invites us to take a "full view" of the sea and appreciate the variety of its sights and sounds. Like Arnold, Auden wrote of the sea breaking against the white cliffs of the land, but the darkness of "Dover Beach" is dispelled by the lightness of "On This Island."

Look, stranger, on this island now
The leaping light for your delight discovers,
Stand stable here
And silent be,
That through the channels of the ear
May wander like a river
The swaying sound of the sea.

Here at the small field's ending pause
Where the chalk wall falls to the foam, and its
    tall ledges
Oppose the pluck

And knock of the tide,
And the shingle scrambles after the suck-
ing surf, and the gull lodges
A moment on its sheer side.

Far off like floating seeds the ships
Diverge on urgent voluntary errands;
And the full view
Indeed may enter
And move in memory as now these clouds do,
That pass the harbour mirror
And all the summer through the water saunter.

The Day the Tide

Philip Booth

Few contemporary poets use the image of the sea more effectively than Philip Booth. His knowledge and reverence of the sea, born of his experiences along the Maine coast, are clearly evident in this poem about the essential meaning of the sea to our lives. The poem is from his collection Margins, *published in 1970.*

The day the tide went out,
and stayed, not just as Mean
Low Water or Spring Ebb,
but out, out all the way
perhaps as far as Spain,
until the bay was empty,

it left us looking down
at what the sea, and our
reflections on it, had
(for generations of
good fish, and wives fair
as vessels) saved us from.

We watched our fishboats ground
themselves, limp-chained in mud;
careened, as we still are
(though they lie far below us)
against this sudden slope
that once looked like a harbor.

We're level, still, with islands,
or what's still left of them
now that treelines invert:
the basin foothills rock
into view like defeated castles,
with green and a flagpole on top.

Awkward as faith itself,
heron still stand on one leg
in trenches the old tide cut;
maybe they know what the moon's
about, working its gravity
off the Atlantic shelf.

Blind as starfish, we
look into our dried reservoir
of disaster: fouled trawls, old
ships hung up on their monument ribs; the skeletons
of which our fathers were master.

We salt such bones down with self-
consolation, left to survive,
if we will, on this emptied slope.
Réunion Radio keeps reporting
how our ebb finally flooded
the terrible Cape of Good Hope.

Dauber

John Masefield

Sharing the drama and excitement of the short story with the rhythmic magic of poetry, this narrative poem by John Masefield reflects his intense experience with the sea. Like the hero of the poem, Masefield could portray the sea "From the inside / By one who really knows." If the sea represented freedom and adventure in "Sea Fever," here it becomes the protagonist, the cruel test of manhood. Despised and mocked by the rest of the crew for being a painter, and a poor one at that, Dauber resolves to win his mates' respect by fighting the storms off Cape Horn along with the other seamen. The poem was published in 1913, as the days of the great sailing ships were slipping into history.

Si talked with Dauber, standing by the side.
"Why did you come to sea, painter?" he said.
"I want to be a painter," he replied,
"And know the sea and ships from A to Z,
And paint great ships at sea before I'm dead;
Ships under skysails running down the Trade—
Ships and the sea; there's nothing finer made.

"But there's so much to learn, with sails and ropes,
And how the sails look, full or being furled,
And how the lights change in the troughs and slopes,
And the sea's colours up and down the world,
And how a storm looks when the sprays are hurled
High as the yard (they say) I want to see;
There's none ashore can teach such things to me.

"And then the men and rigging, and the way
Ships move, running or beating, and the poise
At the roll's end, the checking in the sway—
I want to paint them perfect, short of the noise;
And then the life, the half-decks full of boys,
The fo'c's'les with the men there, dripping wet:
I know the subjects that I want to get.

"It's not been done, the sea, not yet been done,
From the inside, by one who really knows;
I'd give up all if I could be the one,
But art comes dear the way the money goes.
So I have come to sea, and I suppose
Three years will teach me all I want to learn
And make enough to keep me till I earn."

★   ★   ★

And then the thought came: "I'm a failure. All
My life has been a failure. They were right.
It will not matter if I go and fall;
I should be free then from this hell's delight.
I'll never paint. Best let it end to-night.
I'll slip over the side. I've tried and failed."
So in the ice-cold in the night he quailed.

Death would be better, death, than this long hell
Of mockery and surrender and dismay—
This long defeat of doing nothing well,
Playing the part too high for him to play.
"O Death! who hides the sorry thing away,
Take me; I've failed. I cannot play these cards."
There came a thundering from the topsail yards.

And then he bit his lips, clenching his mind,
And staggered out to muster, beating back
The coward frozen self of him that whined.
Come what cards might he meant to play the pack.
"Ai!" screamed the wind; the topsail sheet went clack;
Ice filled the air with spikes; the grey-backs burst.
"Here's Dauber," said the Mate, "on deck the first.

"Why, holy sailor, Dauber, you're a man!
I took you for a soldier. Up now, come!"
Up on the yards already they began
That battle with a gale which strikes men dumb.
The leaping topsail thundered like a drum.
The frozen snow beat in the face like shots.
The wind spun whipping wave-crests into clots.

So up upon the topsail yard again,
In the great tempest's fiercest hour, began
Probation to the Dauber's soul, of pain
Which crowds a century's torment in a span.
For the next month the ocean taught this man,
And he, in that month's torment, while she wested,
Was never warm nor dry, nor full nor rested

But still it blew, or, if it lulled, it rose
Within the hour and blew again; and still
The water as it burst aboard her froze.
The wind blew off an ice-field, raw and chill,
Daunting man's body, tampering with his will;
But after thirty days a ghostly sun
Gave sickly promise that the storms were done.

The Open Boat

Stephen Crane

The sea is once again portrayed as an adversary in "The Open Boat" by Stephen Crane. The story is based on the author's own experiences in the winter of 1896/1897, when, en route to Cuba to report on the rebellion there, he was shipwrecked in a storm. He and three companions survived for fifty hours in a dinghy before they were finally cast upon the shore. His newspaper account of the experience appeared in the New York Press *on January 7, 1897. In transforming the report into a short story, Crane allows the reader to share the meaning of the brute force of the sea with the main character, who becomes a universal figure by being identified simply as "the correspondent."*

A Tale Intended to be after the Fact: Being the Experience of Four Men from the Sunk Steamer *Commodore*

I

None of them knew the color of the sky. Their eyes glanced level, and were fastened upon the waves that swept toward them. These waves were of the hue of slate, save for the tops, which were of foaming white, and all of the men knew the colors of the sea. The horizon narrowed and widened, and dipped and rose, and at all times its edge was jagged with waves that seemed thrust up in points like rocks.

Many a man ought to have a bathtub larger than the boat which here rode upon the sea. These waves were most wrongfully and barbarously abrupt and tall, and each froth-top was a problem in small-boat navigation.

The cook squatted in the bottom, and looked with both eyes at the six inches of gunwale which separated him from the ocean. His sleeves were rolled over his fat forearms, and the two flaps of his unbuttoned vest dangled as he bent to bail out the boat. Often he said, "Gawd! that was a narrow clip." As he remarked it he invariably gazed eastward over the broken sea.

The oiler, steering with one of the two oars in the boat, sometimes raised himself suddenly to keep clear of water that swirled in over the stern. It was a thin little oar, and it seemed often ready to snap.

The correspondent, pulling at the other oar, watched the waves and wondered why he was there.

The injured captain, lying in the bow, was at this time buried in that profound dejection and indifference which comes, temporarily at least, to even the bravest and most enduring when, willy-nilly, the firm fails, the army loses, the ship goes down. The mind of the master of a vessel is rooted deep in the timbers of her, though he command for a day or a decade; and this captain had on him the stern impression of a scene in the grays of dawn of seven turned faces, and later a stump of a topmast with a white ball on it, that slashed to and fro at the waves, went low and lower, and down. Thereafter there was something strange in his voice. Although steady, it was deep with mourning, and of a quality beyond oration or tears.

"Keep'er a little more south, Billie," said he.

"A little more south, sir," said the oiler in the stern.

A seat in this boat was not unlike a seat upon a bucking broncho, and by the same token a broncho is not much smaller. The craft pranced and reared and plunged like an animal. As each wave came, and she rose for it, she seemed like a horse making at a fence outrageously high. The manner of her scramble over these walls of water is a mystic thing, and, moreover, at the top of them were ordinarily these problems in white water, the foam racing down from the summit of each wave requiring a new leap, and a leap from the air. Then, after scornfully bumping a crest, she would slide and race and splash down a long incline, and arrive bobbing and nodding in front of the next menace.

A singular disadvantage of the sea lies in the fact that after successfully surmounting one wave you discover that there is another behind it just as important and just as nervously anxious to do something effective in the way of swamping boats. In a ten-foot dinghy one can get an idea of the resources of the sea in the line of waves that is not probable to the average experience, which is never at sea in a dinghy. As each slaty wall of water approached, it shut all else from the view of the men in the boat, and it was not difficult to imagine that this particular wave was the final outburst of the ocean, the last effort of the grim water. There was a terrible grace in the move of the waves, and they came in silence, save for the snarling of the crests.

In the wan light the faces of the men must have been gray. Their eyes must have glinted in strange ways as they gazed steadily astern. Viewed from a balcony, the whole thing would, doubtless, have been weirdly picturesque. But the men in the boat had no time to see it, and if they had had leisure, there were other things to occupy their minds. The sun swung steadily up the sky, and they knew it was broad day because the color of the sea changed from slate to emerald-green streaked with amber lights, and the foam was like tumbling snow. The process of the breaking day was unknown to them. They were aware only of this effect upon the color of the waves that rolled toward them.

In disjointed sentences the cook and the correspondent argued as to the difference between a life-saving station and a house of refuge. The cook had said: "There's a house of refuge just north of the Mosquito Inlet Light, and as soon as they see us they'll come off in their boat and pick us up."

"As soon as who see us?" said the correspondent.

"The crew," said the cook.

"Houses of refuge don't have crews," said the correspondent. "As I understand them, they are only places where clothes and grub are stored for the benefit of shipwrecked people. They don't carry crews."

"Oh, yes, they do," said the cook.

"No, they don't," said the correspondent.

"Well, we're not there yet, anyhow," said the oiler, in the stern.

"Well," said the cook, "perhaps it's not a house of refuge that I'm thinking of as being near Mosquito Inlet Light; perhaps it's a life-saving station."

"We're not there yet," said the oiler in the stern.

II

As the boat bounced from the top of each wave the wind tore through the hair of the hatless men, and as the craft plopped her stern down again the spray slashed past them. The crest of each of these waves was a hill, from the top of which the men surveyed for a moment a broad tumultuous expanse, shining and wind-riven. It was probably splendid, it was probably glorious, this play of the free sea, wild with lights of emerald and white and amber.

"Bully good thing it's an on-shore wind," said the cook. "If not, where would we be? Wouldn't have a show."

"That's right," said the correspondent.

The busy oiler nodded his assent.

Then the captain, in the bow, chuckled in a way that expressed humor, contempt, tragedy, all in one. "Do you think we've got much of a show now, boys?" said he.

Whereupon the three were silent, save for a trifle of hemming and hawing. To express any particular optimism at this time they felt to be childish and stupid, but they all doubtless possessed this sense of the situation in their minds. A young man thinks doggedly at such times. On the other hand, the ethics of their condition was decidedly against any open suggestion of hopelessness. So they were silent.

"Oh, well," said the captain, soothing his children, "we'll get ashore all right."

But there was that in his tone which made them think; so the oiler quoth, "Yes! if this wind holds."

The cook was bailing. "Yes! if we don't catch hell in the surf."

Canton-flannel gulls flew near and far. Sometimes they sat down on the sea, near patches of brown seaweed that rolled over the waves with a movement like carpets on a line in a gale. The birds sat comfortably in groups, and they were envied by some in the dinghy, for the wrath of the sea was no more to them than it was to a covey of prairie chickens a thousand miles inland. Often they came very close and stared at the men with black bead-like eyes. At these times they were uncanny and sinister in their unblinking scrutiny, and the men hooted angrily at them, telling them to be gone. One came, and evidently decided to alight on the top of the captain's head. The bird flew parallel to the boat and did not circle, but made short sidelong jumps in the air in chicken fashion. His black eyes were wistfully fixed upon the captain's head. "Ugly brute," said the oiler to the bird. "You look as if you were made with a jack-knife." The cook and the correspondent swore darkly at the creature. The captain naturally wished to knock it away with the end of the heavy painter, but he did not dare do it, because anything resembling an emphatic gesture would have capsized this freighted boat; and so, with his open hand, the captain gently and carefully waved the gull away. After it had been discouraged from the pursuit the captain breathed easier on account of his hair, and others breathed easier because the bird struck their minds at this time as being somehow gruesome and ominous.

In the meantime the oiler and the correspondent rowed; and also they rowed. They sat together in the same seat, and each rowed an oar. Then the oiler took both oars; then the correspondent took both oars, then the oiler; then the correspondent. They rowed and they rowed. The very ticklish part of the business was when the time came for the reclining one in the stern to take his turn at the oars. By the

very last star of truth, it is easier to steal eggs from under a hen than it was to change seats in the dinghy. First the man in the stern slid his hand along the thwart and moved with care, as if he were of Sèvres. Then the man in the rowing-seat slid his hand along the other thwart. It was all done with the most extraordinary care. As the two sidled past each other, the whole party kept watchful eyes on the coming wave, and the captain cried; "Look out, now! Steady, there!"

The brown mats of seaweed that appeared from time to time were like islands, bits of earth. They were travelling, apparently, neither one way nor the other. They were, to all intents, stationary. They informed the men in the boat that it was making progress slowly toward the land.

The captain, rearing cautiously in the bow after the dinghy soared on a great swell, said that he had seen the lighthouse at Mosquito Inlet. Presently the cook remarked that he had seen it. The correspondent was at the oars then, and for some reason he too wished to look at the lighthouse; but his back was toward the far shore, and the waves were important, and for some time he could not seize an opportunity to turn his head. But at last there came a wave more gentle than the others, and when at the crest of it he swiftly scoured the western horizon.

"See it?" said the captain.

"No," said the correspondent, slowly; "I didn't see anything."

"Look again," said the captain. He pointed. "It's exactly in that direction."

At the top of another wave the correspondent did as he was bid, and this time his eyes chanced on a small, still thing on the edge of the swaying horizon. It was precisely like the point of a pin. It took an anxious eye to find a lighthouse so tiny.

"Think we'll make it, Captain?"

"If this wind holds and the boat don't swamp, we can't do much else," said the captain.

The little boat, lifted by each towering sea and splashed viciously by the crests, made progress that in the absence of seaweed was not apparent to those in her. She seemed just a wee thing wallowing, miraculously top up, at the mercy of five oceans. Occasionally a great spread of water, like white flames, swarmed into her.

"Bail her, cook," said the captain, serenely.

"All right, Captain," said the cheerful cook.

III

It would be difficult to describe the subtle brotherhood of men that was here established on the seas. No one said that it was so. No one mentioned it. But it dwelt in the boat, and each man felt it warm him. They were a captain, an oiler, a cook, and a correspondent, and they were friends—friends in a more curiously iron-bound degree than may be common. The hurt captain, lying against the water-jar in the bow, spoke always in a low voice and calmly; but he could never command a more ready and swiftly obedient crew than the motley three of the dinghy. It was more than a mere recognition of what was best for the common safety. There was surely in it a quality that was personal and heart-felt. And after this devotion to the commander of the boat, there was this comradeship, that the correspondent, for instance, who had been taught to be cynical of men, knew even at the time was the best experience of his life. But no one said that it was so. No one mentioned it.

"I wish we had a sail," remarked the captain. "We might try my overcoat on the end of an oar, and give you two boys a chance to rest." So the cook and the correspondent held the mast and spread wide the overcoat; the oiler steered; and the little boat made good way with her new rig. Sometimes the oiler had to scull sharply to keep a sea from breaking into the boat, but otherwise sailing was a success.

Meanwhile the lighthouse had been growing slowly larger. It had now almost assumed color, and appeared like a little gray shadow on the sky. The man at the oars could not be prevented from turning his head rather often to try for a glimpse of this little gray shadow.

At last, from the top of each wave, the men in the tossing boat could see land. Even as the lighthouse was an upright shadow on the sky, this land seemed but a long black shadow on the sea. It certainly was thinner than paper. "We must be about opposite New Smyrna," said the cook, who had coasted this shore often in schooners. "Captain, by the way, I believe they abandoned that life-saving station there about a year ago."

"Did they?" said the captain.

The wind slowly died away. The cook and the correspondent were not now obliged to slave in order to hold high the oar. But the waves continued their old impetuous swooping at the dinghy, and the little craft, no longer under way, struggled woundily over them. The oiler or the correspondent took the oars again.

Shipwrecks are *apropos* of nothing. If men could only train for them and have them occur when the men had reached pink condition, there would be less drowning at sea. Of the four in the dinghy none had slept any time worth mentioning for two days and two nights previous to embarking in the dinghy, and in the excitement of clambering about the deck of a foundering ship they had also forgotten to eat heartily.

For these reasons, and for others, neither the oiler nor the correspondent was fond of rowing at this time. The correspondent wondered ingenuously how in the name of all that was sane could there be people who thought it amusing to row a boat. It was not an amusement; it was a diabolical punishment, and even a genius of mental aberrations could never conclude that it was anything but a horror to the muscles and a crime against the back. He mentioned to the boat in general how the amusement of rowing struck him, and the weary-faced oiler smiled in full sympathy. Previously to the foundering, by the way, the oiler had worked a double watch in the engine-room of the ship.

"Take her easy now, boys," said the captain. "Don't spend yourselves. If we have to run a surf you'll need all your strength, because we'll sure have to swim for it. Take your time."

Slowly the land arose from the sea. From a

black line it became a line of black and a line of white—trees and sand. Finally the captain said that he could make out a house on the shore. "That's the house of refuge, sure," said the cook. "They'll see us before long, and come out after us."

The distant lighthouse reared high. "The keeper ought to be able to make us out now, if he's looking through a glass," said the captain. "He'll notify the life-saving people."

"None of those other boats could have got ashore to give word of this wreck," said the oiler, in a low voice, "else the life-boat would be out hunting us."

Slowly and beautifully the land loomed out of the sea. The wind came again. It had veered from the northeast to the southeast. Finally a new sound struck the ears of the men in the boat. It was the low thunder of the surf on the shore. "We'll never be able to make the lighthouse now," said the captain. "Swing her head a little more north, Billie."

"A little more north, sir," said the oiler.

Whereupon the little boat turned her nose once more down the wind, and all but the oarsman watched the shore grow. Under the influence of this expansion doubt and direful apprehension were leaving the minds of the men. The management of the boat was still most absorbing, but it could not prevent a quiet cheerfulness. In an hour, perhaps, they would be ashore.

Their backbones had become thoroughly used to balancing in the boat, and they now rode this wild colt of a dinghy like circus men. The correspondent thought that he had been drenched to the skin, but happening to feel in the top pocket of his coat, he found therein eight cigars. Four of them were soaked with sea-water; four were perfectly scatheless. After a search, somebody produced three dry matches; and thereupon the four waifs rode impudently in their little boat and, with an assurance of an impending rescue shining in their eyes, puffed at the big cigars, and judged well and ill of all men. Everybody took a drink of water.

IV

"Cook," remarked the captain, "there don't seem to be any signs of life about your house of refuge."

"No," replied the cook. "Funny they don't see us!"

A broad stretch of lowly coast lay before the eyes of the men. It was of low dunes topped with dark vegetation. The roar of the surf was plain, and sometimes they could see the white lip of a wave as it spun up the beach. A tiny house was blocked out black upon the sky. Southward, the slim lighthouse lifted its little gray length.

Tide, wind, and waves were swinging the dinghy northward. "Funny they don't see us," said the men.

The surf's roar was here dulled, but its tone was nevertheless thunderous and mighty. As the boat swam over the great rollers the men sat listening to this roar. "We'll swamp sure," said everybody.

It is fair to say here that there was not a life-saving station within twenty miles in either direction; but the men did not know this fact, and in consequence they made dark and opprobrious remarks concerning the eyesight of the nation's life-savers. Four scowling men sat in the dinghy and surpassed records in the invention of epithets.

"Funny they don't see us."

The light-heartedness of a former time had completely faded. To their sharpened minds it was easy to conjure pictures of all kinds of incompetency and blindness and, indeed, cowardice. There was the shore of the populous land, and it was bitter and bitter to them that from it came no sign.

"Well," said the captain, ultimately, "I suppose we'll have to make a try for ourselves. If we stay out here too long, we'll none of us have strength left to swim after the boat swamps."

And so the oiler, who was at the oars, turned the boat straight for the shore. There was a sudden tightening of muscles. There was some thinking.

"If we don't all get ashore," said the captain—"if we don't all get ashore, I suppose you fellows know where to send news of my finish?"

They then briefly exchanged some addresses and admonitions. As for the reflections of the men, there was a great deal of rage in them. Perchance they might be formulated thus: "If I am going to be drowned—if I am going to be drowned—if I am going to be drowned, why, in the name of the seven mad gods who rule the sea, was I allowed to come thus far and contemplate sand and trees? Was I brought here merely to have my nose dragged away as I was about to nibble the sacred cheese of life? It is preposterous. If this old ninny-woman, Fate, cannot do better than this, she should be deprived of the management of men's fortunes. She is an old hen who knows not her intention. If she has decided to drown me, why did she not do it in the beginning and save me all this trouble? The whole affair is absurd. . . . But no; she cannot mean to drown me. She dare not drown me. She cannot drown me. Not after all this work." Afterward the man might have had an impulse to shake his fist at the clouds. "Just you drown me, now, and then hear what I call you!"

The billows that came at this time were more formidable. They seemed always just about to break and roll over the little boat in a turmoil of foam. There was a preparatory and long growl in the speech of them. No mind unused to the sea would have concluded that the dinghy could ascend these sheer heights in time. The shore was still afar. The oiler was a wily surfman. "Boys," he said swiftly, "she won't live three minutes more, and we're too far out to swim. Shall I take her to sea again, Captain?"

"Yes; go ahead!" said the captain.

This oiler, by a series of quick miracles and fast and steady oarsmanship, turned the boat in the middle of the surf and took her safely to sea again.

There was a considerable silence as the boat bumped over the furrowed sea to deeper water.

Then somebody in gloom spoke: "Well, anyhow, they must have seen us from the shore by now."

The gulls went in slanting flight up the wind toward the gray, desolate east. A squall, marked by dingy clouds and clouds brick-red, like smoke from a burning building, appeared from the southeast.

"What do you think of those life-saving people? Ain't they peaches?"

"Funny they haven't seen us."

"Maybe they think we're out for sport! Maybe they think we're fishin'. Maybe they think we're damned fools."

It was a long afternoon. A changed tide tried to force them southward, but wind and wave said northward. Far ahead, where coast-line, sea, and sky formed their mighty angle, there were little dots which seemed to indicate a city on the shore.

"St. Augustine?"

The captain shook his head. "Too near Mosquito Inlet."

And the oiler rowed, and then the correspondent rowed; then the oiler rowed. It was a weary business. The human back can become the seat of more aches and pains than are registered in books for the composite anatomy of a regiment. It is a limited area, but it can become the theater of innumerable muscular conflicts, tangles, wrenches, knots, and other comforts.

"Did you ever like to row, Billie?" asked the correspondent.

"No," said the oiler; "hang it!"

When one exchanged the rowing-seat for a place in the bottom of the boat, he suffered a bodily depression that caused him to be careless of everything save an obligation to wiggle one finger. There was cold sea-water swashing to and fro in the boat, and he lay in it. His head, pillowed on a thwart, was within an inch of the swirl of a wave-crest, and sometimes a particularly obstreperous sea came inboard and drenched him once more. But these matters did not annoy him. It is almost certain that if the boat had capsized he would have tumbled comfortably out upon the ocean as if he felt sure that it was a great soft mattress.

"Look! There's a man on the shore!"

"Where?"

"There! See 'im? See 'im?"

"Yes, sure! He's walking along."

"Now he's stopped. Look! He's facing us!"

"He's waving at us!"

"So he is! By thunder!"

"Ah, now we're all right! Now we're all right! There'll be a boat out here for us in half an hour."

"He's going on. He's running. He's going up to that house there."

The remote beach seemed lower than the sea, and it required a searching glance to discern the little black figure. The captain saw a floating stick, and they rowed to it. A bath towel was by some weird chance in the boat, and, tying this on the stick, the captain waved it. The oarsman did not dare turn his head, so he was obliged to ask questions.

"What's he doing now?"

"He's standing still again. He's looking, I think. . . . There he goes again—toward the house. . . . Now he's stopped again."

"Is he waving at us?"

"No, not now; he was, though."

"Look! There comes another man!"

"He's running."

"Look at him go, would you!"

"Why, he's on a bicycle. Now he's met the other man. They're both waving at us. Look!"

"There comes something up the beach."

"What the devil is that thing?"

"Why, it looks like a boat."

"Why, certainly, it's a boat."

"No; it's on wheels."

"Yes, so it is. Well, that must be the life-boat. They drag them along shore on a wagon."

"That's the life-boat, sure."

"No, by God, it's—it's an omnibus."

"I tell you it's a life-boat."

"It is not! It's an omnibus. I can see it plain. See? One of these big hotel omnibuses."

"By thunder, you're right. It's an omnibus,

sure as fate. What do you suppose they are doing with an omnibus? Maybe they are going around collecting the life-crew, hey?"

"That's it, likely. Look! There's a fellow waving a little black flag. He's standing on the steps of the omnibus. There come those other two fellows. Now they're all talking together. Look at the fellow with the flag. Maybe he ain't waving it!"

"That ain't a flag, is it? That's his coat. Why, certainly, that's his coat."

"So it is; it's his coat. He's taken it off and is waving it around his head. But would you look at him swing it!"

"Oh, say, there isn't any life-saving station there. That's just a winter-resort hotel omnibus that has brought over some of the boarders to see us drown."

"What's that idiot with the coat mean? What's he signalling, anyhow?"

"It looks as if he were trying to tell us to go north. There must be a life-saving station up there."

"No; he thinks we're fishing. Just giving a merry hand. See? Ah, there, Willie!"

"Well, I wish I could make something out of those signals. What do you suppose he means?"

"He don't mean anything; he's just playing."

"Well, if he'd just signal us to try the surf again, or to go to sea and wait, or go north, or go south, or go to hell, there would be some reason in it. But look at him! He just stands there and keeps his coat revolving like a wheel. The ass!"

"There come more people."

"Now there's quite a mob. Look! Isn't that a boat?"

"Where? Oh, I see where you mean. No, that's no boat."

"That fellow is still waving his coat."

"He must think we like to see him do that. Why don't he quit it? It don't mean anything."

"I don't know. I think he is trying to make us go north. It must be that there's a life-saving station there somewhere."

"Say, he ain't tired yet. Look at 'im wave!"

"Wonder how long he can keep that up. He's been revolving his coat ever since he caught sight of us. He's an idiot. Why aren't they getting men to bring a boat out? A fishing-boat—one of those big yawls—could come out here all right. Why don't he do something?"

"Oh, it's all right now."

"They'll have a boat out here for us in less than no time, now that they've seen us."

A faint yellow tone came into the sky over the low land. The shadows on the sea slowly deepened. The wind bore coldness with it, and the men began to shiver.

"Holy smoke!" said one, allowing his voice to express his impious mood, "if we keep on monkeying out here! If we've got to flounder out here all night!"

"Oh, we'll never have to stay here all night! Don't you worry. They've seen us now, and it won't be long before they'll come chasing out after us."

The shore grew dusky. The man waving a coat blended gradually into this gloom, and it swallowed in the same manner the omnibus and the group of people. The spray, when it dashed uproariously over the side, made the voyagers shrink and swear like men who were being branded.

"I'd like to catch the chump who waved the coat. I feel like socking him one, just for luck."

"Why? What did he do?"

"Oh, nothing, but then he seemed so damned cheerful."

In the meantime, the oiler rowed, and then the correspondent rowed, and then the oiler rowed. Gray-faced and bowed forward, they mechanically, turn by turn, plied the leaden oars. The form of the lighthouse had vanished from the southern horizon, but finally a pale star appeared, just lifting from the sea. The streaked saffron in the west passed before the all-merging darkness, and the sea to the east was black. The land had vanished, and was expressed only by the low and drear thunder of the surf.

"If I am going to be drowned—if I am going to be drowned— if I am going to be drowned, why, in the name of the seven mad gods who rule the sea, was I allowed to come thus far and contemplate sand and trees? Was I brought here merely to have my nose dragged away as I was about to nibble the sacred cheese of life?"

The patient captain, drooped over the water-jar, was sometimes obliged to speak to the oarsman.

"Keep her head up! Keep her head up!"

"Keep her head up, sir." The voices were weary and low.

This was surely a quiet evening. All save the oarsman lay heavily and listlessly in the boat's bottom. As for him, his eyes were just capable of noting the tall black waves that swept forward in a most sinister silence, save for an occasional subdued growl of a crest.

The cook's head was on a thwart, and he looked without interest at the water under his nose. He was deep in other scenes. Finally he spoke. "Billie," he murmured, dreamfully, "what kind of pie do you like best?"

V

"Pie!" said the oiler and the correspondent, agitatedly. "Don't talk about those things, blast you!"

"Well," said the cook, "I was just thinking about ham sandwiches, and—"

A night on the sea in an open boat is a long night. As darkness settled finally, the shine of the light, lifting from the sea in the south, changed to full gold. On the northern horizon a new light appeared, a small bluish gleam on the edge of the waters. These two lights were the furniture of the world! Otherwise there was nothing but waves.

Two men huddled in the stern, and distances were so magnificent in the dinghy that the rower was enabled to keep his feet partly warm by thrusting them under his companions. Their legs indeed extended far under the rowing-seat until they touched the feet of the captain forward. Sometimes, despite the efforts of the tired oarsman, a wave came piling into the boat, an icy wave of the night, and the chilling water soaked them anew. They would twist their bodies for a moment and groan, and sleep the dead sleep once more, while the water in the boat gurgled about them as the craft rocked.

The plan of the oiler and the correspondent was for one to row until he lost the ability, and then arouse the other from his sea-water couch in the bottom of the boat.

The oiler plied the oars until his head dropped forward and the overpowering sleep blinded him; and he rowed yet afterward. Then he touched a man in the bottom of the boat, and called his name. "Will you spell me for a little while?" he said meekly.

"Sure, Billie," said the correspondent, awaking and dragging himself to a sitting position. They exchanged places carefully, and the oiler, cuddling down in the sea-water at the cook's side, seemed to go to sleep instantly.

The particular violence of the sea had ceased. The waves came without snarling. The obligation of the man at the oars was to keep the boat headed so that the tilt of the rollers would not capsize her, and to preserve her from filling when the crests rushed past. The black waves were silent and hard to be seen in the darkness. Often one was almost upon the boat before the oarsman was aware.

In a low voice the correspondent addressed the captain. He was not sure that the captain was awake, although this iron man seemed to be always awake. "Captain, shall I keep her making for that light north, sir?"

The same steady voice answered him. "Yes. Keep it about two points off the port bow."

The cook had tied a life-belt around himself in order to get even the warmth which this clumsy cork contrivance could donate, and he seemed almost stove-like when a rower, whose teeth invariably chattered wildly as soon as he ceased his labor, dropped down to sleep.

The correspondent, as he rowed, looked down at the two men sleeping underfoot. The cook's arm was around the oiler's shoulders, and, with their fragmentary clothing and haggard faces, they were the babes of the sea—a grotesque rendering of the old babes in the wood.

Later he must have grown stupid at his work, for suddenly there was a growling of water, and a crest came with a roar and a swash into the boat, and it was a wonder that it did not set the cook afloat in his life-belt. The cook continued to sleep, but the oiler sat up, blinking his eyes and shaking with the new cold.

"Oh, I'm awful sorry, Billie," said the correspondent, contritely.

"That's all right, old boy," said the oiler, and lay down again and was asleep.

Presently it seemed that even the captain dozed, and the correspondent thought that he was the one man afloat on all the ocean. The wind had a voice as it came over the waves, and it was sadder than the end.

There was a long, loud swishing astern of the boat, and a gleaming trail of phosphorescence, like blue flame, was furrowed on the black waters. It might have been made by a monstrous knife.

Then there came a stillness, while the correspondent breathed with open mouth and looked at the sea.

Suddenly there was another swish and another long flash of bluish light, and this time it was alongside the boat, and might almost have been reached with an oar. The correspondent saw an enormous fin speed like a shadow through the water, hurling the crystalline spray and leaving the long glowing trail.

The correspondent looked over his shoulder at the captain. His face was hidden, and he seemed to be asleep. He looked at the babes of the sea. They certainly were asleep. So, being bereft of sympathy, he leaned a little way to one side and swore softly into the sea.

But the thing did not then leave the vicinity of the boat. Ahead or astern, on one side or the other, at intervals long or short, fled the long sparkling streak, and there was to be heard the *whirroo* of the dark fin. The speed and power of the thing was greatly to be admired. It cut

the water like a gigantic and keen projectile.

The presence of this biding thing did not affect the man with the same horror that it would if he had been a picnicker. He simply looked at the sea dully and swore in an undertone.

Nevertheless, it is true that he did not wish to be alone with the thing. He wished one of his companions to awake by chance and keep him company with it. But the captain hung motionless over the water-jar, and the oiler and the cook in the bottom of the boat were plunged in slumber.

VI

"If I am going to be drowned—if I am going to be drowned—if I am going to be drowned, why, in the name of the seven mad gods who rule the sea, was I allowed to come thus far and contemplate sand and trees?"

During this dismal night, it may be remarked that a man would conclude that it was really the intention of the seven mad gods to drown him, despite the abominable injustice of it. For

Winslow Homer, *The Gulf Stream*, 1899
"[He] thought that he was the one man afloat on all the ocean. . . . Ahead or astern, on one side or the other . . . was to be heard the whirroo *of the dark fin."*—Crane

it was certainly an abominable injustice to drown a man who had worked so hard, so hard. The man felt it would be a crime most unnatural. Other people had drowned at sea since galleys swarmed with painted sails, but still—

When it occurs to a man that nature does not regard him as important, and that she feels she would not maim the universe by disposing of him, he at first wishes to throw bricks at the temple, and he hates deeply the fact that there are no bricks and no temples. Any visible expression of nature would surely be pelleted with his jeers.

Then, if there be no tangible thing to hoot, he feels, perhaps, the desire to confront a personification and indulge in pleas, bowed to one knee,

40

and with hands supplicant, saying, "Yes, but I love myself."

A high cold star on a winter's night is the word he feels that she says to him. Thereafter he knows the pathos of his situation.

The men in the dinghy had not discussed these matters, but each had, no doubt, reflected upon them in silence and according to his mind. There was seldom any expression upon their faces save the general one of complete weariness. Speech was devoted to the business of the boat.

To chime the notes of his emotion, a verse mysteriously entered the correspondent's head. He had even forgotten that he had forgotten this verse, but it suddenly was in his mind.

A soldier of the Legion lay dying in Algiers;
There was lack of woman's nursing, there was dearth of woman's tears;
But a comrade stood beside him, and he took that comrade's hand,
And he said, "I never more shall see my own, my native land."

In his childhood the correspondent had been made acquainted with the fact that a soldier of the Legion lay dying in Algiers, but he had never regarded it as important. Myriads of his schoolfellows had informed him of the soldier's plight, but the dinning had naturally ended by making him perfectly indifferent. He had never considered it his affair that a soldier of the Legion lay dying in Algiers, nor had it appeared to him as a matter for sorrow. It was less to him than the breaking of a pencil's point.

Now, however, it quaintly came to him as a human, living thing. It was no longer merely a picture of a few throes in the breast of a poet, meanwhile drinking tea and warming his feet at the grate; it was an actuality—stern, mournful, and fine.

The correspondent plainly saw the soldier. He lay on the sand with his feet out straight and still. While his pale left hand was upon his chest in an attempt to thwart the going of his life, the blood came between his fingers. In the far Algerian distance, a city of low square forms was set against a sky that was faint with the last sunset hues. The correspondent, plying the oars and dreaming of the slow and slower movements of the lips of the soldier, was moved by a profound and perfectly impersonal comprehension. He was sorry for the soldier of the Legion who lay dying in Algiers.

The thing which had followed the boat and waited had evidently grown bored at the delay. There was no longer to be heard the slash of the cutwater, and there was no longer the flame of the long trail. The light in the north still glimmered, but it was apparently no nearer to the boat. Sometimes the boom of the surf rang in the correspondent's ears, and he turned the craft seaward then and rowed harder. Southward, some one had evidently built a watchfire on the beach. It was too low and too far to be seen, but it made a shimmering, roseate reflection upon the bluff in back of it, and this could be discerned from the boat. The wind came stronger, and sometimes a wave suddenly raged out like a mountain-cat, and there was to be seen the sheen and sparkle of a broken crest.

The captain, in the bow, moved on his waterjar and sat erect. "Pretty long night," he observed to the correspondent. He looked at the shore. "Those life-saving people take their time."

"Did you see that shark playing around?"

"Yes, I saw him. He was a big fellow, all right."

"Wish I had known you were awake."

Later the correspondent spoke into the bottom of the boat. "Billie!" There was a slow and gradual disentanglement. "Billie, will you spell me?"

"Sure," said the oiler.

As soon as the correspondent touched the cold, comfortable sea-water in the bottom of the boat and had huddled close to the cook's life-belt he was deep in sleep, despite the fact that his teeth played all the popular airs. This sleep was so good to him that it was but a moment before he heard a voice call his name in a tone that demonstrated the last stages of exhaustion. "Will you spell me?"

41

"Sure, Billie."

The light in the north had mysteriously vanished, but the correspondent took his course from the wide-awake captain.

Later in the night they took the boat farther out to sea, and the captain directed the cook to take one oar at the stern and keep the boat facing the seas. He was to call out if he should hear the thunder of the surf. This plan enabled the oiler and the correspondent to get respite together. "We'll give those boys a chance to get into shape again," said the captain. They curled down and, after a few preliminary chatterings and trembles, slept once more the dead sleep. Neither knew they had bequeathed to the cook the company of another shark, or perhaps the same shark.

As the boat caroused on the waves, spray occasionally bumped over the side and gave them a fresh soaking, but this had no power to break their repose. The ominous slash of the wind and the water affected them as it would have affected mummies.

"Boys," said the cook, with the notes of every reluctance in his voice, "she's drifted in pretty close. I guess one of you had better take her to sea again." The correspondent, aroused, heard the crash of the toppled crests.

As he was rowing, the captain gave him some whiskey-and-water, and this steadied the chills out of him. "If I ever get ashore and anybody shows me even a photograph of an oar—"

At last there was a short conversation.

"Billie! . . . Billie, will you spell me?"

"Sure," said the oiler.

VII

When the correspondent again opened his eyes, the sea and the sky were each of the gray hue of the dawning. Later, carmine and gold was painted upon the waters. The morning appeared finally, in its splendor, with a sky of pure blue, and the sunlight flamed on the tips of the waves.

On the distant dunes were set many little black cottages, and a tall white windmill reared above them. No man, nor dog, nor bicycle appeared on the beach. The cottages might have formed a deserted village.

The voyagers scanned the shore. A conference was held in the boat. "Well," said the captain, "if no help is coming, we might better try a run through the surf right away. If we stay out here much longer we will be too weak to do anything for ourselves at all." The others silently acquiesced in this reasoning. The boat was headed for the beach. The correspondent wondered if none ever ascended the tall wind-tower, and if then they never looked seaward. This tower was a giant, standing with its back to the plight of the ants. It represented in a degree, to the correspondent, the serenity of nature amid the struggles of the individual—nature in the wind, and nature in the vision of men. She did not seem cruel to him then, nor beneficent, nor treacherous, nor wise. But she was indifferent, flatly indifferent. It is, perhaps, plausible that a man in this situation, impressed with the unconcern of the universe, should see the innumerable flaws of his life, and have them taste wickedly in his mind, and wish for another chance. A distinction between right and wrong seems absurdly clear to him, then, in this new ignorance of the grave-edge, and he understands that if he were given another opportunity he would mend his conduct and his words, and be better and brighter during an introduction or at a tea.

"Now, boys," said the captain, "she is going to swamp sure. All we can do is to work her in as far as possible, and then when she swamps, pile out and scramble for the beach. Keep cool now, and don't jump until she swamps sure."

The oiler took the oars. Over his shoulders he scanned the surf. "Captain," he said, "I think I'd better bring her about and keep her head-on to the seas and back her in."

"All right, Billie," said the captain. "Back her in." The oiler swung the boat then, and, seated in the stern, the cook and the correspondent were obliged to look over their shoulders to contemplate the lonely and indifferent shore.

The monstrous inshore rollers heaved the boat high until the men were again enabled to

42

see the white sheets of water scudding up the slanted beach. "We won't get in very close," said the captain. Each time a man could wrest his attention from the rollers, he turned his glance toward the shore, and in the expression of the eyes during this contemplation there was a singular quality. The correspondent, observing the others, knew that they were not afraid, but the full meaning of their glances was shrouded.

As for himself, he was too tired to grapple fundamentally with the fact. He tried to coerce his mind into thinking of it, but the mind was dominated at this time by the muscles, and the muscles said they did not care. It merely occurred to him that if he should drown it would be a shame.

There were no hurried words, no pallor, no plain agitation. The men simply looked at the shore. "Now, remember to get well clear of the boat when you jump," said the captain.

Seaward the crest of a roller suddenly fell with a thunderous crash, and the long white comber came roaring down upon the boat.

"Steady now," said the captain. The men were silent. They turned their eyes from the shore to the comber and waited. The boat slid up the incline, leaped at the furious top, bounced over it, and swung down the long back of the wave. Some water had been shipped, and the cook bailed it out.

But the next crest crashed also. The tumbling, boiling flood of white water caught the boat and whirled it almost perpendicular. Water swarmed in from all sides. The correspondent had his hands on the gunwale at this time, and when the water entered at that place he swiftly withdrew his fingers, as if he objected to wetting them.

The little boat, drunken with this weight of water, reeled and snuggled deeper into the sea.

"Bail her out, cook! Bail her out!" said the captain.

"All right, Captain," said the cook.

"Now, boys, the next one will do for us sure," said the oiler. "Mind to jump clear of the boat."

The third wave moved forward, huge, furi-

43

ous, implacable. It fairly swallowed the dinghy, and almost simultaneously the men tumbled into the sea. A piece of life-belt had lain in the bottom of the boat, and as the correspondent went overboard he held this to his chest with his left hand.

The January water was icy, and reflected immediately that it was colder than he had expected to find it off the coast of Florida. This appeared to his dazed mind as a fact important enough to be noted at the time. The coldness of the water was sad; it was tragic. This fact was somehow mixed and confused with his opinion of his own situation, so that it seemed almost a proper reason for tears. The water was cold.

When he came to the surface he was conscious of little but the noisy water. Afterward he saw his companions in the sea. The oiler was ahead in the race. He was swimming strongly and rapidly. Off to the correspondent's left, the cook's great white and corked back bulged out of the water; and in the rear the captain was hanging with his one good hand to the keel of the overturned dinghy.

There is a certain immovable quality to a shore, and the correspondent wondered at it amid the confusion of the sea.

It seemed also very attractive; but the correspondent knew that it was a long journey, and he paddled leisurely. The piece of life-preserver lay under him, and sometimes he whirled down the incline of a wave as if he were on a hand-sled.

But finally he arrived at a place in the sea where travel was beset with difficulty. He did not pause swimming to inquire what manner of current had caught him, but there his progress ceased. The shore was set before him like a bit of scenery on a stage, and he looked at it and understood with his eyes each detail of it.

As the cook passed, much farther to the left, the captain was calling to him, "Turn over on your back, cook! Turn over on your back and use the oar."

"All right, sir." The cook turned on his back, and, paddling with an oar, went ahead as if he were a canoe.

Presently the boat also passed to the left of

the correspondent, with the captain clinging with one hand to the keel. He would have appeared like a man raising himself to look over a board fence if it were not for the extraordinary gymnastics of the boat. The correspondent marvelled that the captain could still hold to it.

They passed on nearer to shore—the oiler, the cook, the captain—and following them went the water-jar, bouncing gaily over the seas.

The correspondent remained in the grip of this strange new enemy, a current. The shore, with its white slope of sand and its green bluff topped with little silent cottages, was spread like a picture before him. It was very near to him then, but he was impressed as one who, in a gallery, looks at a scene from Brittany or Algiers.

He thought: "I am going to drown? Can it be possible? Can it be possible? Can it be possible?" Perhaps an individual must consider his own death to be the final phenomenon of nature.

But later a wave perhaps whirled him out of this small deadly current, for he found suddenly that he could again make progress toward the shore. Later still he was aware that the captain, clinging with one hand to the keel of the dinghy, had his face turned away from the shore and toward him, and was calling his name. "Come to the boat! Come to the boat!"

In his struggle to reach the captain and the boat, he reflected that when one gets properly wearied drowning must really be a comfortable arrangement—a cessation of hostilities accompanied by a large degree of relief; and he was glad of it, for the main thing in his mind for some moments had been horror of the temporary agony; he did not wish to be hurt.

Presently he saw a man running along the shore. He was undressing with most remarkable speed. Coat, trousers, shirt, everything flew magically off him.

"Come to the boat!" called the captain.

"All right, Captain." As the correspondent paddled, he saw the captain let himself down to bottom and leave the boat. Then the correspondent performed his one little marvel of the voyage. A large wave caught him and flung him with ease and supreme speed completely over the boat and far beyond it. It struck him even then as an event in gymnastics and a true miracle of the sea. An overturned boat in the surf is not a plaything to a swimming man.

The correspondent arrived in water that reached only to his waist, but his condition did not enable him to stand for more than a moment. Each wave knocked him into a heap, and the undertow pulled at him.

Then he saw the man who had been running and undressing, and undressing and running, come bounding into the water. He dragged ashore the cook, and then waded toward the captain; but the captain waved him away and sent him to the correspondent. He was naked—naked as a tree in winter; but a halo was about his head, and he shone like a saint. He gave a strong pull, and a long drag, and a bully heave at the correspondent's hand. The correspondent, schooled in the minor formulae, said, "Thanks, old man." But suddenly the man cried, "What's that?" He pointed a swift finger. The correspondent said, "Go."

In the shallows, face downward, lay the oiler. His forehead touched sand that was periodically, between each wave, clear of the sea.

The correspondent did not know all that transpired afterward. When he achieved safe ground he fell, striking the sand with each particular part of his body. It was as if he had dropped from a roof, but the thud was grateful to him.

It seems that instantly the beach was populated with men with blankets, clothes, and flasks, and women with coffee-pots and all the remedies sacred to their minds. The welcome of the land to the men from the sea was warm and generous; but a still and dripping shape was carried slowly up the beach, and the land's welcome for it could only be the different and sinister hospitality of the grave.

When it came night, the white waves paced to and fro in the moonlight, and the wind brought the sound of the great sea's voice to the men on the shore, and they felt that they could then be interpreters.

Moby-Dick

Herman Melville

A blend of whaling facts, fiction, and symbolism, Herman Melville's story of Captain Ahab's pursuit of the white whale is now regarded as one of the classics of world literature. Critics have long disagreed as to the ultimate meaning of the novel, some regarding it as the struggle between good and evil, others as man's struggle to find meaning and order in an essentially chaotic and uncontrollable universe. Melville's own voyages aboard whalers from 1841 to 1844 provided the rich details for the book, first published in 1851. The loneliness of the sailor's life and of the sea itself—interestingly portrayed as masculine—permeate the novel. As Melville wrote in the opening chapter, the sea represented "the ungraspable phantom of life."

The Symphony

*I*t was a clear steel-blue day. The firmaments of air and sea were hardly separable in that all-pervading azure; only, the pensive air was transparently pure and soft, with a woman's look, and the robust and man-like sea heaved with long, strong, lingering swells, as Samson's chest in his sleep.

Hither, and thither, on high, glided the snow-white wings of small, unspeckled birds; these were the gentle thoughts of the feminine air; but to and fro in the deeps, far down in the bottomless blue, rushed mighty leviathans, sword-fish, and sharks; and these were the strong, troubled, murderous thinkings of the masculine sea.

But though thus contrasting within, the contrast was only in shades and shadows without; those two seemed one; it was only the sex, as it were, that distinguished them.

Aloft, like a royal czar and king, the sun seemed giving this gentle air to this bold and rolling sea; even as bride to groom. And at the girdling line of the horizon, a soft and tremulous motion—most seen here at the equator—denoted the fond, throbbing trust, the loving alarms, with which the poor bride gave her bosom away.

Tied up and twisted; gnarled and knotted with wrinkles; haggardly firm and unyielding; his eyes glowing like coals, that still glow in the ashes of ruin; untottering Ahab stood forth in the clearness of the morn; lifting his splintered helmet of a brow to the fair girl's forehead of heaven.

Oh, immortal infancy, and innocency of the azure! Invisible winged creatures that frolic all round us! Sweet childhood of air and sky! how oblivious were ye of old Ahab's close-coiled woe! But so have I seen little Miriam and Martha, laughing-eyed elves, heedlessly gambol around their old sire; sporting with the circle of singed locks which grew on the marge of that burnt-out crater of his brain.

Slowly crossing the deck from the scuttle, Ahab leaned over the side, and watched how his shadow in the water sank and sank to his gaze, the more and the more that he strove to pierce the profundity. But the lovely aromas in that enchanted air did at last seem to dispel, for a moment, the cankerous thing in his soul. That glad, happy air, that winsome sky, did at last stroke and caress him; the step-mother world, so long cruel—forbidding—now threw affectionate arms round his stubborn neck, and did seem to joyously sob over him, as if over one, that however wilful and erring, she could yet find it in her heart to save and to bless. From beneath his slouched hat Ahab dropped a tear into the sea; nor did all the Pacific contain such wealth as that one wee drop.

Starbuck saw the old man; saw him, how he heavily leaned over the side; and he seemed to hear in his own true heart the measureless sobbing that stole out of the centre of the serenity around. Careful not to touch him, or be noticed by him, he yet drew near to him, and stood there.

Ahab turned.

"Starbuck!"

"Sir."

"Oh, Starbuck! it is a mild, mild wind, and a mild looking sky. On such a day—very much such a sweetness as this—I struck my first whale—a boy-harpooneer of eighteen! Forty—forty—forty years ago!—ago! Forty years of continual whaling! forty years of privation, and peril, and storm-time! forty years on the pitiless sea! for forty years has Ahab forsaken the peaceful land, for forty years to make war on the horrors of the deep! Aye and yes, Starbuck, out of those forty years I have not spent three ashore. When I think of this life I have led; the desolation of solitude it has been; the masoned, walled-town of a Captain's exclusiveness, which admits but small entrance to any sympathy from the green country without—oh, weariness! heaviness! Guinea-coast slavery of solitary command!—when I think of all this; only half-suspected, not so keenly known to me before—and how for forty years I have fed upon dry salted fare—fit emblem of the dry nourishment of my soul!—when the poorest landsman has had fresh fruit to his daily hand, and broken the world's fresh bread to my mouldy crusts—away, whole oceans away, from that young girl-wife I wedded past fifty, and sailed for Cape Horn the next day, leaving but one dent in my marriage pillow— wife? wife?—rather a widow with her husband alive! Aye, I widowed that poor girl when I married her, Starbuck; and then, the madness, the frenzy, the boiling blood and the smoking brow, with which, for a thousand lowerings old Ahab has furiously, foamingly chased his prey—more a demon than a man!—aye, aye! what a forty years' fool—fool—old fool, has old Ahab been! Why this strife of the chase? why weary, and palsy the arm at the oar, and the iron, and the lance? how the richer or better is Ahab now? Behold. Oh, Starbuck! is it not hard, that with this weary load I bear, one poor leg should have been snatched from under me? Here, brush this old hair aside; it blinds me, that I seem to weep. Locks so grey did never grow but from out some ashes! But do I look very old, so very, very old, Starbuck? I feel deadly faint, bowed, and humped, as though I were Adam, staggering beneath the piled centuries since Paradise. God! God! God!—crack my heart!—stave my brain!—mockery! mockery! bitter, biting mockery of grey hairs, have I lived enough joy to wear ye; and seem and feel thus intolerably old? Close! stand close to me, Starbuck; let me look into a human eye; it is better than to gaze into sea or sky; better than to gaze upon God. By the green land; by the bright hearth-stone! this is the magic glass, man; I see my wife and my child in thine eye. No, no; stay on board, on board!—lower not when I do; when branded Ahab gives chase to Moby Dick. That hazard shall not be thine. No, no! not with the far away home I see in that eye!"

"Oh, my Captain! my Captain! noble soul! grand old heart, after all! why should any one give chase to that hated fish! Away with me! let us fly these deadly waters! let us home! Wife and child, too, are Starbuck's—wife and child of his brotherly, sisterly, play-fellow youth; even as thine, sir, are the wife and child of thy loving, longing, paternal old age! Away! let us away!—this instant let me alter the course! How cheerily, how hilariously, O my Captain, would we bowl on our way to see old Nantucket again! I think, sir, they have some such mild blue days, even as this, in Nantucket."

"They have, they have. I have seen them—some summer days in the morning. About this time—yes, it is his noon nap now—the boy vivaciously wakes; sits up in bed; and his mother tells him of me, of cannibal old me; how I am abroad upon the deep, but will yet come back to dance him again."

"'Tis my Mary, my Mary herself! She promised that my boy, every morning, should be carried to the hill to catch the first glimpse of his father's sail! Yes, yes! no more! it is done! we head for Nantucket! Come, my Captain, study out the course, and let us away! See, see! the boy's face from the window! the boy's hand on the hill!"

But Ahab's glance was averted; like a blighted fruit tree he shook, and cast his last, cindered apple to the soil.

"What is it, what nameless, inscrutable, unearthly thing is it; what cozening, hidden lord and master, and cruel, remorseless emperor commands me; that against all natural lovings and longings, I so keep pushing, and crowding, and jamming myself on all the time; recklessly making me ready to do what in my own proper, natural heart, I durst not so much as dare? Is Ahab, Ahab? Is it I, God, or who, that lifts this arm? But if the great sun move not of himself; but is as an errand-boy in heaven; nor one single star can revolve, but by some invisible power; how then can this one small heart beat; this one small brain think thoughts; unless God does that beating, does that thinking, does that living, and not I. By heaven, man, we are turned round and round in this world, like yonder windlass, and Fate is the handspike. And all the time, lo! that smiling sky, and this unsounded sea! Look! see yon Albicore! who put it into him to chase and fang that flying-fish? Where do murderers go, man! Who's to doom, when the judge himself is dragged to the bar? But it is a mild, mild wind, and a mild looking sky; and the airs smells now, as if it blew from a far-away meadow; they have been making hay somewhere under the slopes of the Andes, Starbuck, and the mowers are sleeping among the new-mown hay. Sleeping? Aye, toil we how we may, we all sleep at last on the field. Sleep? Aye, and rust amid greenness; as last year's scythes flung down, and left in the half-cut swaths—Starbuck!"

But blanched to a corpse's hue with despair, the Mate had stolen away.

Ahab crossed the deck to gaze over on the other side; but started at two reflected, fixed eyes in the water there. Fedallah was motionlessly leaning over the same rail.

The Chase—First Day

That night, in the mid-watch, when the old man—as his wont at intervals—stepped forth from the scuttle in which he leaned, and went to his pivot-hole, he suddenly thrust out his face fiercely, snuffing up the sea air as a sagacious ship's dog will, in drawing nigh to some barbarous isle. He declared that a whale must be near. Soon that peculiar odor, sometimes to a great distance given forth by the living sperm whale, was palpable to all the watch; nor was any mariner surprised when, after inspecting the compass, and then the dog-vane,[1] and then ascertaining the precise bearing of the odor as nearly as possible, Ahab rapidly ordered the ship's course to be slightly altered, and the sail to be shortened.

The acute policy dictating these movements was sufficiently vindicated at daybreak, by the sight of a long sleek on the sea directly and lengthwise ahead, smooth as oil, and resembling in the pleated watery wrinkles bordering it, the polished metallic-like marks of some swift tide-rip, at the mouth of a deep, rapid stream.

"Man the mast-heads! Call all hands!"

Thundering with the butts of three clubbed handspikes on the forecastle deck, Daggoo roused the sleepers with such judgment claps that they seemed to exhale from the scuttle, so instantaneously did they appear with their clothes in their hands.

"What d'ye see?" cried Ahab, flattening his face to the sky.

"Nothing, nothing, sir!" was the sound hailing down in reply.

"T'gallant sails—stunsails! alow and aloft, and on both sides!"

All sail being set, he now cast loose the life-line, reserved for swaying him to the main royal-mast head; and in a few moments they were hoisting him thither, when, while but two thirds of the way aloft, and while peering ahead

---

[1] A small vane placed on the weather gunwale to show the direction of the wind.

through the horizontal vacancy between the main-top-sail and top-gallant-sail, he raised a gull-like cry in the air, "There she blows!—there she blows! A hump like a snow-hill! It is Moby Dick!"

Fired by the cry which seemed simultaneously taken up by the three look-outs, the men on deck rushed to the rigging to behold the famous whale they had so long been pursuing. Ahab had now gained his final perch, some feet above the other look-outs, Tashtego standing just beneath him on the cap of the top-gallant-mast, so that the Indian's head was almost on a level with Ahab's heel. From this height the whale was now seen some mile or so ahead, at every roll of the sea revealing his high sparkling hump, and regularly jetting his silent spout into the air. To the credulous mariners it seemed the same silent spout they had so long ago beheld in the moonlit Atlantic and Indian oceans.

"And did none of ye see it before?" cried Ahab, hailing the perched men all around him.

"I saw him almost that same instant, sir, that Captain Ahab did, and I cried out," said Tashtego.

"Not the same instant; not the same—no, the doubloon is mine, Fate reserved the doubloon for me. *I* only; none of ye could have raised the White Whale first. There she blows! there she blows!—there she blows! There again!—there again!" he cried, in long-drawn, lingering, methodic tones, attuned to the gradual prolongings of the whale's visible jets. "He's going to sound! In stunsails! Down top-gallant-sails! Stand by three boats. Mr. Starbuck, remember, stay on board, and keep the ship. Helm there! Luff,

DESIGN FROM SCRIMSHAW

luff a point! So; steady, man, steady! There go flukes! No, no; only black water! All ready the boats there? Stand by, stand by! Lower me, Mr. Starbuck; lower, lower,—quick, quicker!" and he slid through the air to the deck.

"He is heading straight to leeward, sir," cried Stubb, "right away from us; cannot have seen the ship yet."

"Be dumb, man! Stand by the braces! Hard down the helm!—brace up! Shiver her!—shiver her!—So; well that! Boats, boats!"

Soon all the boats but Starbuck's were dropped; all the boat-sails set—all the paddles plying; with rippling swiftness, shooting to leeward; and Ahab heading the onset. A pale, death-glimmer lit up Fedallah's sunken eyes; a hideous motion gnawed his mouth.

Like noiseless nautilus shells, their light prows sped through the sea; but only slowly they neared the foe. As they neared him, the ocean grew still more smooth; seemed drawing a carpet over its waves; seemed a noon-meadow, so serenely it spread. At length the breathless hunter came so nigh his seemingly unsuspecting prey, that his entire dazzling hump was distinctly visible, sliding along the sea as if an isolated thing, and continually set in a revolving ring of finest, fleecy, greenish foam. He saw the vast, involved wrinkles of the slightly projecting head beyond. Before it, far out on the soft Turkish-rugged waters, went the glistening white shadow from his broad, milky forehead, a musical rippling playfully accompanying the shade; and behind, the blue waters interchangeably flowed over into the moving valley of his steady wake; and on either hand bright bubbles arose and danced by his side. But these were broken again by the light toes of hundreds of gay fowl softly feathering the sea, alternate with

their fitful flight; and like to some flag-staff rising from the painted hull of an argosy, the tall but shattered pole of a recent lance projected from the white whale's back; and at intervals one of the cloud of soft-toed fowls hovering, and to and fro skimming like a canopy over the fish, silently perched and rocked on this pole, the long tail feathers streaming like pennons.

A gentle joyousness—a mighty mildness of repose in swiftness, invested the gliding whale. Not the white bull Jupiter swimming away with ravished Europa clinging to his graceful horns; his lovely, leering eyes sideways intent upon the maid; with smooth bewitching fleetness, rippling straight for the nuptial bower in Crete; not Jove, not that great majesty Supreme! did surpass the glorified White Whale as he so divinely swam.

On each soft side—coincident with the parted swell, that but once leaving him, then flowed so wide away—on each bright side, the whale shed off enticings. No wonder there had been some among the hunters who namelessly transported and allured by all this serenity, had ventured to assail it; but had fatally found that quietude but the vesture of tornadoes. Yet calm, enticing calm, oh, whale! thou glidest on, to all who for the first time eye thee, no matter how many in that same way thou may'st have bejuggled and destroyed before.

And thus, through the serene tranquillities of the tropical sea, among waves whose hand-clappings were suspended by exceeding rapture,

Moby Dick moved on, still withholding from sight the full terrors of his submerged trunk, entirely hiding the wrenched hideousness of his jaw. But soon the fore part of him slowly rose from the water; for an instant his whole marbleized body formed a high arch, like Virginia's Natural Bridge, and warningly waving his bannered flukes in the air, the grand god revealed himself, sounded, and went out of sight. Hoveringly halting, and dipping on the wing, the white sea-fowls longingly lingered over the agitated pool that he left.

With oars apeak, and paddles down, the sheets of their sails adrift, the three boats now stilly floated, awaiting Moby Dick's reappearance.

"An hour," said Ahab, standing rooted in his boat's stern; and he gazed beyond the whale's place, towards the dim blue spaces and wide wooing vacancies to leeward. It was only an instant; for again his eyes seemed whirling round in his head as he swept the watery circle. The breeze now freshened; the sea began to swell.

"The birds!—the birds!" cried Tashtego.

In long Indian file, as when herons take wing, the white birds were now all flying towards Ahab's boat; and when within a few yards began fluttering over the water there, wheeling round and round, with joyous, expectant cries. Their vision was keener than man's; Ahab could discover no sign in the sea. But suddenly as he peered down and down into its depths, he pro-

THE WHALING MUSEUM, NEW BEDFORD, MASSACHUSETTS

foundly saw a white living spot no bigger than a white weasel, with wonderful celerity uprising, and magnifying as it rose, till it turned, and then there were plainly revealed two long crooked rows of white, glistening teeth, floating up from the undiscoverable bottom. It was Moby Dick's open mouth and scrolled jaw; his vast, shadowed bulk still half blending with the blue of the sea. The glittering mouth yawned beneath the boat like an open-doored marble tomb; and giving one sidelong sweep with his steering oar, Ahab whirled the craft aside from this tremendous apparition. Then, calling upon Fedallah to change places with him, went forward to the bows, and seizing Perth's harpoon, commanded his crew to grasp their oars and stand by to stern.

Now, by reason of this timely spinning round the boat upon its axis, its bow, by anticipation, was made to face the whale's head while yet under water. But as if perceiving this stratagem, Moby Dick, with that malicious intelligence ascribed to him, sidelingly transplanted himself, as it were, in an instant, shooting his pleated head lengthwise beneath the boat.

Through and through; through every plank and each rib, it thrilled for an instant, the whale obliquely lying on his back, in the manner of a biting shark, slowly and feelingly taking its bows full within his mouth, so that the long, narrow, scrolled lower jaw curled high up into the open air, and one of the teeth caught in a row-lock. The bluish pearl-white of the inside of the jaw was within six inches of Ahab's head, and reached higher than that. In this attitude the White Whale now shook the slight cedar as a mildly cruel cat her mouse. With unastonished eyes Fedallah gazed, and crossed his arms; but the tiger-yellow crew were tumbling over each other's heads to gain the uttermost stern.

And now, while both elastic gunwales were springing in and out, as the whale dallied with the doomed craft in this devilish way; and from his body being submerged beneath the boat, he could not be darted at from the bows, for the bows were almost inside of him, as it were; and while the other boats involuntarily paused, as

before a quick crisis impossible to withstand, then it was that monomaniac Ahab, furious with this tantalizing vicinity of his foe, which placed him all alive and helpless in the very jaws he hated; frenzied with all this, he seized the long bone with his naked hands, and wildly strove to wrench it from its gripe. As now he thus vainly strove, the jaw slipped from him; the frail gunwales bent in, collapsed, and snapped, as both jaws, like an enormous shears, sliding further aft, bit the craft completely in twain, and locked themselves fast again in the sea, midway between the two floating wrecks. These floated aside, the broken ends drooping, the crew at the stern-wreck clinging to the gunwales, and striving to hold fast to the oars to lash them across.

At that preluding moment, ere the boat was yet snapped, Ahab, the first to perceive the whale's intent, by the crafty upraising of his head, a movement that loosed his hold for the time; at that moment his hand had made one final effort to push the boat out of the bite. But only slipping further into the whale's mouth, and tilting over sideways as it slipped, the boat had shaken off his hold on the jaw; spilled him out of it, as he leaned to the push; and so he fell flat-faced upon the sea.

Ripplingly withdrawing from his prey, Moby Dick now lay at a little distance, vertically thrusting his oblong white head up and down in the billows; and at the same time slowly revolving his whole spindled body; so that when his vast wrinkled forehead rose—some twenty or more feet out of the water—the now rising swells, with all their confluent waves, dazzlingly broke against it; vindictively tossing their shivered spray still higher into the air.[2] So, in a gale, the but half baffled Channel billows only recoil from the base of the Eddystone, triumphantly to overleap its summit with their scud.

---

[2] This motion is peculiar to the sperm whale. It receives its designation (pitchpoling) from its being likened to that preliminary up-and-down poise of the whale-lance, in the exercise called pitchpoling. . . . By this motion the whale must best and most comprehensively view whatever objects may be encircling him. [Melville's note.]

But soon resuming his horizontal attitude, Moby Dick swam swiftly round and round the wrecked crew; sideways churning the water in his vengeful wake, as if lashing himself up to still another and more deadly assault. The sight of the splintered boat seemed to madden him, as the blood of grapes and mulberries cast before Antiochus's elephants in the book of Maccabees.[3] Meanwhile Ahab half smothered in the foam of the whale's insolent tail, and too much of a cripple to swim,—though he could still keep afloat, even in the heart of such a whirl-pool as that; helpless Ahab's head was seen, like a tossed bubble which the least chance shock might burst. From the boat's fragmentary stern, Fedallah incuriously and mildly eyed him; the clinging crew, at the other drifting end, could not succor him; more than enough was it for them to look to themselves. For so revolvingly appalling was the White Whale's aspect, and so planetarily swift the ever-contracting circles he made, that he seemed horizontally swooping upon them. And though the other boats, un-harmed, still hovered hard by; still they dared not pull into the eddy to strike, lest that should be the signal for the instant destruction of the jeopardized castaways, Ahab and all; nor in that case could they themselves hope to escape. With straining eyes, then, they remained on the outer edge of the direful zone, whose centre had now become the old man's head.

Meantime, from the beginning all this had been descried from the ship's mast heads; and squaring her yards, she had borne down upon the scene; and was now so nigh, that Ahab in the water hailed her—"Sail on the"—but that moment a breaking sea dashed on him from Moby Dick, and whelmed him for the time. But struggling out of it again, and chancing to rise on a towering crest, he shouted,—"Sail on the whale!—Drive him off!"

The Pequod's prows were pointed; and break-ing up the charmed circle, she effectually parted the white whale from his victim. As he sullenly swam off, the boats flew to the rescue.

Dragged into Stubb's boat with blood-shot, blinded eyes, the white brine caking in his wrin-kles; the long tension of Ahab's bodily strength did crack, and helplessly he yielded to his body's doom for a time, lying all crushed in the bottom of Stubb's boat, like one trodden under foot of herds of elephants. Far inland, nameless wails came from him, as desolate sounds from out ravines.

But this intensity of his physical prostration did but so much the more abbreviate it. In an instant's compass, great hearts sometimes con-dense to one deep pang, the sum total of those shallow pains kindly diffused through feebler men's whole lives. And so, such hearts, though summary in each one suffering; still, if the gods decree it, in their life-time aggregate a whole age of woe, wholly made up of instantaneous in-tensities; for even in their pointless centres, those noble natures contain the entire circum-ferences of inferior souls.

"The harpoon," said Ahab, half way rising, and draggingly leaning on one bended arm—"is it safe?"

"Aye, sir, for it was not darted; this is it," said Stubb, showing it.

"Lay it before me;—any missing men?"

"One, two, three, four, five;—there were five oars, sir, and here are five men."

"That's good.—Help me, man; I wish to stand. So, so, I see him! there! there! going to leeward still; what a leaping spout!—Hands off from me! The eternal sap runs up in Ahab's bones again! Set the sail; out oars; the helm!"

It is often the case that when a boat is stove, its crew, being picked up by another boat, help to work that second boat, and the chase is thus continued with what is called double-banked oars. It was thus now. But the added power of the boat did not equal the added power of the whale, for he seemed to have treble-banked his every fin; swimming with a velocity which plainly showed, that if now, under these circum-stances, pushed on, the chase would prove an

---

[3] I Maccabees vi:34. The sight of the wine was apparently supposed to arouse and madden the elephants, as if it were blood, "that they might prepare them for battle."

indefinitely prolonged, if not a hopeless one; nor could any crew endure for so long a period, such an unintermitted, intense straining at the oar; a thing barely tolerable only in some one brief vicissitude. The ship itself, then, as it sometimes happens, offered the most promising intermediate means of overtaking the chase. Accordingly, the boats now made for her, and were soon swayed up to their cranes—the two parts of the wrecked boats having been previously secured by her—and then hoisting everything to her side, and stacking her canvas high up, and sideways outstretching it with stun-sails, like the double-jointed wings of an albatross; the Pequod bore down in the leeward wake of Moby Dick. At the well known, methodic intervals, the whale's glittering spout was regularly announced from the manned mast-heads; and when he would be reported as just gone down, Ahab would take the time, and then pacing the deck, binnacle-watch in hand, so soon as the last second of the allotted hour expired, his voice was heard.—"Whose is the doubloon now? D'ye see him?" and if the reply was No, sir! straightway he commanded them to lift him to his perch. In this way the day wore on; Ahab, now aloft and motionless; anon, unrestingly pacing the planks.

As he was thus walking, uttering no sound, except to hail the men aloft, or to bid them hoist a sail still higher, or to spread one to a still greater breadth—thus to and fro pacing, beneath his slouched hat, at every turn he passed his own wrecked boat, which had been dropped upon the quarter-deck, and lay there reversed; broken bow to shattered stern. At last he paused before it; and as in an already over-clouded sky fresh troops of clouds will sometimes sail across, so over the old man's face there now stole some such added gloom as this.

Stubb saw him pause; and perhaps intending, not vainly, though, to evince his own unabated fortitude, and thus keep up a valiant place in his Captain's mind, he advanced, and eyeing the wreck exclaimed—"The thistle the ass refused; it pricked his mouth too keenly, sir, ha! ha!"

"What soulless thing is this that laughs before a wreck? Man, man! did I not know thee brave as fearless fire (and as mechanical) I could swear thou wert a poltroon. Groan nor laugh should be heard before a wreck."

"Aye, sir," said Starbuck drawing near, "'tis a solemn sight; an omen, and an ill one."

"Omen? omen?—the dictionary! If the gods think to speak outright to man, they will honorably speak outright; not shake their heads, and give an old wives' darkling hint.—Begone! Ye two are the opposite poles of one thing; Starbuck is Stubb reversed, and Stubb is Starbuck; and ye two are all mankind; and Ahab stands alone among the millions of the peopled earth, nor gods nor men his neighbors! Cold, cold—I shiver!—How now? Aloft there! D'ye see him? Sing out for every spout, though he spout ten times a second!"

The day was nearly done; only the hem of his golden robe was rustling. Soon it was almost dark, but the lookout men still remained unset.

"Can't see the spout now, sir;—too dark"—cried a voice from the air.

"How heading when last seen?"

"As before, sir,—straight to leeward."

"Good! he will travel slower now 'tis night. Down royals and top-gallant stun-sails, Mr. Starbuck. We must not run over him before morning; he's making a passage now, and may heave-to a while. Helm there! keep her full before the wind!—Aloft! come down!—Mr. Stubb, send a fresh hand to the fore-mast head, and see it manned till morning."—Then advancing towards the doubloon in the main-mast—"Men, this gold is mine, for I earned it; but I shall let it abide here till the White Whale is dead; and then, whosoever of ye first raises him, upon the day he shall be killed, this gold is that man's; and if on that day I shall again raise him, then, ten times its sum shall be divided among all of ye! Away now! the deck is thine, sir."

And so saying, he placed himself half way within the scuttle, and slouching his hat, stood there till dawn, except when at intervals rousing himself to see how the night wore on.

The Poulps

Jules Verne

The mystery of the sea was captured in one of the earliest and most popular of all science-fiction tales Twenty Thousand Leagues Under the Sea, *published in 1870. While working on the book, Jules Verne wrote to his father, "Whatever one man is capable of imagining, other men will prove themselves capable of realizing." Blending fact with fantasy and forecasting with remarkable accuracy the technological developments of a future time, Verne relates the voyage of the* Nautilus, *an underwater vessel that is at once a scientific laboratory for studying the infinite variety of ocean life and a refuge from the rest of mankind for its captain, Nemo. For Nemo, "The sea is everything. Its breath is pure and healthy. Here man is never lonely, for on all sides he feels life astir. The sea does not belong to despots. Upon its surface men can still make unjust laws, fight, tear one another to pieces, wage wars of terrestrial horror. But at thirty feet below their reign ceases, their influence is quenched, and their power disappears. . . . There alone . . . I am free!" Walking the ocean floor in diving suits or cruising through the ocean's depths in their windowed submarine, the crew encounters many marvelous creatures, including those described below in one of the classics of all "monster literature."*

*F*or several days the *Nautilus* kept off from the American coast. Evidently it did not wish to risk the tides of the Gulf of Mexico, or of the sea of the Antilles. April 16, we sighted Martinique and Guadaloupe from a distance of about thirty miles. I saw their tall peaks for an instant. The Canadian, who counted on carrying out his projects in the Gulf, by either landing, or hailing one of the numerous boats that coast from one island to another, was quite disheartened. Flight would have been quite practicable, if Ned Land had been able to take possession of the boat without the captain's knowledge. But in the open sea it could not be thought of. The Canadian, Conseil, and I had a long conversation on this subject. For six months we had beeen prisoners on board the *Nautilus*. We had travelled 17,000 leagues; and, as Ned Land said, there was no reason why it should not come to an end. We could hope nothing from the captain of the *Nautilus,* but only from ourselves. Besides, for some time past he had become graver, more retired, less sociable. He seemed to shun me. I met him rarely. Formerly, he was pleased to explain the submarine marvels to me; now, he left me to my studies, and came no more to the saloon. What change had come over him? For what cause?

For my part, I did not wish to bury with me my curious and novel studies. I had now the power to write the true book of the sea; and this book, sooner or later, I wished to see daylight. Then again, in the water by the Antilles, ten yards below the surface of the waters, by the open panels, what interesting products I had to enter on my daily notes! There were, among other zoöphytes, those known under the name of physalis pelagica, a sort of large oblong bladder, with mother-of-pearl rays, holding out their membranes to the wind, and letting their blue tentacles float like threads of silk; charming medusae to the eye, real nettles to the touch, that distil a corrosive fluid. There were also annelides, a yard and a half long, furnished with a pink horn, and with 1,700 locomotive organs, that wind through the waters, and throw out in passing all the light of the solar spectrum. There were, in the fish category, some Malabar rays, enormous gristly things, ten feet long, weighing 600 pounds, the pectoral fin triangular in the midst of a slightly humped back, the eyes fixed in the extremities of the face, beyond the head, and which floated like weft, and looked sometimes like an opaque shutter on our glass window. There were American balistae, which nature has only dressed in black and white; gobies, with yellow fins and prominent jaw; mackerel sixteen feet long, with short-pointed teeth, covered with small scales, belonging to the albacore species. Then, in swarms, appeared gray mullet, covered with stripes of gold from the head to the tail, beating their resplendent fins, like masterpieces of jewelry, consecrated formerly to Diana, particularly sought after by rich Romans, and of which the proverb says, "Whoever takes them does not eat them."

Lastly, pomacanthe dorees, ornamented with emerald bands, dressed in velvet and silk, passed before our eyes like Veronese lords; spurred spari passed with their pectoral fins; clupano-dons fifteen inches long, enveloped in their phosphorescent light; mullet beat the sea with their large jagged tail; red vendaces seemed to mow the waves with their showy pectoral fins; and silvery selenes, worthy of their name, rose on the horizon of the waters like so many moons with whitish rays. April 20, we had risen to a mean height of 1,500 yards. The land nearest us then was the archipelago of the Bahamas. There rose high submarine cliffs covered with large weeds, giant laminariae and fuci, a perfect espalier of hydrophytes worthy of a Titan world. It was about eleven o'clock when Ned Land drew my attention to a formidable prick-ing, like a sting of an ant, which was produced by means of large seaweeds.

"Well," I said, "these are proper caverns for poulps, and I should not be astonished to see some of these monsters."

"What!" said Conseil; "cuttlefish, real cuttle-fish, of the cephalopod class?"

"No," I said; "poulps of huge dimensions."

"I will never believe that such animals exist," said Ned.

"Well," said Conseil, with the most serious air in the world; "I remember perfectly to have seen a large vessel drawn under the waves by a cephalopod's arm."

"You saw that?" said the Canadian.

"Yes, Ned."

"With your own eyes?"

"With my own eyes."

"Where, pray, might that be?"

"At St. Malo," answered Conseil.

"In the port?" said Ned, ironically.

"No; in a church," replied Conseil.

"In a church?" cried the Canadian.

"Yes; friend Ned. In a picture representing the poulp in question."

"Good!" said Ned Land, bursting out laughing.

"He is quite right," I said. "I have heard of this picture; but the subject represented is taken from a legend, and you know what to think of legends in the matter of natural history. Besides, when it is a question of monsters, the imagination is apt to run wild. Not only is it supposed that these poulps can draw down vessels, but a certain Olaüs Magnus speaks of a cephalopod a mile long, that is more like an island than an animal. It is also said that the Bishop of Nidros was building an altar on an immense rock. Mass finished, the rock began to walk, and returned to the sea. The rock was a poulp. Another bishop, Pontoppidan, speaks also of a poulp on which a regiment of cavalry could maneuver. Lastly, the ancient naturalists speak of monsters whose mouths were like gulfs, and which were too large to pass through the Straits of Gibraltar."

"But how much is true of these stories?" asked Conseil.

"Nothing, my friends; at least of that which passes the limit of truth to get to fable or legend. Nevertheless, there must be some ground for the imagination of the story-tellers. One cannot deny that poulps and cuttlefish exist of a large species, inferior, however, to the cetaceans. Aristotle had stated the dimensions of a cuttle-fish as five cubits, or nine feet two inches. Our fishermen frequently see some that are more than four feet long. Some skeletons of poulps are preserved in the museums of Trieste and Montpellier, that measure two yards in length. Besides, according to the calculations of some naturalists, one of these animals, only six feet

A HOLLYWOOD KRAKEN
Model of the cephalopod used in Walt Disney's film version of Twenty Thousand Leagues Under the Sea.

long, would have tentacles twenty-seven feet long. That would suffice to make a formidable monster."

"Do they fish for them in these days?" asked Ned.

"If they do not fish for them, sailors see them at least. One of my friends, Captain Paul Bos of Havre, has often affirmed that he met one of these monsters, of colossal dimensions, in the Indian seas. But the most astonishing fact, and which does not permit of the denial of the existence of these gigantic animals, happened some years ago, in 1861."

"What is the fact?" asked Ned Land.

"This is it. In 1861, to the north-east of Teneriffe, very nearly in the same latitude we are in now, the crew of the despatch-boat *Alector* perceived a monstrous cuttlefish swimming in the waters. Captain Bouguer went near to the animal, and attacked it with harpoons and guns, without much success, for balls and harpoons glided over the soft flesh. After several fruitless attempts, the crew tried to pass a slip-knot round the body of the mollusk. The noose slipped as far as the caudal fins, and there stopped. They tried then to haul it on board, but its weight was so considerable that the tightness of the cord separated the tail from the body, and, deprived of this ornament, he disappeared under the water."

"Indeed! is that a fact?"

"An indisputable fact, my good Ned. They proposed to name this poulp 'Bouguer's cuttlefish.'"

"What length was it?" asked the Canadian.

"Did it not measure about six yards?" said Conseil, who, posted at the window, was examining again the irregular windings of the cliff.

"Precisely," I replied.

"Its head," rejoined Conseil, "was it not crowned with eight tenacles, that beat the water like a nest of serpents?"

"Precisely."

"Had not its eyes, placed at the back of its head, considerable development?"

"Yes, Conseil."

"And was not its mouth like a parrot's beak?"

"Exactly, Conseil."

"Very well! no offence to master," he replied, quietly; "if this is not Bouguer's cuttlefish, it is, at least, one of its brothers."

I looked at Conseil. Ned Land hurried to the window.

"What a horrible beast!" he cried.

I looked in my turn, and could not repress a gesture of disgust. Before my eyes was a horrible monster, worthy to figure in the legends of the marvelous. It was an immense cuttlefish, being eight yards long. It swam crossways in the direction of the Nautilus with great speed, watching us with its enormous staring green eyes. Its eight arms, or rather feet, fixed to its head, that have given the name of cephalopod to these animals, were twice as long as its body, and were twisted like the furies hair. One could see the 250 air-holes on the inner side of the tentacles. The monster's mouth, a horned beak like a parrot's, opened and shut vertically. Its tongue, a horned substance, furnished with several rows of pointed teeth, came out quivering from this veritable pair of shears.

What a freak of nature, a bird's beak on a mollusk! Its spindle-like body formed a fleshy mass that might weigh 4,000 to 5,000 lbs.; the varying color changing with great rapidity, according to the irritation of the animal, passed successively from livid gray to reddish brown. What irritated this mollusk? No doubt the presence of the Nautilus, more formidable than itself, and on which its suckers or its jaws had no

hold. Yet, what monsters these poulps are! what vitality the Creator has given them! what vigor in their movements! and they possess three hearts! Chance had brought us in the presence of this cuttlefish, and I did not wish to lose the opportunity of carefully studying this specimen of cephalopods. I overcame the horror that inspired me; and, taking a pencil, began to draw it.

"Perhaps this is the same which the Alecto saw," said Conseil.

"No," replied the Canadian; "for this is whole, and the other had lost its tail."

"That is no reason," I replied. "The arms and tails of these animals are reformed by redintegration; and, in seven years, the tail of Bouguer's cuttlefish has no doubt had time to grow."

By this time other poulps appeared at the port light. I counted seven. They formed a procession after the Nautilus, and I heard their beaks gnashing against the iron hull. I continued my work. These monsters kept in the water with such precision, that they seemed immovable. Suddenly the Nautilus stopped. A shock made it tremble in every plate.

"Have we struck anything?" I asked.

"In any case," replied the Canadian, "we shall be free, for we are floating."

The Nautilus was floating, no doubt, but it did not move. A minute passed. Captain Nemo, followed by his lieutenant, entered the drawing-room. I had not seen him for some time. He seemed dull. Without noticing or speaking to us, he went to the panel, looked at the poulps, and said something to his lieutenant. The latter went out. Soon the panels were shut. The ceiling was lighted. I went towards the Captain.

"A curious collection of poulps?" I said.

"Yes, indeed, Mr. Naturalist," he replied; "and we are going to fight them, man to beast."

I looked at him. I thought I had not heard aright.

"Man to beast?" I repeated.

"Yes, Sir. The screw is stopped. I think that the horny jaws of one of the cuttlefish are entangled in the blades. That is what prevents our moving."

56

"What are you going to do?"

"Rise to the surface, and slaughter this vermin."

"A difficult enterprise."

"Yes, indeed. The electric bullets are powerless against the soft flesh, where they do not find resistance enough to go off. But we shall attack them with the hatchet."

"And the harpoon, Sir," said the Canadian, "if you do not refuse my help."

"I will accept it, Master Land."

"We will follow you," I said, and following Captain Nemo, we went towards the central staircase.

There, about ten men with boarding hatchets were ready for the attack. Conseil and I took two hatchets; Ned Land seized a harpoon. The *Nautilus* had then risen to the surface. One of the sailors, posted on the top ladder-step, unscrewed the bolts of the panels. But hardly were the screws loosed, when the panel rose with great violence, evidently drawn by the suckers of a poulp's arm. Immediately one of these arms slid like a serpent down the opening, and twenty others were above. With one blow of the axe, Captain Nemo cut this formidable tentacle, that slid wriggling down the ladder. Just as we were pressing one on the other to reach the platform, two other arms, lashing the air, came down on the seaman placed before Captain Nemo, and lifted him up with irresistible power. Captain Nemo uttered a cry, and rushed out. We hurried after him.

What a scene! The unhappy man, seized by the tentacle, and fixed to the suckers, was balanced in the air at the caprice of this enormous trunk. He rattled in his throat, he was stifled, he cried, "Help! help!" These words, *spoken in French,* startled me! I had a fellowcountryman on board, perhaps several! That heartrending cry! I shall hear it all my life. The unfortunate man was lost. Who could rescue him from that powerful pressure? However, Captain Nemo had rushed to the poulp, and with one blow of the axe had cut through one arm. His lieutenant struggled furiously against other monsters that crept on the flanks of the *Nautilus*. The crew fought with their axes. The Canadian, Conseil, and I buried our weapons in the fleshy masses; a strong smell of musk penetrated the atmosphere. It was horrible!

For one instant, I thought the unhappy man, entangled with the poulp, would be torn from its powerful suction. Seven of the eight arms had been cut off. One only wriggled in the air, brandishing the victim like a feather. But just as Captain Nemo and his lieutenant threw themselves on it, the animal ejected a stream of black liquid. We were blinded with it. When the cloud dispersed, the cuttlefish had disappeared, and my unfortunate countryman with it. Ten or twelve poulps now invaded the platform and sides of the *Nautilus*. We rolled pell-mell into the midst of this nest of serpents, that wriggled on the platform in the waves of blood and ink. It seemed as though these slimy tentacles sprang up like the hydra's heads. Ned Land's harpoon, at each stroke, was plunged into the staring eyes of the cuttlefish. But my bold companion was suddenly overturned by the tentacles of a monster he had not been able to avoid.

Ah! how my heart beat with emotion and horror! The formidable beak of a cuttlefish was open over Ned Land. The unhappy man would be cut in two. I rushed to his succor. But Captain Nemo was before me; his axe disappeared between the two enormous jaws, and miraculously saved the Canadian, rising, plunged his harpoon deep into the triple heart of the poulp.

"I owed myself this revenge!" said the captain to the Canadian.

Ned bowed without replying. The combat had lasted a quarter of an hour. The monsters, vanquished and mutilated, left us at last, and disappeared under the waves. Captain Nemo, covered with blood, nearly exhausted, gazed upon the sea that had swallowed up one of his companions, and great tears gathered in his eyes.

Kraken

Frank W. Lane

Accounts of sea monsters such as the one depicted by Jules Verne go back at least to the days of the Greek poet Homer, who described the many-headed, many-footed Scylla more than 2,500 years ago. Were these monsters merely the inventions of fertile imaginations? Were they gross exaggerations of actual creatures? Or did they really exist? There is no doubt that many people firmly believed in the existence of such monsters until well into the sixteenth century, and maps of that period are replete with drawings of the monsters and indications of the best routes for avoiding them. A more skeptical age cast doubt upon their existence, until scientific evidence could no longer be denied. In the following excerpt from Kingdom of the Octopus *(1960), Frank W. Lane, author of several books on natural history, relates how the existence of the kraken, or giant squid, was finally established. His account gives us cause to question whether we may have too readily dismissed reports of other monsters of the deep.*

Stories of a giant many-armed sea creature occur in the ancient literature of several maritime nations . . . [but] the first definite references to the kraken are contained in Norwegian literature. In fact the name is derived from the Norwegian dialect word *krake,* the *n* denoting the definite article.

The earliest identifiable reference appears to be by Olaüs Magnus who, in a history of the northern nations published in 1555, refers to "monstrous fish on the Coasts or Sea of Norway," and says: "Their Forms are horrible, their Heads square, all set with prickles, and they have sharp and long horns round about, like a Tree rooted up by the Roots[1] . . . one of these Sea-Monsters will drown easily many great ships provided with many strong Mariners." Magnus also refers to the beast's huge eyes and the unusual size of the head compared with the rest of the body. Apart from obvious exaggerations the "monstrous fish" is clearly a huge cephalopod.

Magnus was the last archbishop of the Swedish Roman Catholic Church, and was for years Archbishop of Upsala. He was an assiduous but uncritical collector of other men's tales. His account of the kraken seems to have come from men who had seen it—and had been badly frightened.

Two hundred years after Magnus another ecclesiastic, the Protestant Bishop Erik Pontoppidan, wrote at length about the kraken in his *Natural History of Norway,* the English edition of which was published in 1755. In fact he has been falsely accused of inventing the monster. He recounts various tales which, while containing gross exaggeration, obviously refer to giant cephalopods. . . .

In 1735 Carl von Linnaeus published the first edition of his *Systema Naturae,* the pioneer work which introduced system into the nomenclature of the animal world. In this work he included the kraken, and gave it the specific name *Sepia microcosmos.* But he omitted it from subsequent editions.

The scene now shifts to the France of the early 19th century where Pierre Denys de Montfort played the kraken story for all it was worth—and a lot more. Denys de Montfort worked in the Museum of Natural History in Paris, and wrote a book on systematic conchology. But the book which made him notorious was his *Histoire Naturelle générale et particulière des Mollusques,* published in six volumes between 1802 and 1805.

[1] One of the meanings of *krake* is "a stunted tree."

A COLOSSAL KRAKEN

Engraving by Denys de Montfort, early nineteenth century

In this work Denys de Montfort shamelessly mixed science with fantasy. He invented a *Poulpe colossal* and said that it wrapped its tentacles round the masts of a vessel and was on the point of dragging it to the bottom when the crew saved themselves by cutting off its immense arms with cutlasses and hatchets. Not content with that, which may, as will be seen later, have had *some* foundation in fact, he then asserted that a horde of giant cephalopods sank ten men-o'-war in one night!

In view of the wild stories which always seem to have accompanied accounts of the kraken, it is hardly surprising that scientists did not take it seriously. As has happened before in science, the pendulum swung too far the other way. Rather than be taken in by a vulgar fraud, most scientists closed their minds even to the possibility that the legendary kraken could have any basis in fact.

Then in the middle of the 19th century Japetus Steenstrup in Denmark and Pieter Harting in Holland studied old records of captures and strandings of giant cephalopods, and reexamined some of their remains, preserved and long forgotten, in museums. Their studies showed that there was evidence of the occasional capture or stranding of large cephalopods from the middle of the 16th century to contemporary times. Moreover, the "spirit vaults and bottle departments" of great museums had long preserved physical evidence of the kraken's existence. In the basement of the British Museum of Natural History in London there was, at this time, a tall glass jar containing one of the arms of a giant cephalopod of unknown origin. This arm was 9 ft. long and 11 in. in circumference at the base, and bore some 300 suckers.

Despite the foregoing evidence, the scientific world was still skeptical of the kraken. Then in 1861 something happened which, to say the least, required a lot of explaining away—a French warship engaged a kraken off the Canary Islands. The ship was the steam despatch boat *Alecton* commanded by Lieutenant Frédéric-Marie Bouyer. . . .

Sabin Berthelot, the French Consul in the Canary Islands, . . . spoke to Bouyer and some of his officers when the *Alecton* anchored there. They estimated the weight of the beast at more than 2,000 kilograms, or about two tons. It had enormous eyes, which were flush with its head. Its mouth was like a parrot's beak and it could open it to almost half a meter, or about 1½ ft.

The "battle" lasted over three hours, during which the kraken was hit by several harpoons, but they failed to hold. Berthelot was shown the only part of the animal which the crew were able to get aboard [part of a fin weighing almost 20 kilograms or 44 pounds] and says: "I have myself questioned old Canary Island fishermen, who have assured me that they have seen several times, out at sea, great reddish Calmars [squids] two meters [6½ ft.] and more in length, which they did not dare to capture."

It is said that this remarkable incident prompted Jules Verne to include in his *Twenty Thousand Leagues Under the Sea* the famous battle between the submarine *Nautilus* and a kraken. This has been made into a fantastic sequence in Walt Disney's film of Verne's novel—complete with two-ton rubber kraken which it took 24 men to operate with electronics, compressed air, hydraulics and remote controls!

The documentary evidence of strandings, the parts of cephalopods preserved in museum basements, the occasional sightings, and now the *Alecton's* evidence presented to the French Academy of Sciences—all these together made it difficult to doubt that there was a definite basis in fact for the legend of the kraken. But in the 1860s it was still possible for the skeptics to fall back on the well-tried scientific maxim "No body, no animal." It was, or course, true that up to that time no scientist had examined and described a kraken in reasonably complete condition, but in the next few years the seas provided all the evidence that even the most hardened skeptic could require. In fact the 1870s supplied more bodies—and therefore irrefutable evidence—than any decade since.

On October 26, 1873, two fishermen, Theophilus Piccot and Daniel Squires, and Piccot's twelve-year-old son Tom, were fishing for herring off Portugal Cove, Newfoundland. They saw a large object floating on the surface and, thinking it was a piece of wreckage, rowed over to it. One of the men struck it with a boathook. Instantly the supposed dead mass reared up, and the fishermen saw what they had attacked—a kraken.

With its huge eyes flashing, the animal lunged towards them and struck the gunwale with its horny beak. A long thin tentacle shot out and instantly coiled round the boat. A shorter but thicker arm followed, and held it fast. The body of the kraken sank beneath the surface, and began to drag the boat with it. (H.J. Squires tells me that the boat would be 15–20 ft. long and 4–5 ft. across at the widest part.)

The fishermen were almost paralyzed with fear. Water was pouring into the boat as it settled in the water and they thought it was only a matter of seconds before they would all drown. But if the men were resigned to die young Tom Piccot was not. He picked up a small tomahawk from the bottom of the boat and smashed at the arm and tentacle. He severed them just in time. The boat righted itself but the kraken was still alongside. It discharged pints of ink which darkened the water all round them. But it did not renew the fight. The beast's huge bulk seemed to "slide off" and disappear. The last the men saw of it were the huge fins at the end of its body which, they said, measured six or seven feet.

Fearing pursuit, the fishermen rowed for the shore with all their strength. Tom Piccot, the boy-hero to whom the men owed their lives, clutched the arm and tentacle and brought them ashore as his battle trophies. The arm he threw on the ground where it was soon eaten by dogs. The long thin tentacle, however, he carefully preserved.

The local clergyman at Portugal Cove heard of the affair and suggested to Tom that he should take the long tentacle to the Rev. Moses Harvey who lived in nearby St. John's, and was a keen naturalist with a special interest in kraken lore. Remember that even at this time, 1873, although marine zoologists were greatly interested in the stories of sightings, encounters with krakens and their occasional strandings, conclusive evidence was still missing. Harvey's enthusiasm when he saw young Tom's prize knew no bounds. He listened eagerly to his story and paid him handsomely for the tentacle. This was 19 ft. long but was only part of the total length. It was 3½ in. in circumference, and was "exceedingly tough and strong."

The thrill that Harvey felt at that moment, the thrill of the true naturalist when he feels he is about to solve a long-debated mystery of natural history, can still be felt in this passage, published in 1899:

"I was now the possessor of one of the rarest curiosities in the whole animal kingdom—the veritable [tentacle] of the hitherto mythical devil fish, about whose existence naturalists had been disputing for centuries. I knew that I held in my hand the key of the great mystery, and that a new chapter would now be added to Natural History."

Harvey hastened to interrogate the men to see what details they could add to the boy's story.

"I found the two fishermen but partially recovered from the terror of the scene through which they had passed. They still shuddered as they spoke of it. What most impressed them was the huge green eyes gleaming with indescribable fury, and the parrot-like beak that suddenly leaped from a cavity in the middle of the head, as if eager to rend them."

Harvey now knew beyond doubt that krakens existed, although the all-important *body* was still missing.

In November, less than a month after the Portugal Cove encounter, four fishermen were hauling in a large herring-net in nearby Logy Bay. The net was unusually heavy, and as it neared the surface they were startled by its behavior. It was moving violently and they were afraid their catch might either burst the net or carry it away. It took the strength of all four men to raise the net, and when it broke surface they almost let it go. They had caught a kraken!

The net was a mass of writhing arms, and in the center were two large gleaming eyes. The two tentacles shot out through a rent in the net, reaching for the boat. They quivered for a moment in the air, seeking their prey, but the distance was too great and they shot back again.

The fishermen were so alarmed they were tempted to cut the net free rather than risk battle with such a foe. But one of the men drew his sharp fish-splitting knife and, waiting his chance, plunged it in behind the kraken's eyes and rapidly severed the head from the rest of the body. One of the fishermen told Harvey afterwards that he had had such a bad half-hour that no amount of money would induce him to take part in capturing another of the beasts.

The moment Harvey heard of the capture he went to Logy Bay. A quarter of a century later he wrote:

"I remember to this day how I stood on the shore of Logy Bay, gazing on the dead giant and 'rolling as a sweet morsel under my tongue' the thought of how I would astonish the savants, and confound the naturalists, and startle the world at large. I resolved that only the interests of science should be considered. I speedily completed a bargain with the fishermen, whom I astonished by offering 10 dols. to deliver the beast carefully at my house. . . ."

This specimen proved beyond doubt that krakens exist, that they are giant squids, and that they are the largest invertebrates on earth. . . .

What are the enemies of giant squids? It is possible that large sharks sometimes attack them, but their only proven foe, apart from man, is the sperm whale, or cachalot (*Physeter catodon*). This is the largest of the toothed whales and the males, which grow to about twice the length of the females, sometimes measure 60 ft. and weigh over 50 tons. The old-time whalers claimed to have caught some cachalots measuring 85 ft. . . .

Moby Dick was a sperm whale and his character was in keeping with what is known of the breed. When roused it can be a terrible foe. Ships have been sunk when charged by enraged sperm whales—they can charge at 20 miles per hour. The most famous sinking occurred on November 20, 1820, when the 238-ton whaleship *Essex* was twice rammed by a huge sperm whale which inflicted such damage that she sank. This was the tragedy—most of the crew never reached land—that prompted Herman Melville to end his novel with the whale sinking the ship. . . .

Such is the foe of the giant squid, and the titanic struggles that take place between the two must be among the most awesome in Nature. The bodies of sperm whales bear some evidence of what occurs when kraken meets leviathan. The heads of these whales are nearly always scarred by the sucker-rings of giant squids. Sometimes the wounds are an inch deep in the blubber, and the head bears huge sores where the tooth-rimmed suckers have gouged and torn the skin. Often the rounded imprint of the suckers is branded into the flesh. The struggle sometimes continues inside the whale, for occasionally squids are swallowed alive, and sucker marks have been found on the walls of the whale's stomach. . . .

Surprising as it sounds, sperm whales sometimes swallow giant squids whole, although I know of no record of a 50- or 60-footer being thus devoured intact.

DIVER WITH A GIANT OCTOPUS
"The octopus, far from being the aggressive monster so dear to fiction writers and filmmakers, is a shy, retiring creature."—Jacques Cousteau

Sharkproof!

Eugenie Clark

Perhaps the most fearsome monster of the deep—one whose existence was never questioned—is the shark. The ominous shark appears in countless stories of man's confrontation with the sea, including those by Melville and Crane. Fortunately, of the 250 or so species of shark, only about a dozen are known killers; of these, the great white shark, featured in Peter Benchley's best-selling book Jaws, *is the greatest threat to humans. But even this most rapacious of predators may soon be rendered harmless as scientists learn the secret of an unassuming, sharkproof sole. In the following article from* National Geographic *(1974), Eugenie Clark, noted shark expert and professor of zoology at the University of Maryland, tells of her experiments that may lead to the development of the first truly effective shark repellent.*

*W*e watched in disbelief as the big shark swam toward the tethered baitfish, opened wide its saw-toothed mouth, half enveloped its prey, and then—instant retreat!

The deadly predator jerked away, its jaws "frozen" open. Vigorously shaking its head from side to side, the shark dashed around the pool in the marine laboratory at Elat, Israel, before at last succeeding in closing its mouth. The captive baitfish kept up its easy undulating motion as if nothing had happened.

Of ordinary appearance—it looks very like a flounder you might buy in a supermarket—the flatfish that repelled the shark is a species of sole known to scientists as *Pardachirus marmoratus.* What makes it remarkable is the lethally toxic, milky poison secreted by glands along its dorsal and anal fins. Nevertheless, local people along the Red Sea eat the fish—after they cook it, of course—and consider it quite tasty.

Pardachirus is a translucent white on the side facing downward as it swims about or lies in the sea-floor sand, but speckled on the top. Israelis call it the "Moses sole." When Moses parted the Red Sea, the story goes, a fish, caught in the middle, was split; the halves became our sole.

The Moses sole swims quite abundantly in the crystalline waters of the Gulf of Aqaba, the northeastern arm of the Red Sea, where for many years I have been studying garden eels and other marine fish. In small aquariums at Hebrew University's Heinz Steinitz Marine Biology Laboratory at Elat, I had been subjecting a variety of marine creatures to the toxic "milk" secreted by *Pardachirus.*

Echinoderms—sea urchins and starfish—had died from contact with the poison, even when it was highly diluted in seawater. Small reef fish, too, had succumbed. Could the toxin also kill, or repel, larger marine animals?

Two reef whitetip sharks, *Triaenodon obesus,* caught three years earlier, were more pets than experimental animals at the laboratory. Everyone on the staff tossed fish into their aquarium to watch them be gobbled up.

"There's no fish those sharks won't eat," the laboratory people assured me.

But I consulted undersea naturalist David Fridman, the laboratory's aquarist and collector. The sharks were his special charge.

"Sure, go ahead and try *Pardachirus* on them," said David. "Can't believe such a little fish could hurt them; they're too tough."

Midget Morsel Holds Hunters at Bay

I had been winding up my summer work at the laboratory. Left over was one live eight-inch female *Pardachirus,* already partly milked during echinoderm experiments. I was sure, though, she had some poison left.

We tied her with a string through the gill openings and lowered her into the shark pool. As soon as the wriggling sole touched the water, the two sharks swept in toward her, with the surprising result I've described.

Both sharks kept circling back toward the sole, attracted by vibrations from her movements. But each time a shark attempted to take a bite, the outcome was the same: jaws agape, as if unable to close.

During a six-hour vigil, we never saw the fish release her paralyzing milk, though we were sure the sharks were reacting to invisible amounts of it.

What kind of substance was this that could foil a shark in the process of taking a bite? David and I were dumbfounded. I almost missed my homeward plane.

"You better come back, Genie," was David's farewell injunction. "We have a lot more to learn about *Pardachirus.*"

In the summer of 1973, sponsored by the National Geographic Society, I returned to the Red Sea to make further studies of the Moses sole. . . .

We performed some of our experiments in the sea. Catching a *Pardachirus,* we placed it in a plastic bag, which we inverted over a small branched coral, the home of many hidden fish and invertebrates. Then we squeezed the poisonous milk from the *Pardachirus,* and timed what happened. Knowing the volume of the bag and the amount of milk that we could squeeze out (about a thimbleful), we could estimate our dilution.

One part milk in five thousand parts seawater killed every small fish tested in a matter of minutes, even hardy damselfish. We also verified the initial attraction and ultimate repellent effect of a *Pardachirus* on dangerous animals in the sea. I placed a wriggling Moses sole outside the den of a moray eel. Soon the moray eased out and tried to bite the fish. It came as no surprise that the eel could not lay a tooth on *Pardachirus,* but instead writhed backward in a hurried escape.

David Doubilet's experience was more exciting. After photographing a large sole on the sea floor, he grabbed it to bring it up.

"Just then," David later recalled, "a barracuda appeared. I could have counted its teeth as it charged straight toward the *Pardachirus* in my hand. But at the last possible moment it stopped cold, shook its head, and shot away. Its teeth never touched us!"

I first came upon this mysterious sole in 1960 while diving near Elat collecting sea horses and

pipefish. Netting the *Pardachirus,* I had been surprised to see milky fluid flowing out of pores along the fins. The milk felt slippery and caused a tightening sensation in my fingers. I suspected it might be poisonous.

I preserved the fish and later read about it in Professor Steinitz's library in Jerusalem. The milk from *Pardachirus* had first been reported by Dr. C.B. Klunzinger in 1871. My interest continued, and in 1971 one of my students, Stella Chao of the University of Maryland, ran tests that proved the milk, indeed, has toxic effects. But no chemical experimentation was undertaken until 1972, when I was a visiting professor at Hebrew University in Jerusalem. One of my graduate students, Naftali Primor, was working on the biochemistry of animal toxins.

After hearing me lecture about my early observations of the Red Sea sole, Naftali pleaded, "I want to study that *Pardachirus* toxin." He dropped almost everything else and devoted himself to this work.

Poison Ravages Blood Cells

In his laboratory Naftali injected a mouse with a fifth of a milliliter of the undiluted secretion. It was a massive dose, immediately sending the rodent into violent convulsions. Within two minutes it was dead. Naftali found that the organs in the mouse's body cavity had hemorrhaged. From his description of the creature's last moments, it seemed that the milk had attacked nerves as well as blood, indicating both neurotoxic and hemotoxic effects.

More research is needed to find out if the sole's milk really does affect the nervous system. But, thanks to Naftali's later experiments, we now know for certain that the poison destroys red blood cells.

Dr. Eliahu Zlotkin, toxicologist at Hebrew University, assured me that Naftali's results were reliable. The graduate-student researcher probed deeper. He separated the milk into three protein components, and found he could isolate an inhibitor of the hemotoxin—a substance mixed into the poison that can prevent it from taking effect! I think this inhibitor could eventually prove to be the explanation for the sole's seeming imperviousness to its own deadly poison.

Naftali's work went even further. He discovered that *Pardachirus*'s toxin inhibitor also counteracts the hemotoxic effects of venom from scorpions, bees, and elapid snakes such as corals, cobras, and mambas.

"Because of these findings," Dr. Zlotkin said, "your Moses sole has promising and exciting medical possibilities."

Research even now is continuing in Israel and the U.S. toward utilizing this inhibitor

FISH FIGURE, SOUTHERN SOLOMON ISLANDS
Wood with mother-of-pearl inlay, 80¾" long. Solomon Islanders believe that shark spirits keep them from harm.

ingredient. And *Pardachirus* toxin itself is being evaluated as a possible shark repellent for use by swimmers and divers. None of four species of sharks exposed to the Moses sole have attacked it. Flies, too, seem to be repelled by its poison.

Humans Try a Taste Test

What about the sole's poison as a hazard to humans? From our experiments with rats—most of them recovered from moderate injections of the poison—I was confident that it would take a strong dose of *Pardachirus* poison placed directly into a person's bloodstream to do any damage. As with snake venom, you probably could drink *Pardachirus*'s potent milk without any ill effect if you have no cut in your mouth or ulcer in your stomach.

We didn't go that far, but we tasted the milk from a live *Pardachirus*. I dipped a finger in the poison and touched it to my tongue. Then Gail and, finally, reluctantly, the boys followed suit. We all experienced not only the highly unpleasant bitterness and the strong taste, but also the fast-acting astringent action caused by the tiniest drop. We wiggled our tongues uncomfortably for 20 minutes before the sensation went away.

Now we could better understand the violent reactions of our marine test animals: Why sea anemones contracted and everted their stomachs; why the pulsating soft coral, *Xenia*, became discombobulated and distorted; why the feet of mollusks fragmented into shrunken pieces; and why fish shuddered and turned belly up when put in water to which the milk had been added. Brittle stars would either coil up in a ball and die, or suddenly stiffen straight out, their slender arms like five radiating exclamation points expressing their horror.

Dissecting *Pardachirus*, we found that the poison glands, an average of 240 per sole, are located in pairs at the base of all dorsal and anal fin rays except the first and last. Each gland opens through a tiny pore, and from the pores

the milk runs out into the membranous grooves between the stiff, gristly rays that give rigidity to the fins.

During our experiments with the sharks in the Steinitz laboratory tank, only in two instances did *Pardachirus* release a visible amount of milk; the poison came out as two or three white threadlike wisps that dissolved within a few moments. I believe *Pardachirus* constantly exudes very minute amounts. Our experiments showed that a halo of protection several inches in radius surrounds the sole.

Ordinary fish suspended in the pool immediately adjacent to *Pardachirus* were shunned by the sharks. If we let a dead fish dangle from the tail of *Pardachirus,* so part of it was three inches away, the sharks would eventually bite the end farthest from *Pardachirus,* but only after many minutes.

When we wiped the skin of a live *Pardachirus* with alcohol, removing the mucus-poison mix, and at once dropped the fish in the tank, it was inside a shark's stomach in a flash.

Back to the Sea for a Final Test

For our concluding fieldwork, we wanted to see how free-swimming sharks react to live, tethered *Pardachirus*. On the southern tip of the Sinai Peninsula, at Ras Muhammad, we found a magnificent spot to set shark test lines. . . .

Our experience at Ras Muhammad confirmed the observations at the tank in Elat: *Pardachirus* could hang for hours in the sea and repel all comers among the finny predators of dawn and dusk.

I look forward to the day when research on *Pardachirus* and its potent toxin has advanced to allow this scenario: I get into my wet suit and spray it with the synthesized poison of the Moses sole. Then I dive in and swim at ease among the sharks, exempt from concern that these old friends may make me an item in their diet.

Shark Attack Against Man
H. David Baldridge

Reports about unprovoked shark attacks have come from around the world, with the exception of extreme northern and southern latitudes. In an attempt to learn more about shark behavior and what precipitates their attacks, scientists have recently enlisted the aid of the computer. Supported by the United States Navy, which has long been searching for ways to reduce the risks of shark attacks on its personnel, zoologist H. David Baldridge, undertook a computer analysis of 1,165 documented shark attacks in various parts of the world over a period of more than twenty years. The occurrence of the attacks was correlated with such factors as water temperature and depth, distance from shore, and time of day. But perhaps the most startling findings in the report, which was completed in 1973, regard the characteristics of the victims. These findings are reprinted below.

There are for shark attack victims a series of considerations which at best can be described as open-ended or totally uncontrolled. That is, in the absence of corresponding data on non-victims, there are very few conclusions that can be drawn which have any true significance, statistical or otherwise. This, however, has not stilled the hands of an assemblage of earlier writers, both popular and scientific, from drawing a number of often-quoted associations and assigning to them the status of cause-effect relationships. . . .

Sex of Victims

Almost all of the victims of coded attacks were identifiable as to sex. A surprising 93.1% (1,080 out of 1,160) were male. Perhaps even more meaningful is the fact that this predominance of male victims held true when race was also considered: white, 93.0% of 725 victims were male; black, 94.3% of 176; brown/yellow, 92.4% of 105; and those of unknown race, 92.9% of 154. On the average, there were 13.5 male victims for each female.

The first thought that came to mind was that more men than women are engaged in occupations which offer greater opportunity for shark attack, that is, fishermen, mariners, etc. So, the computer was restricted to only those cases associated with recreational activities at or near

beaches. Again, shark attack victims were found to be very predominantly male: white, 89.7% of 341; black, 91.4% of 58; yellow/brown, 93.9% of 33; and race unknown, 87.5% of 32. The overall ratio of male to female victims at beaches was 9.1 to 1.

Counts of people in the water at typical bathing beaches showed a slight predominance of man, but nothing consistent with an attack ratio of over 9-to-1. Our observations, however, did confirm suspicions that males, in general, are more active when in the water. Activity, and the associated sonic and visual stimuli provided to sharks, has long been recognized as one of the prime triggers of aggressive shark behavior. In studies where laboratory-bred rats were used to simulate survivors at sea, sharks repeatedly struck struggling, splashing rodents while showing little or no interest in either the same animals or others nearby when they were relatively motionless in the water. Perhaps males present to sharks significantly different olfactory profiles than do females. It has been often suggested that the danger of shark attack to a female might be affected by the stage of her menstrual cycle, but, in the absence of definitive data, this remains but conjecture. Very little is known about the chemical interchange which occurs between the body of a man or woman and the water in which they may be totally immersed.

Maybe there is in this interchange some substance, hormonal perhaps, that is more peculiar to males and which sharks interpret as indicating some form of threat to which they respond with aggressive behavior. This begins to make sense when, as discussed later, it is realized that a considerable percentage of shark attacks do not appear to have been motivated by hunger. In any event, we have in this about 10-to-1 male-to-female ratio in shark attack victims something that can not be explained off as being consistent with observed patterns in beach populations. Clearly, there is here a need for further basic research. . . .

Depth of Victim in Water at Time of Attack

In popular writings, it has been hypothesized that shark attack is primarily a hazard of surface swimmers, and that divers, being beneath the surface, are in a less hazardous position. In support of this, it is generally stated that a diver disturbs the environment far less than a swimmer and he also is in a better position to see an aggressive shark in time to counter its charge. There are 881 cases in the SAF [Shark Attack File] where there were data on how deep the victims were in the water when the attacks occurred. By far the majority (797 cases or 90%) happened either at the surface or no deeper than 5 feet from the surface. Sixty-two attacks occurred within 6–30 feet of the surface (23 at 6–10 feet, 27 at 11–20 feet, and 12 at 21–30 feet). Deeper attacks were as follows: 8 at 31–40 feet, 2 at 41–50 feet, 2 at 51–60 feet, 3 at 61–70 feet, 4 at 71–80 feet, 1 at 101–110 feet, 1 at 111–120 feet, and 1 at over 150 feet. Great care must be taken in interpreting these data in the absence of control information on non-victims. It is true that only about 10% of the shark attacks on file were directed against subsurface victims. These figures appear to confirm the idea of relative immunity for divers. Yet, intuition tells us that it is surely not likely for any period of time on a worldwide basis that 10% of the people in the water, either at beaches or off-

shore, are to be found more than five feet below the surface. Considering the great masses of people who swim at beaches in the warm water regions of the world, divers of all categories would be expected to constitute only a very small percentage (certainly far less than 10%) of persons exposed to shark attack. And consider further the fact that skindiving, SCUBA diving, spearfishing, and other related subsurface water sports have only recently become immensely popular. While the data in the SAF can not be taken to conclusively indicate a greater hazard for divers as compared to surface swimmers, it certainly does not appear to support any status of immunity for divers. That divers are actually more likely to be attacked than surface swimmers would be consistent with . . . [other] conclusions regarding distance from shore. In order to dive to any appreciable depth, one would generally have to move farther out from shore than for ordinary surface swimming. To move farther out from shore is to make encounter with a shark possibly more likely. So, the fact that the person is either at the surface or well below it may not matter at all. There simply is not enough information to separate the effects of attack depth and distance from shore.

It is extremely interesting that of the 65 cases of attacks on females where depths of the attack sites were reported, *not a single case is on record of a female being attacked by a shark below 0–5 feet from the surface.* Thus, it may be that not only do females not venture away from shore as far as males, but when they do, as in the case of female skindivers, perhaps they also do not do some of the things done by male skindivers which are likely to invite the shark attack. Could it be that female skindivers swim with a basically different movement than males—a movement far less exciting to sharks? Is it possible that sharks differentiate chemically between male and female humans even when the divers are encased in full wetsuits? Or maybe female skindivers are not as taken to such ac-

John Singleton Copley, *Watson and the Shark,* 1778
*Depicting an actual shark attack on Brook Watson in
Havana Harbor, this is the first significant marine painting
by an American artist.*

tivities as spearfishing, or wearing black wetsuits where sharks feed on seals, or prying
abalone from rocks in waters inhabited by large
sharks, or tying bleeding fish to their belts,
or any number of other things reported in association with shark attacks on male skindivers.

Nature of Wounds

It was this part of the analysis that first led to
the conclusion that many shark attacks have not
been efforts on the part of the attackers to devour
the victims. After a high percentage of attacks,
there was no significant amount of flesh missing
from the victims—a fact clearly inconsistent
with the known effectiveness of sharks as predators. Furthermore, photographs and descriptions of many of the lacerations indicated that
they were not as likely to have been caused by
a bite in the usual sense as they were by some
other application of the teeth such as open-
mouth raking—most likely with the flat, ser-
rated, very sharp edges of the upper teeth—
resulting in severe cuts and slashes without
significant loss of tissue.

The data . . . indicate that the primary medical problems associated with shark attack on an emergency basis would be loss of blood from massive wounds and shock, and that these factors, complicated by the possibility of drowning, would be the most likely causes of death among victims. It was relatively rare that a victim received wounds which could not have been survived if treated promptly and definitively. There were those cases, of course, where the location of the wound on the body overrode the nature of the injury in determining survival. As to inherently fatal wounds, the body cavities of victims were opened in only 42 (5.0%) known cases, the trunk was actually severed in only 9 out of 959 cases considered, and essentially total removal of flesh from the victims' skeletons was reported only 7 times out of 978 cases where sufficient information was available for judgment. And, the classically envisioned picture of a shark swallowing its victim whole was thought to have happened in only 6 cases on record.

Considering these observations on wound characteristics, it is not too surprising that about three-fourths of all present day victims survive being attacked by sharks. And I repeat, the array of wounds . . . is not at all consistent with determined efforts of an otherwise very successful predator to generally attack, kill, and devour man as a prey of choice.

Seascape and the American Imagination

Roger B. Stein

From prehistoric times to the present, the sea has been an important theme in the art of all maritime nations. Reflecting both their individual experiences and the nation's involvement with the sea, American artists have produced a rich and varied tradition of seascape painting. In the following excerpt from Seascape and the American Imagination *(1975), Roger B. Stein, who assembled an exhibition of American marine paintings for the Whitney Museum of American Art, discusses both American seascapes in particular and the relationship through the centuries of all seascape artists to the sea.*

*F*rom the discoveries of Columbus in 1492 to the first settlements at Jamestown and Plymouth at the outset of the seventeenth century and for three hundred years thereafter, American experience has been inextricably linked to the sea. The sea has been the barrier to be crossed on the way to the New World, which dislocated the immigrant from all that was familiar in the old. It has been the source of cod and whale oil to feed and light the lamps of Europe and America, a nautical highway for burgeoning commerce and industry, and a testing ground for American skill and prowess. In a multitude of ways American culture has been defined and shaped profoundly by its experience with the sea. Our powerful thrust into the western frontier has been counterbalanced by a thrust eastward into the equally vast and untamed Atlantic Ocean, and our art has recognized and recorded that fact. We have heard much about the artistic recorders of our landscape from the Hudson River, the Catskills, and Niagara Falls to the Rocky Mountains and beyond, but the imaginative record of American experience with the sea is less well known. And yet, historically speaking, seascape art has a prior claim upon our attention, for the sea engaged the American imagination long before Americans fell in love with their land. The history of American seascape should thus document the range and variety of our encounters with the maritime world.

It can do more than that, however. Seascape is an artistic mode, an organization of expressive gestures that can give permanence to our visual awareness of the sea, the coast, and the voyage. As such it is as much a record of our thoughts and feelings about that outer fluid space and our attempts to map it and to impose upon it a conceptual grid as it is a documentary record of that space itself. The works of art that comprise the tradition of American seascape art demand to be read, thus, not only as elements in the history of a particular genre called "marine painting" but also as ways in which Americans have understood themselves as a people in the New World. Because the expressive language is artistic, we must inevitably confront how an individual artist constructs his pictorial world, for each artist's gesture is his own and unique, though it may later be transformed and multiplied by a Currier & Ives print, rearranged and stamped in red on a muslin souvenir kerchief, incised as scrimshaw on a whale's tooth, or reproduced in some other way, in whole or in part, by another artist. Yet because seascape taps such an important shared cultural experience, these individual works become particular configurations of that experience, forms of cultural self-definition. By exploring the range and variety of our seascape tradition, we can achieve a richer sense of our artistic heritage and perhaps come to understand better the nature of American experience: our aspirations and our fears, our achievements and our failures, our triumphs and our tragedies.

At the outset of our study, there is a paradox, or at least a tension, inherent in the very notion of seascape; it implies the imposition of form and order, of a perspective and a point of view, upon that which is by definition formless and shapeless. The sea has embodied through the centuries ideas of the vast and the untamed, the All, the Universal Oneness, or even Chaos. In the Judeo-Christian tradition the first seascape artist is Deity himself, in the opening chapter of Genesis, when He separates the heavens from the earth and delineates a horizon between sea and sky. As if to emphasize the tenuous nature of the act, the later chapters on Noah and the flood reinforce the point: because of human beings' moral failure, that stay against chaos, that line of separation which differentiates and makes space measurable, may be withdrawn and the land world of the human community swallowed up by the Deluge. Often with a rich sense of this threat implicit in our creation myth, the seascape artist over the centuries has attempted to further differentiate the original space, to make humanly understandable, if not ultimately humanly habitable, the endless and intrinsically terrifying openness of the undifferentiated space of the sea.

Winslow Homer, *The Life Line,* 1884

The Art of the Seaman

Gerhard Timmermann and
Helen L. Winslow

Much of the best literature and art of the sea reflects the personal experience of the writer or artist with the ocean. Few artists have ever been in a better position to interpret the sea than those whose lives are inextricably bound to it—the seamen themselves. Throughout the nineteenth century, these folk artists, unschooled but highly skilled, took advantage of leisure time at sea to turn whale teeth and whalebone into works of art. Scrimshaw and such other handicrafts of the sailor as painted sea chests and model ships provide a rich and beautiful record of the life of the sailor in the days of sail. These crafts and their craftsmen are described in the following article by Gerhard Timmermann, a German expert on the history of shipbuilding and marine architecture, and Helen L. Winslow, who has written on scrimshaw for the Historical Society at Nantucket, Massachusetts, one of the great whaling centers of the nineteenth century. The article originally appeared in a collection titled Art and the Seafarer, *edited by Hans Jürgen Hansen and published in 1968.*

*I*t has already been established in connection with ship portraiture and model building that many works in these fields were done by amateurs—by seamen, in fact. But the skill of seamen was by no means confined to these industries, and it is worth taking an overall look at the whole of their artistic repertoire.

All of their products are characterized by the fact that, whilst being what is generally called folk art, they were international in their style, their subject matter and the materials from which they were made, because those who made them belonged to every seafaring nation in the world. The products of this amateur art have found an increasing number of admirers in recent times, and without doubt represent the most original and interesting collectors' pieces.

Seamen are condemned to spend a part of their life in a very restricted place of work which they cannot leave for long periods at a time. In the days of the sailing ships such a stay on board, could, depending on circumstances, last weeks, months or even years. It was hard work in bad weather and there was very little free time. Everybody was only too pleased when such a voyage was over. But in fine weather the crew of a sailing ship had enough time during the

watch below to busy themselves with their hobbies. Reading was not a common occupation then because very few seamen could read. So there was nothing else to do but to pass the time with some kind of handicraft.

Some of the crew drew or painted, often with primitive materials, others carved pictures in ebony, bone, ivory, copper or whalebone. Others borrowed tools from the ship's carpenter, the "chippie," if they did not possess their own, and made little chests, cupboards, benches and stools and other small items of furniture, either to be used on board or for their relatives back home. Sewing cases and tobacco boxes were favourite presents. All along the North Sea coast of Germany and also in Denmark richly decorated mangle boards, mostly from the 18th century, carved by seamen during their long voyages and then brought back home, are very common. Other skilfully ornamented objects made by sailors included writing stands with or without ink-pots, dressing-table tops with mirrors and tools for every conceivable purpose.

A particularly favoured material were the bones or teeth of walruses or sperm whales which were caught on the voyages to Greenland and in the South Sea fishing grounds. In particular the members of the so-called South Sea fishing expeditions (which were not fishing but whaling expeditions and took place north of the equator as well as in the South Seas) made large numbers of "scrimshaws" from ivory, engraved with attractive patterns, whaling scenes or scenes from the Bible, occasionally even romantic scenes. They are reminiscent in their execution of man's earliest pictorial efforts, the engraved figures of animals on reindeer bones from the early Stone Age Ahrensburg culture, and at the same time of the motifs seamen had tattooed on their skin.

But these rough men also knew of subtler ways of working the ivory: many cut proper bas-reliefs and very delicate pictures from this durable material. Without the help of a lathe their large hands shaped fragile little boxes and needle-holders which were scarcely inferior to the work of an ivory turner. There were also rulers, prettily decorated, clothes pegs, fishing rods, clothes hangers, pastry cutters, paper knives, butter knives, dividers, corset stiffeners and many other useful objects. All these things are to be found in large numbers in American maritime museums, because the Americans made the longest whaling trips. The late President Kennedy collected some very fine pieces. Apart from bones and ivory the baleen whale, which was first hunted in the Arctic and later in the Antarctic, also yielded baleen, popularly called whalebone, which was ideally suited for making boxes and caskets. Elliptical hat boxes could be shaped very easily from the elastic and flexible whalebone; they were often decorated with engraved ornaments and painted into the bargain.

It is worth going into the art of scrimshaw in some detail—this most remarkable of folk arts. In the middle of the last century Herman Melville wrote in *Moby Dick*: "Throughout the Pacific, and also in Nantucket, and New Bedford, and Sag Harbor, you will come across lively sketches of whales and whaling scenes, graven by the fishermen themselves on Sperm Whale teeth, or ladies' busks wrought out of the Right Whale bone, and other like skrimshander articles, as the whalemen call the numerous little ingenious contrivances they elaborately carve out of the rough material, in their hours of ocean leisure. Some of them have little boxes of dentistical-looking implements, specially intended for the skrimshandering business. But, in general, they toil with their jack-knives alone; and with that almost omnipotent tool of the sailor, they will turn you out anything you please, in the way of a mariner's fancy.

"Long exile from Christendom and civilisation inevitably restores a man to that condition in which God placed him, i.e. what is called savagery. Your true whale-hunter is as much a savage as an Iroquois. I myself am a savage, owing no allegiance but to the King of the Cannibals; and ready at any moment to rebel against him. Now, one of the peculiar characteristics of the savage in his domestic hours, is his wonderful patience of industry. An ancient Hawaiian war-club or spear-paddle, in its full multiplicity and elaboration of carving, is as great a trophy of human perseverance as a Latin lexicon. For, with but a bit of broken sea-shell or a shark's tooth, that miraculous intricacy of wooden network has been achieved; and it has cost steady years of steady application. As with the Hawaiian savage, so with the white sailor-savage. With the same marvellous patience, and with the same single shark's tooth, or his one poor jack-knife, he will carve you a bit of bone sculpture, not quite as workmanlike, but as close packed in its maziness of design, as the Greek savage, Achilles's shield; and full of barbaric spirit and suggestiveness, as the prints of that fine old Dutch savage, Albert Dürer."

The art of scrimshaw has been considered the only important indigenous folk art, except that of the Indians, which has ever developed in America. At the same time it has, of course, always been international. The term once applied to all forms of carving or decorating of whales' teeth, walrus' tusks, or bone, but nowadays usually refers specifically to the teeth engraved by the whalemen. The subject has been given

scant notice by literate whalemen and the historians of whaling, except for Herman Melville. In recent years, however, an increase in antique collecting has focussed attention on these miracles of whale ivory sculpture achieved with the crude, scanty tools available to the sailor of the whaling era.

The origin of the word "scrimshaw" is obscure. Several dictionaries attempt to derive it from the surname "Scrimshaw." This is doubtless in error, since Scrimshaw is the more recent of the several forms of the name. Others, tracing the word to Nantucket, surmise it to be of Indian origin. Another theory, advanced on the basis of "skimshander" or "skrimshander" offers a possible analogy between these forms and the words "skimp" and "scrimp," meaning "scant," or, in verb form, "to economize." There was always a dearth of good material, for large pieces of whale ivory were rare. Thus the nature of the material commonly made scrimping necessary. However, considering the term "scrimshant," an early form, others believe that scrimshaw comes from an old word "scrimshander" or "scrimshanker," an idle, worthless fellow. Gradually the term may have come to mean the artistic results of a sailor's idle hours at sea. This would seem to be the most satisfactory derivation, but it is no more conclusive.

The earliest reference to the art of scrimshaw, by name, occurs in the logbook of the brig *By Chance* of Dartmouth, Massachusetts, preserved in the collection of the New Bedford Whaling Museum. Under the date, 20 May 1826, it reads: "All these 24 hours small breezes and thick foggy weather, made no sale [*sic*]. So ends this day, all hands employed Scrimshanting." However, the true beginnings of the art must be found in the 18th century.

Soon after Nantucket's first sperm whaling venture in 1712, longer voyages left considerable time at the disposal of the whalemen. To Nantucketers indoctrinated with the belief that

SCRIMSHAW BUSK

idleness was a "most heinous sin" and trained in the cooper's trade whatever their future occupation might be, wood-carving was second nature. . . .

Although at first wood was used more than ivory, whalemen must soon have discovered the choicer material. Certainly, despite its lack of publicity, the art of scrimshaw played a major role in the everyday life of the whalemen. On some ships every man from captain to cabin boy had an article of scrimshaw under way. Men swapped tobacco—the universal currency aboard whalers—washed clothes or did other menial tasks in order to gain coveted pieces of ivory. New Bedford owners once fiercely debated whether the engrossing interest of the whaleman was not seriously detrimental to the success of voyages. Men had even been known to sight whales and then fail to report them rather than interrupt some particularly fascinating stage of their artistry. For captains to forbid scrimshaw altogether was unusual but not unknown. On some vessels scrimshaw was limited to the forecastle and was subject to confiscation if brought on deck. Scrimshaw was so widespread that it may be said to have become universal.

Clifford Ashley, a foremost student of the technique, summarized the requirements of the ambitious fashioner of scrimshaw as including proficiency alike in joinery, turning, carving, inlay, coopering and engraving. That the sailor's tools in no way met the most meagre requirements of any of these trades is further tribute to his artistry. We have already quoted Melville on the tools used. Sometimes sail knives were used instead of jack-knives. Improvized files, converted chisels and gimlets fashioned from nails were also employed. Green whalebone was soft enough to be planed or otherwise worked, but the harder whale tooth required sharp cutting tools.

Some ships had jigsaws or home-made turning lathes—most often owned by mates in their more spacious quarters—to attain lace-like effects in the pieces of more intricate design.

However, the majority of the examples of fine turning and execution resembling scroll-saw work were actually produced by the use of makeshift files. For the American whaleman pioneered a form of art which required that he must first fashion the very tools he would use.

The sailor's needle was the most versatile element in the paraphernalia of scrimshaw. From it were contrived a variety of files, fine saws, and the piercing and boring instruments used in executing the openwork patterns found in rings, bracelets, brooches, needle cases, etc. These awl-like tools were also used to trace the outline of a future design by a series of pricked or punched holes. Ivory or bone parts entered into a large proportion of the hand-made implements. Bone handles were standard. It would seem that the tools themselves could be placed in the category of scrimshaw.

The polishing process has been given credit for much of the mastery of the pioneers. This is due no less to the preliminary grinding, filing, smoothing, or sandpapering necessary to reduce the ribbed whale's tooth to a more workable surface than to the finishing done with wood ash and laborious hand polishing. The more skilful, on occasion, even used the skin of the shark, in lieu of sandpaper, as a smoothing agent. Ash, pumice, and whiting were usually employed for conditioning the whale teeth. In this, as in every other stage of scrimshaw evolution, patience was the price of perfection. Whalemen put the fresh teeth in brine to soften them, for, as they grew older, the teeth became harder and correspondingly more difficult to work. Even so, it was possible by the use of hot water to get a surface that would respond to their tools.

The scrimshaw creators were cautioned by the knowledge that a slip of the knife, or a cut too deep with the bodkin, would force them to discard a literal labour of love which had occupied months or even years. Some of the men became every expert in carving and decorating the teeth and produced marvellously delicate and beautiful handiwork despite the crudeness of their tools. The inking or colouring of these scrimshaw etchings or engravings was a detail that is in no small degree responsible for the elusive characteristics of old scrimshaw. Indeed, it is the inability of most of the present practitioners of the art to simulate this subtle quality which has been responsible for the criticism that contemporary pieces look "too new." The supposition has been that India ink alone was used to emphasize the engraved designs, while in actuality ink was not always as available as paint, tar, or even soot from the try-works. In most cases black pigment was depended upon to obtain contrast for the incised lines. Occasionally red was introduced in conjunction with the black. In rare instances other colours, notably green, blue or orange, were used in combination. By this laborious and crude method results equalling the finest steel engravings were often produced, although the majority of scrimshaw teeth show little evidence of artistic talent.

There is little reminiscent of other arts in scrimshaw. Despite attempts of writers to seek its origins in the primitive arts of the Eskimo or of the South-Sea islander, it is well established that the sources of inspiration were to be found in the environment of the whaleman—either in the home surroundings he had left or in his life at sea. Certainly a trophy of the whale hunt, the symbol of the whalemen's success—a huge tooth taken from the gigantic sea mammal whose capture held his life in constant danger—was a most natural gift to a distant friend or loved one: a gift made even more meaningful by its carefully etched pictures of the ship under full sail, incidents of the chase and capture of whales or other maritime scenes.

SCRIMSHAW JAGGING WHEEL OR PASTRY CRIMPER

In his choice of scrimshaw designs the sailor expressed great individuality, although, on occasion, a man of artistic talent might influence the output of a whole ship. Life around him constituted his dictionary of ornament—the knots in the rigging; the stars in the heavens above him; the figure-head and the stern board of his ship; the fish of the sea; whales, birds, sails, boats, casks, bells; the wheel, the anchor and similar symbols. Some of the best and most elaborate work was traced or transferred from books, magazines or illustrated papers, which found their way into the forecastle of the whaling ship. Pictures of women were frankly copied, primarily from *Godey's Book*. In all the seaman's work, however, there is a marked and sturdy originality in selection.

The earliest dated piece of scrimshaw in the collection of the Nantucket Historical Association is a tooth decorated "off the coast of Japan" on the first voyage of the ship *Susan* of Nantucket in 1829. On the reverse one can read the following couplet:

"Death to the Living, Long live the Killers, Success to Sailors' Wives, and Greasy Luck to Whalers."

There is also a whaling scene and the name of the master, Captain Frederick Swain.

Although his graphics on whale teeth were the most familiar fruit of his craft, it was in the busk, made of planed whalebone, that the whaleman etched his most inspired pictures and waxed most sentimental. To the uninformed, a busk was a flat ruler-like "stay" about two inches wide, which milady of the 19th century thrust into an open slit at the front of her corset. It has been said, as much in truth as in jest, that any woman so fortified was bound to remain true to her sailor.

Frequently these functional ornaments bore appropriate and tender verses. One could ill afford to omit this oft-quoted tribute to the charms of a loved one and to the lure of the sea:

> "Accept, dear Girl, this busk from me;
> Carved by my humble hand.
> I took it from a Sperm Whale's Jaw,
> One thousand miles from land!
> In many a gale,
> Has been the Whale,
> In which this bone did rest,
> His time is past,
> His bone at last
> Must now support thy breast."

More ambitious and skilful whalemen fashioned articles from the teeth as delicately carved, as well finished, and as intricate in design as any work of the Orient. The majority of these items were designed for human adornment; still others provided recreation after, as well as at the time of, their execution; while a few must be classified purely as objets d'art.

Among the extensive scrimshaw collection of the Nantucket Historical Association one can find numerous items which were commonly found in island kitchens—chopping knives, a corkscrew, butter stamps and butter paddles, dippers, dish mops, spoons and forks, a corn skewer, lamp picks, napkin rings, and rolling pins. More numerous than all such articles are the odd "jagging wheels," elaborate implements for cutting, piercing and crimping the edges of pies. No one seems to know just why the whalemen were so fond of making these. Perhaps it was the challenge they offered in craftsmanship. Perhaps they were a natural

THE PEABODY MUSEUM OF SALEM

78

tribute to the delicious pies baked in New England kitchens, which for years at a time were but memories of the past or expectations of the future. Whatever the reason, these carved pastry wheels, produced in vast numbers, were ingeniously constructed and most beautifully wrought. . . .

Nor did the sailor ignore the housewife's responsibilities for the family wardrobe and linen. In his leisure time he fashioned clothes pins, spool racks, thimble and needle cases. . . .

Other subjects of the whaleman's craft were door knobs, hooks, boxes, baskets, birdcages, paper knives, yardsticks and rulers, seals and stilettos. The latter items frequently found their way into the captain's desk on board ship. Still others served practical purposes in the equipment of navigation, as in ropemaking. Handles were also made for gimlets, hammers, knives, and so on.

Common recreational activities of the 19th century were games of cribbage, checkers, chess and dominoes. To these leisure time pursuits the craftsmanship of the sailor contributed cribbage boards of whale ivory or walrus tusks, checker and chess boards, chessmen and dominoes. Strangely enough few ship models were made. Perhaps quarters were too cramped or perhaps the whaler failed to offer the artistic inspiration of the faster clipper ship. Thus the whalebone model of the *Lagoda* in the Nantucket Whaling Museum is indeed a rarity.

SCRIMSHAW WHALE
TOOTH: SHIP *ELIZABETH*
OF SALEM

79

Another branch of scrimshaw art includes articles made for adornment. For himself the sailor carved collar buttons, cuff links and rings; for his wife, decorative combs, beads, earrings, pins and brooches. Outstanding in this phase of scrimshaw were the handsomely wrought canes and cane heads which the whaleman treated with great individuality, selecting his designs from all areas of his nautical experience. The collection of whale bone and ivory canes in the Nantucket Whaling Museum provides ample illustration of the variety in subjects—innumerable geometric designs, clenched fists, Turk's head knots, sea bird heads, dogs, snakes, and antique editions of pin-up girls in high-buttoned boots.

The sailor benefited from one great advantage: in foreign ports he could get hold of rare species of wood which were easy to work and looked most attractive. From them he usually built his sea chest, the most important piece of equipment on board besides his berth. The crew's quarters on the old sailing ships were exceedingly primitive. They often had no windows or only skylights; along the walls were bunks and in the middle a table and perhaps two benches. The seaman had to stow his private belongings in his sea chest, which often served as a seat as well. Cupboards or wardrobes were unheard of. The sea chest, called *Sviptikista,* was already known in Viking times, and literary references to it are supported by the fact that no kind of seating has ever been found in the remains of Viking ships. Sea chests with all their contents were also found in the *Vasa.* The old sea chests had sloping sides, which made their bases wider than their tops, which gave them a firm stand when the ship was rolling, even with men sitting on them. In some only the two longer sides were sloping, since this was enough to give them the necessary stability. Most chests were about 3 ft. long, 18 in. wide and 18 in. high. Chests from the last days of the sailing ships usually had straight sides and very often were made from inferior wood compared to the old chests. They were all very strongly built to withstand the rough treatment they received, and were often painted on the outside, usually brown, sometimes rust or a handsome bright green. The lid sometimes overlapped and was closed with two strong iron hinges. The sailors were unaccustomed to fitting locks; a simple hasp and staple were enough. Theft was not, and still is not, tolerated on board. The lid was covered with sailcloth and generally painted black. The two rope handles, one at either end, were much easier to grip than the screwed-on iron or brass handles which were used later. These rope handles were not just simple rope, knotted on, but they were "grafted" and decorated with Turks' heads, and their owners took great pride in them.

A sailor who could afford to went in for teak wood and got the ship's carpenter to build him a chest, which was not only extremely strong but needed no painting on the outside. The inside of the lid, however, was usually painted white and then decorated with all kinds of ornaments, flowers and garlands, flags, symbols of faith, love and hope, or at least the owner's initials adorned with countless flourishes. The most handsome chests are probably those decorated by their owners with paintings of ships; although similar to ship portraits or votive pictures they are much more primitive, because the paint on board was of inferior quality. Besides, sailors knew more about painting their own ship than they did about painting pictures. But the loving care with which a sailor built and decorated his sea chest shows how attached he was to it. It was the only piece of furniture which actually belonged to him during the voyage; everything else was the property of the ship-owner. In order to provide a tidy stowage place for buttons, tapes and other small objects the chest was often fitted with small compartments, some with a folding or sliding lid. These lids were occasionally decorated with coloured inlays. All in all the various sea chests were real ornaments in the otherwise stark quarters. Bones and ivory, which we mentioned in connection with scrimshaw, were also used in building ship models because they were easy to work with file and saw. Some of the smallest models were made from bone. Another material that was popular for its workability was mahogany. It is doubtful whether planked models were ever made at sea, but whole and half models made from solid blocks of wood were extremely common. Many of them were far from perfect, but those that turned out particularly misshapen probably did so not because their makers suffered from lack of taste but because they were so awkward with their hands. Even then it is surprising how the rough, gnarled hands managed the delicate task of rigging the models and reeving thin thread through small, block-like pieces of wood. The yards usually turned out too thick, but the standing and running rigging was correct in every detail. Every stay was where it belonged and every sheet was led in exactly the same way as on the ship in which the man crossed the oceans and which he wanted to remember.

The little ships were either put into a glass case or into a wooden one with a painted background which might portray a stormy sea, the white cliffs of Dover or Naples with the smoking cone of Vesuvius. But the water with its magnificent waves had to be made from coloured putty. Sometimes a small pilot cutter would be sailing in front of the ship. Finally a glass front was put on. If the sides of the box were angled obliquely and the scene was framed, as many of them were, it looked like a three-dimensional oil painting. But why build a whole ship if only one side could be seen? So in due course the seamen left out half the ship. They gave them sails, too; not of cloth or paper but of wood, which had to be very thin so that the sails could be made to belly out in the wind. They were usually close-hauled to show them in their full width, but some models had them squared off and a mirror put in as background to give the impression of a complete model. Such "panoramic models" can be found on the coasts of all countries, frequently hanging on the wall above the door of a captain's house.

Visions of the Sea

BRONZE AGE SHIPS SET SAIL FROM THE ROCKS IN NORTH BORNHOLM, DENMARK

Bronze Age rock engravings from the second millennium B.C. reflect the sea-faring culture of the Scandinavian people. The high-prowed ship, which held together a trading community that was separated by seas and lakes, was a universal symbol throughout Scandinavia.

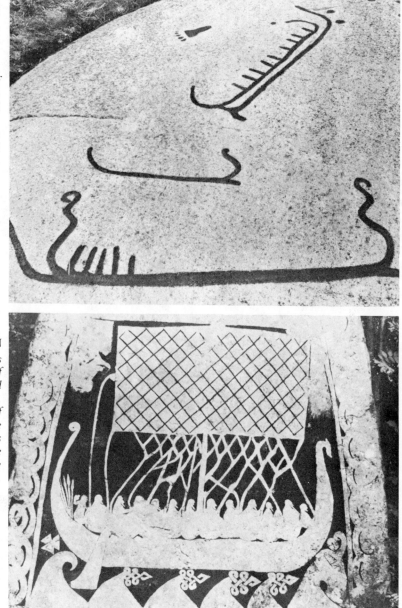

VIKING SHIP ON PICTURE STONE, GOTLAND, SWEDEN

The sailing ships of the Vikings ventured along the coasts of Europe, deep into Russia, and across the Atlantic to America. Figureheads, in the form of fierce-looking animals, were designed to frighten evil spirits and were removed near shore so as not to frighten the friendly land spirits.

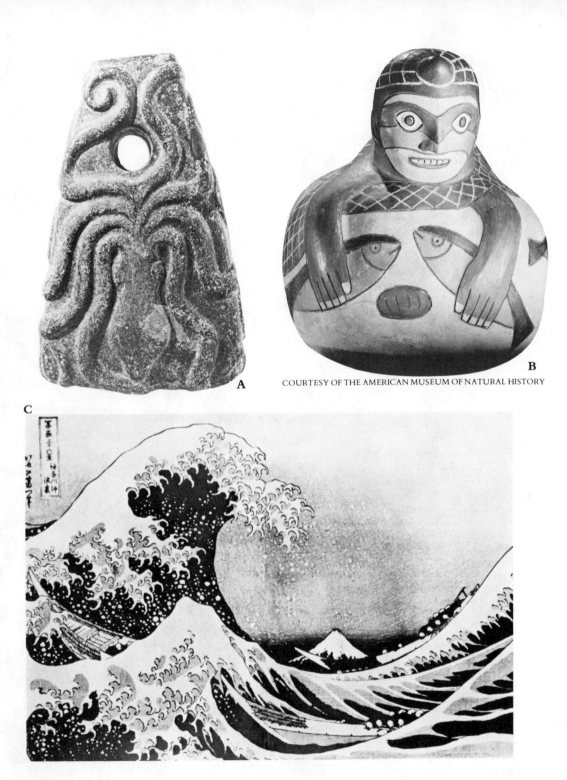

COURTESY OF THE AMERICAN MUSEUM OF NATURAL HISTORY

A

B

C

D

A. OCTOPUS: ORNAMENTAL PORPHYRY WEIGHT, KNOSSOS, CRETE
Circa 2000 B.C.

B. SWIMMING FISHERMAN CONTAINER, PERU
Early Nazca Period, second or third century A.D.

C. Katsushika Hokusai, *The Crest of the Great Wave off Kanagawa,* circa 1829
One of a series of thirty-six views of Mt. Fiji, this Japanese print has immortalized the "tsunami." These towering waves are caused by underwater earthquakes or volcanic eruptions and have nothing to do with tides.

D. Winslow Homer, *Breezing Up,* 1876

E. Alexander Calder, *Oceans,* 1976

E

H.M.S. *Challenger* off Tristan de Cunha, painted by Benjamin Shepherd, a member of the crew.

TRISTAN. DE. CUNHA. OCT. 14TH 1873.

Unit Three

Few people have the breadth to see the sea as a whole. Most of us are like the blind men of the fable who thought the whole elephant was like the small part each was touching. This is particularly true of scientists, who are required by profession to discover new things and to examine them very, very carefully. Most of what scientists discover is difficult to explain at first just because it is, by definition, new. It can be communicated only to specialists and advanced students. It is usually even harder to explain the point of scientific work or why taxpayers should pay for it—as they do.

Now and then, perhaps once in two or three generations of scientists, a scientific revolution occurs. Some of the endless array of new, obscure, seemingly pointless facts simply cannot be explained by the theories that the scientists have accepted since they were students. This happened when Copernicus showed that the earth moves around the sun and when Darwin demonstrated evolution.

Such a scientific revolution has occurred during the past two decades. Scientists have found that the continents move and have measured the rates at which they spread apart. The proof of their theories came from the sea and, so, is of interest here. Fortunately, when an idea is revolutionary it is easy to grasp—hard

Science and Myth

to believe, perhaps, but understandable by the nonspecialist because there are no specialists talking jargon to each other.

This unit of the Reader is concerned with scientists, science, and myths. It begins with several selections related to how scientists study the oceans—by going to sea, taking their laboratories with them on oceanographic expeditions.

The most important expedition of all was one of the first, the voyage of the British ship *Challenger* from 1872 to 1876. It was supported by the Royal Navy and carried out by the Royal Society of London for rather practical reasons related to submarine telegraphic cables and the location of fisheries. We have selected part of the original correspondence between the navy and the Royal Society that sets forth the objectives and planning. To illustrate the general public interest and the attitude of some members of Parliament we include a poem from the humor magazine *Punch* dated December 14, 1872. "But, pray, how about cost?" says John Bull. He is assured it will be nominal and says in that case, go ahead. An economic analysis a century later showed that the *Challenger* Expedition turned out to be the most expensive oceanographic expedition ever undertaken.

A selection from *Anatomy of an Expedition* by H. William Menard carries the motivations, organization, and cost of expeditions into our times. They have changed surprisingly little in many ways, although the scientific problems under investigation keep changing.

From *Seven Miles Down* by Jacques Piccard and Robert S. Dietz we have a personal account of the first dive in a bathyscaph to the bottom of the Challenger Deep—the deepest spot in the ocean. (This deep spot was discovered not by the great *Challenger* but a worthy successor of the same name.) This was not so much a scientific expedition as an effort of the human spirit—doing it, as Mallory said of climbing Mt. Everest, "because it's there."

In the selection by William Wertenbaker we see scientists in the act of discovery. We focus on a few people at the Lamont-Doherty Geological Observatory of Columbia University in New York. It is one of the leading oceanographic centers in the world and has existed for hardly a quarter of a century. So rapidly has science been applied to the sea!

These scientists recount their discoveries and how they looked at the world and saw it changed, as in Ariel's song, "into something rich and strange." We see it now in terms of "plate tectonics"—the movements of enormous rigid pieces of the earth's crust. These may contain a whole continent like a ship in an ice flow.

In "Minerals and Plate Tectonics" by Allen L. Hammond we pass from the excitement of the first discoveries of plate motions to the application of the new ideas. The mineral deposits of the land are now seen in new ways, and this may help us find badly needed resources.

In the last three selections we return to ideas, to history and prehistory, to the scientific tests of myth. We have already seen the happy art of Crete and read Homer's brief description of the kingdom. In J. V. Luce's *The End of Atlantis* we read Plato's original account of the lost continent and its destruction. Then we follow a bookish trail of interpretations by scholars during recent centuries. Where was it? What happened? Finally we come to the application of modern science and a new solution to the problem. It isn't lost—you can take a cruise to Atlantis.

From *Deep Water, Ancient Ships* by Willard Bascom we have a few short descriptions of some of the most important underwater archeological sites near Atlantis. From the preserving cold of the deep we are recovering a profusion of objects that reveal the ancient world in intimate detail. The amphorae of commerce, the marble statuary, the ships themselves show a lively world that has vanished. Most remarkable of all is the clockwork "computer" from Antikythera. Professor Derek Price of Yale University has found that the purpose of the machine was to grind out astronomical data such as the positions of the stars. Ancient Greece had what we now call "high technology." Is it a lesson for our times that what survived was not the technology but the art?

What outlives all are ideas, and the most ancient we call "myths" or "gospel" according to our wont. One of the most pervasive of ancient ideas, existing in cultures world-wide, is the idea of an ancient flood. In our culture it is the biblical flood on which Noah floated in the ark.

In the 1960s several symposia were held, in the words of Paul Zimmerman, "For the man who in the last third of the 20th century continues to be concerned about the relationship of the truths of the Bible to the truths of science." "Geology and the Flood" by Paul C. Tychsen is a selection from the book *Rock Strata and the Bible Record,* which summarizes the results of the meetings.

When glaciers covered much of the far north, water was withdrawn from the sea, and sea level dropped about 300 feet. Humans all over the world took their ancient fishhooks and boats and followed the receding shoreline. When the ice began to melt for the most recent time 18,000 years ago, sea level began to rise. For 12,000 years coastal families and tribes abandoned homes to an inexorably advancing sea. Occasionally the advance was catastrophic when the rising sea overtopped a low basin on the land, but it always continued.

It is little wonder that so many cultures preserve an ancient idea about a great flood. Can the geological record confirm or deny the particular flood recorded in the Bible? Read the selection and make up your mind.

The Voyage of the H. M. S. Challenger

In December 1872 a converted British man-of-war, the H.M.S. Challenger, left Portsmouth, England on a three-and-one-half-year voyage of oceanographic exploration. This date is frequently cited as the beginning of the era of modern oceanography. Efforts to discover what lies beneath the shimmering surface of the sea date back at least to the ancient Greeks in the fourth century B.C., *but it was not until 2,000 years later that scientists began systematically to seek the answers to the age-old questions: How deep is the sea, and how deep can animal life be found in it?*

Some deep-sea soundings and dredgings were made in the late 1700s and early 1800s, largely as a fringe benefit of geographic exploration. By the middle of the nineteenth century, however, the expansion of merchant shipping, the need of telegraph cable companies to survey routes under entire oceans, and the controversy aroused by Darwin's theory of evolution combined to create a new interest in the physical and biological characteristics of the deep oceans. At that time, most people still believed that any form of life was untenable in the perpetual darkness, high pressure, and low temperatures of the greatest depths of the ocean. But scientific explorations in both the United States and Britain in the 1860s produced considerable evidence that life did occur even in what had been thought of as a lifeless zone and that a system of oceanic circulation was responsible for markedly different temperatures of water in close proximity.

Spurred on by these findings, Charles Wyville Thomson, a natural scientist at the University of Edinburgh, and William B. Carpenter, professor at the University of London and vice-president of the Royal Society of London, determined to ask the British Admiralty to support an exploration of the deep oceans on a global scale. The correspondence between the Admiralty and the Royal Society that led to the Challenger *Expedition is reprinted below from the official* Challenger Reports, *prepared by Thomson, who headed the scientific staff of the expedition, and by John Murray, one of the naturalists on board.*

Also reprinted is a humorous poem from Punch *magazine that indicates the popular mood toward the expedition.*

The expedition covered 68,890 miles, and many of the samples taken on the voyage are still being studied by scientists. The reports of the Challenger *Expedition fill fifty volumes, written over twenty years. Among the most important findings of the trawling, sounding, and dredging operations were that animal life exists throughout the oceans, regardless of depth and that the ocean sediments are of two distinct kinds, one coming from nearby continents or islands, the other coming from the surface waters and falling as a continuous rain upon the sea bed.*

The vast ocean lay scientifically unexplored. All the efforts of the previous decade had been directed to the strips of water round the coast and to enclosed or partially enclosed seas; great things had certainly been done there, but as certainly far greater things remained to be done beyond. This consideration led to the conception of the idea of a great exploring expedition which should circumnavigate the globe, find out the most profound abysses of the ocean, and extract from them some sign of what went on at the greatest depths.

The following correspondence extracted from the Minutes of Council of the Royal Society giving expression to this idea, and tracing the progress of its realisation, will best show how all the difficulties in the way of inaugurating an undertaking of such magnitude and novelty were successfully surmounted; . . . high as were the hopes entertained by the promoters of the Expedition, the performance was even greater than had been anticipated.

June 29th, 1871.

Read the following Letter from Dr. Carpenter:—

University of London, Burlington Gardens, W.
June 15, 1871

Dear Prof. Stokes,—The information we have lately received as to the activity with which other nations are now entering upon the Physical and Biological Exploration of the Deep Sea, makes it appear to my colleagues and myself that the time is now come for bringing before our own Government the importance of initiating a more complete and systematic course of research than we have yet had the means of prosecuting.

The accompanying slip from last week's "Nature" will make known to the Council what is going on elsewhere, and the feeling entertained on the subjects alike in the scientific world and (as I have good reason to believe) by the public generally.[1]

For adequately carrying out any extensive plan of research, it would be requisite that special provision should be made; and as the Estimates for next year will have to be framed before the end of the present year, no time ought now to be lost, if the matter is to be taken up at all.

In order that the various departments of Science to which these researches are related should be adequately represented,—so that any Application made to Government should be on the broadest basis possible,—I should suggest that the Council of the Royal Society, as the promoters of all that has been already done in the matter, should take the initiative; and should appoint a Committee to consider a Scheme, in conjunction with the President of the British Association, and the Presidents of the Chemical, Geographical, Geological, Linnean, and Zoological Societies. Such a Committee might meet before the Recess, and decide upon some general plan; and this would be then considered as to its details by the Members representing different departments of Scientific Enquiry, so that they might be able to report to the Council, and enable it to lay that Scheme (if approved) before the Government by the end of November.

Believe me, dear Prof. Stokes,
Yours faithfully,
William B. Carpenter.

Prof. Stokes.

Resolved,—That the subject of Dr. Carpenter's Letter be taken into consideration at an early Meeting of the Council after the Recess.

---

[1]*Nature,* Vol. IV (1871) p. 107. The paragraph states that the Governments of Germany, Sweden, and the United States were preparing to despatch ships to various parts of the ocean expressly fitted for deep-sea exploration.

In reference to the subject of Dr. Carpenter's Letter of the 15th June, read at the last Meeting of Council, the Secretary stated that he had received a subsequent Letter from him, dated Malta, 29th Sept., which was now read. In this Letter Dr. Carpenter urges the expediency of making arrangements for the proposed circumnavigating Expedition without delay, and communicates a correspondence with the First Lord of the Admiralty, from which it appears that H.M. Government will be prepared to give the requisite aid in furtherance of such an Expedition on receipt of a formal Application from the Royal Society; and in consequence of this information, Dr. Carpenter now suggests a modification in the composition of the Committee to which, in his former Letter, he had proposed that the matter should be referred.

Resolved,—That a Committee be appointed to consider the plan of operations it would be advisable to follow in the proposed Expedition, the staff of scientific superintendents and assistants to be employed, and the different provisions and arrangements to be made, with an estimate of the probable expense, and to submit to the Council for approval a scheme which might be laid before H.M. Government, if the Council see fit, at as early a period as may be convenient. The Committee to consist of the President and Officers of the Royal Society, Dr. Carpenter, Dr. Frankland, Dr. Hooker, Professor Huxley, the Hydrographer of the Admiralty, Mr. Gwyn Jeffreys, Mr. Siemens, Sir William Thomson, Dr. Wyville Thomson, and Dr. Williamson, with power to add to their number.

The Report of the Committee on the subject of a Scientific Circumnavigation Voyage, received at the last Meeting, having been taken into consideration, it was

Resolved,—That application be made to Her Majesty's Government, as recommended by the Committee, and that the following Draft of a Letter to be addressed by the Secretary to the Secretary of the Admiralty be approved:—

FROM H.M.S. *CHALLENGER* REPORTS

To the Secretary of the Admiralty.

THE ROYAL SOCIETY, BURLINGTON HOUSE,
December 8th, 1871.

SIR,—I am directed by the President and Council of the Royal Society to request that you will represent to the Lords Commissioners of the Admiralty that the experience of the recent scientific investigations of the deep sea, carried on in European waters by the Admiralty at the instance of the Royal Society (Reports of which will be found in their "Proceedings" herewith enclosed), has led them to the conviction that advantages of great importance to Science and to Navigation would accrue from the extension of such investigations to the great oceanic regions of the Globe. The President and Council therefore venture to submit to their Lordships' favourable consideration a proposal for fitting out an Expedition commensurate to the objects in view; which objects are briefly as follows:—

(1) The Physical conditions of the deep sea throughout all the great Ocean-basins.

(2) The chemical constitution of the water at various depths from the surface to the bottom.

(3) The physical and chemical characters of the deposits.

(4) The distribution of organic life throughout the areas explored.

For effectively carrying out these researches there would, in the opinion of the President and Council, be required—

(1) A ship of sufficient size to afford accommodation and storage-room for sea-voyages of considerable length and for a probable absence of four years.

(2) A staff of scientific men qualified to take charge of the several branches of investigation.

(3) A supply of everything necessary for the collection of the objects of research, for the prosecution of the physical and chemical investigations, and for the study and preservation of the specimens of organic life.

The President and Council hope that, in the event of their recommendation being adopted, it may be possible for the Expedition to leave England some time in the year 1872; and they would suggest that as its organization will require much time and labour, no time should be lost in the commencement of preparations.

The President and Council desire to take this opportunity of expressing their readiness to render every assistance in their power to such an undertaking; to advise upon (1) the route which might be followed by the Expedition, (2) the scientific equipment, (3) the composition of the scientific staff, (4) the instructions for that staff; as well as upon any matter connected with the Expedition upon which their Lordships might desire their opinion.

The President and Council have abstained from any allusion to geographical discovery or hydrographical investigations, for which the proposed Expedition will doubtless afford abundant opportunity, because their Lordships will doubtless be better judges of what may be conveniently undertaken in these respects, without departing materially from the primary objects of the voyage; and they would only add their hope that in accordance with the precedents followed by this and other countries under somewhat similar circumstances, a full account of the voyage and its scientific results may be published under the auspices of the Government as soon after its return as convenient, the necessary expense being defrayed by a grant from the Treasury.

The President and Council desire, in conclusion, to express their willingness to assist in the preparation for such publication of the scientific results.

I remain, &c.

91

March 21st, 1872.

Read the following communication from the Admiralty:—

ADMIRALTY, *2nd March 1872.*

SIR,—In reply to your Letter of the 8th of December 1871 conveying a representation from the President and Council of the Royal Society that advantages of great importance of Science and Navigation would result from equipping an Expedition for the Examination of the Physical Conditions of the Deep Sea throughout all the Great Oceanic Basins, and for other special objects therein named,—

2. I am commanded by my Lords Commissioners of the Admiralty to acquaint you, for the information of the President and Council, that they have had the subject under their consideration, and have decided to fit out one of Her Majesty's ships to leave England on a Voyage of Circumnavigation towards the close of the present year, in prosecution of the objects specified in your letter.

3. I am further desired to inform you that their Lordships will be prepared to receive from the President and Council of the Royal Society any suggestions that they may desire to make on the Scientific Equipment of the Vessel, the Composition of the Civilian Scientific Staff, or any other Scientific matter connected with the Expedition upon which that body may desire to offer their opinion.

I am, Sir,
Your obedient Servant,
THOS. WOLLEY.

The Secretary to the Royal Society.

June 20th, 1872.

In reference to the arrangements to be made for the Circumnavigatory Expedition, for which H.M.S. Challenger has now been put in commission, the Committee presented the following Report to the Council, viz.:—

The Committee suggest that the President and Council should direct a Letter to be written to the Secretary of the Admiralty to the following effect:—That it appears desirable that the Scientific gentlemen who are to accompany the Challenger Expedition should be selected at an early date and their salaries decided on, in order that they may be enabled to make the necessary arrangements for an extended absence from England.

The President and Council of the Royal Society therefore recommend as a fit and proper person, to superintend and be at the head of the Civilian Scientific Staff of the Expedition, Wyville Thomson, LL.D., F.R.S., &c., Regius Professor of Natural History in the University of Edinburgh; and that, as Professor Thomson will have to give up his position, with its emoluments, at Edinburgh for the time he is absent, the President and Council are of opinion that a less sum than £1000 per annum cannot properly be offered to him.

They propose that the other members of the Staff and their Salaries should be as follows:—

Mr. John James Wild, as Secretary to the Director and Artist,.............. £ 400
Mr. John Young Buchanan, M.A. (Glas.), Principal Laboratory Assistant
    in the University of Edinburgh, as Chemist and Physicist,.............. 200
Mr. Henry Nottidge Moseley, B.A. (Oxon.), Radcliffe Travelling Fellow
    of Oxford University, as Naturalist,.. 200
Mr. William Stirling, D.Sc. (Edin.), M.B., Falconer Fellow of the
    University of Edinburgh, as Naturalist,..................................... 200
Mr. John Murray, as Naturalist, ... 200

The Committee further report that Professor Wyville Thomson informed them that he had gone with Admiral Richards to Sheerness to examine the Challenger, and that the arrangements appeared to be satisfactory in every respect.

Resolved,—That the Report of the Circumnavigation Committee be adopted, and that a communication be made to the Admiralty in terms of their recommendation.

November 14th, 1872.

The Council proceeded to consider the Report of the Circumnavigation Committee. The following is the Letter from the Admiralty to which the Report refers:—

ADMIRALTY, *August 22nd, 1872.*

Sir,—With reference to my letter of the 6th instant, and to previous correspondence on the subject of the intended deep-sea exploratory Expedition, I am commanded by My Lords Commissioners of the Admiralty to acquaint you that H.M.S. Challenger will probably be ready to leave this country about the end of November; and their Lordships will be glad to learn what are the precise objects of research which the President and Council of the Royal Society have in view, and in what particular portions of the Ocean such investigations may, in their opinion, be carried out with the greatest advantage to science and the best probability of success.

2. The object of their Lordships is to frame their instructions to the Officer in command of the Challengr, so far as may be possible, to meet the recommendations of the President and Council of the Royal Society.

I am, Sir,
Your obedient Servant,
VERNON LUSHINGTON.

W. Sharpey, Esq., M.D., &c.
Secretary of the Royal Society, Burlington House.

The Report having been considered, was adopted as follows:—

  The Circumnavigation Committee have had before them the Letter from the Admiralty to the Royal Society, dated August 22, 1872, and as the Council were not in Session and the matter was pressing, they have thought it best to treat the letter as having been referred to them by the Council. They beg leave to recommend to the Council that an answer be returned to the Admiralty to the following effect:—

    Resolved,—That the Report of the Circumnavigation Committee, now adopted by the Council, be transmitted by the Secretary of the Admiralty. . . .

November 30th, 1872.

Read the following Letter:—

ADMIRALTY, *27th November 1872.*

SIR,—I am commanded by the Lords Commissioners of the Admiralty to thank you for your communication of the 22nd instant, in regard to the objects of research which the Royal Society have in view with reference to the intended voyage of H.M.S. Challenger, and to acquaint you that they are desirous of affording to the President and Council of the Royal Society, as well as the Members of the Circumnavigation Committee, an opportunity of inspecting the ship, and the arrangements made with a view to her equipment for the service she is intended to perform.

2. My Lords therefore invite those gentlemen to proceed to Sheerness on the 6th proximo for the purpose of visiting the Challenger; and a saloon carriage will be ordered to be in readiness to convey them to that port by the 10.30 A.M. train from Victoria Station.

3. The visitors will be able to return by the 5.10 P.M. train from Sheerness, and free railway passes will be provided for them both ways. They will also be met by their Lordships' Hydrographer on the occasion.

4. I am to request you will inform me, as soon as may be convenient, of the number of the gentlemen who will avail themselves of their Lordships' invitation, in order that the proper number of tickets may be procured.

I am, Sir,
Your obedient Servant,
ROBERT HALL.

The Secretary of the Royal Society.

Before the Expedition left England, Dr. William Stirling resigned his appointment as Naturalist; and Dr. Rudolph von Willemoes Suhm, Privat-Docent in Zoology in the University of Munich, was appointed by the Admiralty in his place, on the recommendation of the Council of the Royal Society.

The Challenger Her Challenge

I'M a spar-decked corvette, built of wood not of iron,
  I am good under steam, under sail:
No Sheffield-plate dead-weights my topsides environ,
  So I ride like a duck through a gale.
By my Lords I'm about to be put in commission,
  For a cruise of three years, if not four;
And for all I'm short-handed, I carry provision
  Such as corvette ne'er victualled before.

Mine's no cruise to train officers, boys, or blue-jackets,
  Or BRITANNIA's old flag to display;
To observe and report South American rackets,
  Or enjoy life in Naples' blue bay:
To practise manoeuvres, or study steam-tactics,
  Hunt down pirate-junk or slave-dhow;
The *Challenger* now aims at higher didactics,
  And on different quests sets her prow!

Her task's to sound Ocean, smooth humours or rough in,
  To examine old NEP's deep-sea bed;
Dredge up samples precise of his mattress's stuffing,
  And the bolsters that pillow his head:
To study the dip and the dance of the needle;
  Test the currents of ocean and air—
In a word, all her secrets from Nature to wheedle,
  And the great freight of facts homeward bear.

And by way of a treat—when the *Fauna* and *Flora*
  Of all lands and all seas I've run through,
And learnt if the Austral Antarctic Aurora
  Our Boreal in beauty outdo—
In the Isle of Kerguelen, with nothing between us
  But the thinnest of clouds—O what fun!—
I'm to lurk and look on at the transit of Venus,
  Across the broad blush of the sun!

For this I bear science to seamanship plighted,
  In THOMPSON and NARE and MACLEAR,
While from highest to lowest aboard all united,
  To serve both alike volunteer.
Broadside guns have made room to ship batteries magnetic,
  Apparatus turns out ammunition,
From main-deck to ground-tier I'm a peripatetic
  Polytechnic marine exhibition.

"Mighty fine!" says JOHN BULL. "But, pray, how about cost?
     Cash soon makes ducks and drakes in the Ocean."
Treasury leave was asked first: prayer, of course, aside tost,
     Till LOWE★ went to figures with GÖSCHEN.★★
When they found that the outlay for all this provision,
     To question the land, and the sea,
Would be no more than keeping my hull in commission,
     With nothing to show for 't, would be!

Said LOWE, laughing, "To pay by results is my plan;
     For results here'll be nothing to pay.
Let the *Challenger* go: and I'll challenge the man,
     Be it RYLANDS† himself, who'll gainsay;
For he, like myself, though he's not been to college,
     And 's a shallowish sort of a snob,
Has, at bottom, I'm sure, no objection to knowledge,
     So long as it don't cost a bob."

And so I'm to sail on my grand cruise of science,
     And a prouder ship ne'er put to sea;
In the good of my mission high souls have reliance,
     Whatever the LOWE view may be.
Of the axiom that "nothing of nothing can come,"
     I'm the Challenger. How is it true?
When 'tis clear to BOB LOWE, as a rule-of-three sum,
     Good for nothing I'm not, 'cause I do.

★ *Ed. note:* Robert Lowe, Viscount Sherbrooke, chancellor
  of the exchequer from 1868–1873.

★★*Ed. note:* George Joachim Goschen, 1st Viscount
  Goschen, was First Lord of the Admiralty.

† *Ed. note:* John Rylands, wealthy textile manufacturer
  and merchant.

Anatomy of an Expedition

H. William Menard

No oceanographic expedition has matched the Challenger's three-and-one-half year record of continuous exploration. Today's expeditions are of shorter duration, with more specific objectives and more sophisticated equipment. The sounding techniques of 100 years ago, using weights and hemp line, have been replaced by sonar techniques in which a pulse of sound is bounced off the ocean floor in a matter of seconds. The dredge and the trawl are still used, but now they are supplemented with underwater cameras, devices to remove cores or samples of the ocean floor, and magnetometers. Despite these advances, many of the problems of organization, planning, and fund-raising remain remarkably similar to those of a century ago, as indicated in the following excerpt from Anatomy of an Expedition *by H. William Menard of the Scripps Institution of Oceanography.*

One of the most important results of the expedition was the quite unexpected discovery that the sea floor spreads slowly on the continental side of deep oceanic trenches. This discovery has been investigated by several subsequent expeditions, including some using deep-sea drilling techniques. Perhaps the success of an expedition should be measured not by the problems it solves but by the questions it identifies.

During the last two decades the deep sea floor—two-thirds of the world—has been explored for the first time. For a hundred years a few scientists had known it as a dark, cold, and utterly alien world. Now it is becoming familiar to schoolchildren who have maps of the sea floor next to the maps of the moon on their classroom walls. How did this come about? A small group of deep sea oceanographers of all nations simply went out into the ocean basins again and again and discovered the mountains, the trenches, the volcanoes, and the rifts one by one. The exploration began near the great scientific ship bases: Woods Hole, Vladivostok, Palisades, Plymouth, La Jolla, Sevastopol. Then it spread through the Atlantic and Pacific and finally the Indian Oceans, and now it goes round and round the world. I believe that it is realistic to say that the geographical discoveries have been on a scale unmatched since the sixteenth century, when the continents were disclosed to Western eyes.

I wonder if any of the people involved have thought of the oceanic exploration of the last two decades in any such grand terms? I doubt it very much; they have been and continue to be too busy. This [account] . . . is an attempt to show some of the things that keep ocean-ographers busy, how at a certain place and time a group of them went to sea and why. . . .

An oceanographic expedition is the crystallization of an idea; it begins and ends in the mind with a little sailing around in between. The idea that became the *Nova* Expedition began to take form in the fall of 1965. At that time, I had not been to sea for fourteen months, which was my longest period on land since I became an ocean-ographer in 1949, and I started to think about another expedition. It is now March 1967 and the expedition is about to begin, and at this point I cannot remember exactly when or where the subject first came under discussion. After fifteen expeditions the preliminaries get rather vague. Most expeditions, and I suppose this was no exception, are conceived late in the evening after enough oceanographers have become sufficiently relaxed to be optimistic about work at sea. The reason for this is that everyone knows that the eventual output is going to require a lot of hard labor and discomfort. The conversation can be generalized as follows:

"Ed, why don't we go drill an atoll and find out why it sank?"

"Say that's great. I need a sequence of reef lime-stones to test this new idea. . . ."

"Which atoll? What about Fakarava?"

97

"Where's Fakarava?"

"You remember. Where we stopped on the *Downwind* Expedition in the central Tuamotus."

"Yeah, that was nice, and we can get fuel and plane service at Tahiti. When?"

"Well, there is a gap in the ship schedule for *Argo* next July because Sam canceled out. He can't get delivery on his equipment. We could take some students along."

"I was going to visit Cambridge that month on my way back from the meeting in Stockholm, but I could go there before the meeting."

"We'll need to know the crustal structure in order to pick the best spot to drill. I wonder if George would be interested. Where is George? Hey, George how would you like to go to Fakarava next July?"

Perhaps one out of five of these evening expeditions eventually puts to sea, and then it is at Majuro because the Navy will provide free air service and in August because the ship needs an overhaul in July.

Considering that it is the taxpayer's money, it may sound a little casual, but oceanographers constantly get new ideas to test, and they have to go to sea at one time or another, and there is no virtue in going to an unpleasant atoll if a beautiful one has the same geology. In any event, there is stringent product control. If you don't produce results, you can't get money for your next expedition. You may not get the results you expected—this is research, not bridge building—but you have to get significant results of some sort.

An expedition needs an idea, scientists who are interested in it, a ship capable of the work, and money. Bringing them together takes a lot of talking, adjustment, writing, modification, time, flexibility, and work. The steps have not changed in the last 150 years. The first United States Exploring Expedition in 1838★ and the first great oceanographic expedition on the

Challenger in 1873 followed the same path of organization as the *Nova* Expedition in 1967. Indeed, these and the intervening expeditions had similar problems, used about the same size ships, and cost similar amounts. The great changes have been in the ideas under investigation and the equipment on the ships. All this historical precedent, however, is little consolation as the painful process is repeated yet again.

The basic idea of *Nova* is to try to determine the development and geological history of the peculiar Melanesian region in the southwestern Pacific where the sea floor seems to be part continent and part ocean basin. . . . Because the region is on the opposite side of the Pacific from Scripps Institution of Oceanography, the ships will incidentally be available for other work coming and going. Several scientists at Scripps were immediately interested in the idea, and we decided to try to get a commitment for some time on one or more of our oceanographic ships. The ships belong to Scripps, and a committee of our oceanographers schedules their use. Even so, ships are hard to come by because so many scientists want them. On my first inquiry in the fall of 1965, I found that the next long period definitely available on one of our ships was in August of 1968, and then only if I had any plans for the area between Mauritius and Cape Town in the Indian Ocean. Fortunately some of my interested colleagues had time allotted in 1967, and by juggling the schedule with other scientists we put several months of ship time into one continuous block and still had everyone as happy as before. The next step was to look for money.

It may be useful at this point to consider how oceanographic expeditions are financed and what they cost. Captain Cook's first voyage in 1768–1770 was supported by King George III with a grant plus a bark, *Endeavour*. The total cost for the ship and outfitting exceeded £8235, and it appears that the voyage cost at least £10,000 or in round terms about $28,000. This sounds very modest, but for comparison with modern expeditions we must allow for the

---

★ *Ed. note:* Led by Charles Wilkes, this naval scientific expedition, with six ships, sailed around the world between 1838 and 1842. The best-known achievement of the expedition was Wilkes' claim to have discovered that Antarctica was a continent.

decreasing value of money. At that time, for example, a man with an income of $1200 a year was able to "keep five or six servants indoors and out, to look well after his relations, to travel freely, and to exercise a generous hospitality to rich and poor."

A reasonable approach is to assume that the purchasing power of a given amount of money has decreased by two percent per year. Accordingly, it takes seven times as much money to buy something at the end of a century as it does at the start. Consequently, it cost King George about 1,400,000 modern dollars for Cook's voyage. In addition, the scientist, Banks, whose tax-free income in modern terms was about $1,000,000 per year, put in a considerable amount of his own money. The expedition lasted almost three years and in modern money cost a minimum of $1300 per day.

The cost of the first United States Exploring Expedition (Wilkes Expedition) in 1838–1842 is buried beyond recall in government records. However, in 1836 Congress appropriated $300,000 for the expedition, and the Navy reported that it was all spent in 1837, a year before the ships put to sea. The expedition involved four ships for four years and cost perhaps $10,000,000 in modern money. Each ship cost more than $2000 per day even though all were very small and some did not last for four years. A century and a quarter later, during *Nova*, we expect to land on the Exploring Islands, named after and surveyed by this expedition.

The value of the *Challenger* Expedition of 1873–1876 was enormous and so was its cost. The British government apparently paid out a sum equivalent to the estimated expense for the recent Mohole Project to drill a hole in the bottom of the sea. The modern project would have been so expensive that the American Congress in 1966 decided that the richest country in the history of the world could not afford it.

So, on historical precedent we needed to locate some rich and generous patrons if the *Nova* Expedition were ever going to sea. Fortunately for the success of American oceanog-

raphy some far-sighted men established just such funding organizations fifteen to twenty years ago. No longer must appeals for support for each expedition be directed to king or Congress. Indeed, it would be quite impossible to have 50 to 100 American oceanographic expeditions per year without the existence and support of federal agencies. The most important for the past two decades has been the Office of Naval Research, which produced modern oceanography by supporting the expansion of marine institutions and fleets of research vessels. Somewhat later the National Science Foundation entered the picture and now provices about half the support.

The main cost of oceanography is for ship time. A research ship without support for operating funds is an albatross around the neck of an oceanographic institution. Not too long ago the cost of a ship had to be justified on the scientific merits of each expedition. Consequently, oceanographic institutions were in about the position of a business which rents out ships but cannot make a profit to cover the expenses while the ship is idle. The daily cost of a ship was determined at the end of a year by dividing the annual cost by the number of days of use. It was a gigantic gamble to use an unpopular ship for a week at the start of the fiscal year. An oceanographer might find that he was expected to find funds for fifty-one idle weeks. This insane system seemed necessary because of the federal system of appropriating money for only one year. Finally, a few years ago officials in the Navy and the National Science Foundation figured out how to fund ships on an annual basis, and life became much simpler.

The total cost of *Nova* was initially estimated as follows:

ARGO, ship time at $3,000/day	$ 720,000
HORIZON, ship time at $2,500/day	400,000
Direct support of expedition	270,000
Indirect costs of research	100,000
Scientific and technical salaries	200,000
	$1,690,000

This is certainly a staggering amount, but at $4000 per day per ship it is roughly what king

DEEP SEA DRILLING PROJECT, SCRIPPS INSTITUTION OF OCEANOGRAPHY

THE *GLOMAR CHALLENGER*

A modern oceanographic ship, the deep-sea drilling vessel Glomar Challenger *is drilling and coring for ocean sediment in all oceans of the world.*

or Parliament or Congress has always paid for marine expeditions during the last 200 years. It would have been much cheaper not too long ago. On our first Scripps deep sea expedition in 1950, *Horizon* cost only $450 per day, but that was before we loaded it with special equipment.

Due to the recent advances in federal money handling we did not need to ask for money to support the ships (which includes wages for the crew). Moreover, most of the salaries for scientists and technicians are paid by the University of California or by long-term state and federal research contracts. The indirect costs of research are those connected with this expedition but which are paid out of existing contracts with some broad objective. Thus the National Science Foundation supports a global study of carbon dioxide in the atmosphere, and some of this money will be used to make measurements during this particular expedition. In sum, in order to put this costly enterprise under way, we estimated that we needed $270,000, or about 15 percent of the total.

My colleagues, Harmon Craig, Edward Goldberg, Edward Winterer, Robert Fisher, and Manuel Bass, plus Charles Helsley of the Graduate Center of the Southwest, joined me in submitting a request to the National Science Foundation for this amount on 13 May 1966.

100

As in all such proposals, we said what we want to do and why it is worth doing, who we are, what we have done that makes it likely that we can do what we are asking money to do, the capabilities of our university to carry out the proposed work, and what it will cost. This proposal went through the customary elaborate review. It was sent to scientists doing related work at other institutions, and they graded it according to the quality of the scientific proposal and the abilities of the people and university. Then a geophysical review panel of government and academic scientists compared our evaluated proposal with those from other scientists and universities and arranged them in order of merit with due consideration of the cost. It is easy to identify one of these panel members if you ride on the airplane he is taking to a meeting. He has with him three brown-covered volumes of proposals with a total thickness of about eight inches, and he is busily reading the last one while other people watch the movie. He does this three times a year and thus evaluates about two feet of proposals requesting perhaps $20,000,000. It is hard work. I was on the panel for three years and for the same reason as the other members. I never met anyone who likes it, but we all agree that if scientists won't evaluate the merits of scientific proposals then it will have to be done by nonscientists who may have to judge proposals by criteria other than scientific merit.

This evaluation process takes three to six months, and meanwhile planning for an expedition has to continue with the hope that it will not be wasted. My colleagues and I talked to other scientists who might be interested in problems in the southwestern Pacific. Gradually more of them decided to participate in *Nova,* and technicians and special equipment were organized and integrated into as comprehensive and efficient a whole as possible. We did not need to know which ports we would use for fueling when we submitted the request for funds. However, we did need such information before sailing so we examined various permutations of ports, distances, ship speeds, fuel con-

sumption, water supplies, explosive handling, airplane connections, and general logistics. Some matters require a long lead time. . . .

Early in August I got a phone call from Joe Creager, an oceanographer from the University of Washington who was doing his public service by working for a year with the National Science Foundation. He said that our proposal was approved on merit but that it cost more money than was available. Could we get along with less? I said we could, although with less I did not think that we could do everything we planned. How much less? Joe said "Half," and another time of troubles began.

What to cut? The categories in our proposal included salaries, assistantships for graduate students, supplies, equipment, travel, and a 42 percent markup to cover University overhead on salaries and wages. Almost all graduate students in oceanography receive substantial support, and most of it comes from federal contracts. This is not unreasonable, because the students help collect and analyze the results. On average they are the cheapest, highest-quality, and most eager workers in the world. We cut support for four students out of the budget, but we knew we would have to find the money somewhere else.

The next item was supplies and expenses. I quote from Sir Wyville Thomson concerning the outfitting of the *Challenger* Expedition in 1873:

> It is almost inconceivable how difficult it is to keep instruments, particularly those which are necessarily made of steel, in working order on board a ship; or how rapidly even with the greatest care they become destroyed or lost. For this reason it is necessary to have an almost unlimited supply of those in most frequent use, such as scissors, forceps, and scalpels of all sizes.

We have the same problems and must constantly restock everything from pencils to expendable equipment which is used only once but is cheaper than the ship time necessary to recover it. We chopped our request for supplies mercilessly but again with the certain knowledge that we would have to ask for more for the next expedition. We did have one windfall. George Shor, who had decided to measure crustal thickness in the southwestern Pacific, needed a large supply of explosives for the purpose. This was in the *Nova* budget, but meanwhile a long-standing request for explosives from the Navy was at last granted and we did not need more from the National Science Foundation. That item alone saved $12,000.

Next we came to heavier equipment which we also wear out or lose in the normal course of oceanographic work. We are not the only ones. A Russian oceanographer, Gleb Udintsev, once learned that we were planning to dredge in the world's deepest water in the Marianas Trench. He told me that if we dredged up a deep sea camera it was his because he had lost one there. The camera cost about $10,000 and we did not find it—in fact we lost one ourselves. What can you expect when you attach a camera to six miles of wire and try to photograph the rocky sea floor from a drifting ship? We eliminated the least urgent items and hoped for good fortune at sea and future replacements even though they would be of no use on this voyage.

Finally we came to a whopping request for $76,400 for travel by air to various ports that the ships would touch. We got our Marine Facilities department which operates the ships to agree to pay for rotating the crew, and that cut the request in half. Of course, it also increased the ship operating cost, but that was covered by another budget. We could have eliminated all this air travel by merely returning the ships to San Diego every few months, but that would have been a waste of money—which is a very different thing from tailoring requests for money to fit the amounts available in different contracts.

After many a painful session we cut the budget request in half and resubmitted the proposal on 10 August. Creager knew how short the planning time was growing and soon phoned that the money was on the way. He also said

that he had hoped that we might reduce the amount of ship time required because the Science Foundation was running short of money for ships. After our juggling this seemed a little ironical. . . .

Next, or meanwhile, we needed to make base maps for our work. If you want a map to do something like hike in the Sierras or navigate a ship into Suva harbor you just buy one from the appropriate government agency. Not so in oceanography, where the maps have to be changed after each expedition and commonly are made on board in order to plan the next day's work. We needed to know about every unpublished geophysical observation anyone had taken in the southwestern Pacific, in order to avoid duplication and for making detailed plans. The only way to get such information is to write to or visit every person who might have it. There are both formal and informal worldwide information networks whereby oceanographers keep track of their colleagues' work. I had already acquired a large amount of information by making the usual carefully respected promises not to publish other people's unpublished data and by agreeing to make our own observations similarly available. There remained the problem of the most recent data collected by oceanographers in Australia, New Zealand, and New Caledonia. A lot is always known which has not even been processed to the point where it can be sent to anyone else. I needed that information but getting it would require work by other people whose time I did not want to waste. The only practical solution was to visit the laboratories and explain exactly what we hoped to do and what information we needed. The trip would also provide an opportunity to interest oceanographers in these countries in participating in *Nova* and thereby getting more scientific results from the same effort at sea. In any event, it is not very courteous to carry out a lengthy expedition in waters near other oceanographic laboratories without asking the scientists there if they would like to join in. Finally, the trip would give me a chance to

Artist's diagram shows Glomar
Challenger *using "dynamic
positioning" to hold its station
above a sonar sound source
placed on the ocean bottom. Worn-
out drill bits can be replaced
by using scanning sonar to locate
the drill reentry cone.*

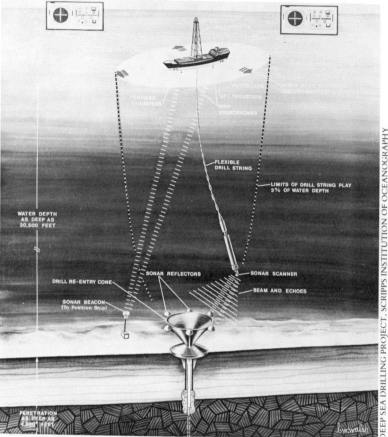

FLEXIBLE
DRILL STRING

LIMITS OF DRILL STRING PLAY
3% OF WATER DEPTH

FORWARD
THRUSTERS

AFT THRUSTERS

SHIP
HYDROPHONES

WATER DEPTH
AS DEEP AS
20,500 FEET

DRILL RE-ENTRY CONE

SONAR REFLECTORS

SONAR SCANNER

BEAM AND ECHOES

SONAR BEACON
(To Position Ship)

PENETRATION
AS DEEP AS
4,000 FEET

DEEP SEA DRILLING PROJECT, SCRIPPS INSTITUTION OF OCEANOGRAPHY

inspect prospective ports once again and see
whether the surrounding countryside would
be suitable for geological field trips for the
students. . . .

[M]y associates Stuart Smith, Tom Chase,
and Isabel Taylor plunged into the compilation
of the new data for our base charts. Tom and
Stu were each to lead one of the early legs of
the expedition and had a vital need for this information. I also had the assistance of several graduate students who would join in the work at
sea and might eventually use some of the results
for Ph.D. theses.

My own effort was concentrated on further
planning. The possible permutations in an
enterprise like this are beyond belief. Somehow
all the senior scientists have to be satisfied that

their work can be done, equipment and supplies
have to be directed to the right places at the right
times, and the budget has to be balanced. It
always reminds me of planning an amphibious
assault—as I did in the same region many years
ago—except that money is important.

I have mentioned the explosives before; the
further problems connected with this item alone
may illustrate something of what is involved.
George Shor had about 12 tons of TNT stock-
piled in the Navy ammunition depot in Pearl
Harbor. Another similar amount would provide
all we needed, and it was available in Southern
California. However, *Horizon* could not carry
its half of the explosives because its magazine
had to take supplies needed for other work
before we took it over in Kwajalein in April

1967. It could have returned to Pearl Harbor, but that would have reduced time in the southwestern Pacific. We hit upon an elaborate scheme. *Argo* was scheduled to conduct oceanographic work from San Diego to Pago Pago to Pearl Harbor and then back south to Suva, where it would be joined by *Horizon*. We could load the local TNT on *Argo,* unload and store it at Pago Pago, and then reload the ship with the Pearl Harbor TNT when it reached there. *Horizon,* which by then would be half empty, would make a short trip to Pago Pago and back to Suva collecting useful although less important data on the way. But could we store explosives at Pago Pago? Appropriate inquiries indicated that we could. Could both ships enter Suva harbor with explosives? We knew from previous expeditions that we had done so, but had the port regulations changed? Our ship agent in Suva said that they had and we would have to unload everything onto a barge in midstream before coming to the dock for fuel. So a little time might be lost.

At about this time a new development entered the explosives problem. Charles Helsley, of the Graduate Center of the Southwest, would also measure crustal thickness and structure on *Nova*. He flew to La Jolla, and he and George Shor developed some intricate negotiations whereby the TNT in California became an equal quantity of Nitramon, which is commercial fertilizer except when it is triggered by an explosive fuse. This safe stuff could be stored on *Argo* outside the magazine and would not have to be off-loaded at Pago Pago before the magazine was filled with TNT at Pearl Harbor. Consequently, *Horizon* need not go to Pago Pago and would be available for more important work. Moreover, Helsley would shoot off all the TNT between Suva and Brisbane, and we should have no problem entering the latter port, which had not yet replied to our inquiries about bringing in explosives. Then Brisbane said it would not let us enter even with the quantity of explosive fuses needed to set off the Nitramon.

So why were we going to Brisbane? Originally we planned to stop at that time in Noumea because it is more centrally located in the southwestern Pacific and it costs less to fly there from California than to Brisbane. However, Helsley had to ship some of his gear on to South Africa to join another oceanographic expedition. To meet this other sailing date he had to have the gear in a port with frequent sailings instead of Noumea, and so we put all other considerations aside. With the news from Brisbane about explosives, the port question was raised again. It turned out that a ship sails on an irregular schedule from Noumea to Sydney about the time we should arrive. By radioing from sea for a final sailing date we could be sure to arrive in time to load Helsley's gear. I phoned him in Texas to ask if this port change was agreeable to him considering the new information about freighter sailings. He agreed and said, moreover, the expedition off Africa was now looking questionable and he was no longer in such a hurry about sending his equipment there. We decided to stop at Noumea, where explosives are brought in frequently for the mines.

The above is a simplified version of the explosives problem. Helsley also shipped some Nitramon from Texas which was supposed to go by commercial truck. A trucking strike stopped that. The Nitramon came by truck driven by a graduate student and his wife. The truck broke down but was repaired and arrived just before sailing time.

There were also the mysterious and complicated affairs of the different air fares to Pago Pago, the missing magnetometers, the carbon dioxide program, and the perplexing problem of who would be working in the Coral Sea.

Finally, in April 1967, *Horizon* sailed south from Kwajalein and *Argo* sailed west from San Diego. The *Nova* Expedition was under way, seventeen months after it was conceived. Not a bad record, really; the *Challenger* Expedition took eighteen months and the United States Exploring Expedition took more than ten years to do the same thing.

Seven Miles Down

Jacques Piccard and
Robert S. Dietz

Although the sounding and sampling techniques of the twentieth century were providing oceanographers with an increased knowledge of the ocean depths, instruments alone could not satisfy a human desire to observe firsthand the "inner space" of the oceans. The first bathysphere—a two-and-one-half-ton steel ball with observation windows—was built in 1930 by William Beebe and Otis Barton. Lowered from a barge by cable, it was dependent on its mother ship and could reach depths of slightly more than half a mile. But it remained for Auguste Piccard of Switzerland, a physicist who had first explored the stratosphere by balloon in 1931, to develop a bathyscaph capable of exploring the greatest depths of the ocean.

The bathyscaph works in water much as a balloon works in the air. The observation cabin, which must withstand crushing water pressure, is heavier than water; the hull to which it is attached is filled with gasoline and is lighter than water, thus serving as a float. To descend, the diver permits sea water to enter the hull, equalizing the pressure inside the hull with that of the surrounding water. The weight of the cabin, plus that of water that enters air chambers in the hull, then causes the bathyscaph to descend. To rise, the diver lightens the bathyscaph by releasing iron pellets, or ballast, from special sections in the hull. Two small screw propellers enable the bathyscaph to move horizontally. The first bathyscaph was tested in 1948. An improved model, the Trieste, *was launched in 1953 and purchased by the United States Navy five years later.*

In 1960 the Trieste *carried Jacques Piccard, son of the inventor, and Lieutenant Don Walsh of the United States Navy 35,800 feet down to the deepest known spot in the ocean—the Challenger Deep, a local depression in a crescent-shape trench in the Pacific near Guam. The following firsthand account of the historic dive is from* Seven Miles Down *by Jacques Piccard and Robert S. Dietz, an oceanographer who was instrumental in persuading the navy to purchase the bathyscaph. Like Cousteau's account of "menfish," this narrative reveals the wonder and excitement of the pioneer on a new frontier.*

*B*uono clasped hands with me and then Walsh. He wished us *buona fortuna.*

"*Grazie,*" I answered. "*Arrivederci.*"

We swung closed the heavy steel door. We turned up the bolt that would seal us securely in our vault. The air was cool and dry in the sphere thanks to the silica gel. But the seas were buffeting the float, swinging our little spherical cell to and fro. I had but one thing on my mind: to dive as quickly as possible into the serene depths.

I checked my watch. It was now 0815. Quickly, I reviewed every essential detail. We were diving without some important instruments. I would have only my watch and my depth gauge with which to calculate our rate of descent. In a descent everything depends on buoyancy control; on conserving and expending ballast with a delicate and knowledgeable touch.

We had a total of 1,440 seconds of ballast; that is to say, I could jettison 25 pounds of iron shot per second for a total of, roughly 16 long tons. That amount of ballast would do it, with a lot to spare.

Through the after porthole I could see water flooding the antechamber. I opened the oxygen valves and checked the air purification system. We could expect to dive momentarily. Had our tachometer★ been operative, the very instant of descent would have been apparent. As it was, we simply had to wait to see the pressure gauge moving.

★*Ed. note:* An instrument used to determine speed.

I wanted to log that instant. My eyes were on my watch. Suddenly, at 0823, the rocking ceased, the sphere became calm. I glanced at the depth gauge. The needle was quivering.

We were on our way down. We were *entre deux eaux,* to use a French expression—in midwater. The *grande plongée* had begun. I looked over at Walsh. We both sighed in relief. . . .

There was, perhaps, a mile of water still beneath us, but the possibility of collision with the trench wall was still on my mind. I pushed the ballast button, slowing us down to two feet per second; then, to one foot per second, as decided before the dive.

. . . 1200—31,000 feet. I flipped on the echo sounder and sought for an echo to record on its 600-foot scale. No echo returned; the bottom, presumably, was still beyond 100 fathoms. Trying moments were ahead. We were venturing beyond the tested capabilities of the *Trieste.* On paper she could descend safely to ten miles and the sphere alone much more. I had confidence in those calculations. She was a complex of nuts and bolts, metal, plastic, and wire. But a dead thing? No. To me she was a living creature with a will to resist the seizing pressure. Above me, in the float, icy water was streaming in as the gasoline contracted, making the craft ever heavier and heavier. It was as if this icy water were coursing through my own veins.

The UQC⋆ was quiet now. The slow, silent descent was disturbed only by the hiss of oxygen escaping and the background hum of electronic instruments. I peered out the window looking for bottom—and then back to the echo sounder. But there is only water and more water. Perhaps the sounder wasn't working. On the last dive we were 120 feet from the bottom before we picked it up. But now I could see noise recording. It made a long smudge on the graph from the iron ballast just released from the tub. "*Très bien,* it is working fine."

Then, an uneasy thought. What would the bottom be like? Clearly, we were in the axis of the trench and most probably we would miss the rocky walls. Dietz who was aboard the *Lewis* had advised me that there was an outside chance that the bottom sediment would be a flocculent and unconsolidated "soup" of recently deposited turbidity current beds. Could we sink and disappear into this material before being aware that we had contacted the bottom? Russian scientists aboard the *Vityaz* reportedly had tried many times, unsuccessfully, to lower a camera and snap pictures in trenches. But each time blank negatives came up. It appeared that the camera had entered a thick, soupy bottom before finally being triggered.

On the other hand, Dietz had emphasized, the *HMS Challenger,* in 1951, had recovered a bottom sample not far distant from our diving site. It was diatomaceous⋆ ooze, composed almost entirely of siliceous remains of tropical diatom *Ethmodiscus rex.* These diatoms live in the surface water and their dead husks settle to the bottom. This would provide a firm bottom for landing. I could only hope that we would land on *Challenger* bottom.

. . . 1206—32,400 feet. A strong, muffled explosion! The sphere shook as though in a small earthquake. I caught Walsh's eye; he was watching me anxiously but calmly. "Have we touched bottom?" Walsh asked. "I do not believe so," I replied. We were waiting for something to happen. Nothing did. I wondered if the light case over the forward port had imploded. I tried to switch it on and it didn't light up. This could be the trouble but I wasn't satisfied. The noise we heard wasn't the high-pitched pop that had accompanied previous implosions. I studied the dials and switched off the UQC to silence the sphere. Still nothing happened. Our equilibrium seemed unaffected—we were not, apparently, losing gas. Our descent continued exactly as before. Without formal discussion, we agreed to continue down.

Thirty-four thousand feet—no bottom . . . 35,000 feet, only water and more water . . .

⋆*Ed. note:* An underwater telephone that permitted the divers to maintain voice contact with the mother ship.

⋆*Ed. note:* Diatoms are minute plants.

THE BATHYSCAPH *TRIESTE*

"Indifferent to the nearly 200,000 tons of pressure clamped on her metal sphere, the Trieste *. . . [made] token claim, in the name of science and humanity, to the ultimate depths in all our oceans."—Piccard and Dietz*

36,000 feet, descending smoothly at 60 feet per minute. Now we were at the supposed depth of the Challenger Deep. Had we found a new hole or was our depth gauge in error? Then a wry thought—perhaps we'd missed the bottom!

. . . 1256, Walsh's eyes were glued to the echo sounder. I was watching alternately through the port and at the fathometer. Suddenly, we saw black echoes on the graph. "There it is, Jacques! It looks like we have found it!" Yes, we had finally found it, just 42 fathoms down.

While I peered through the port preparing to touch-down, Walsh called off the soundings. "Thirty-six fathoms, echo coming in weakly—32—28—25—24—now we are getting a nice trace. Twenty-two fathoms—still going down—yes, this is it! Twenty—18—15—10—makes a nice trace now. Going right down. Six fathoms—we're slowing up, very slowly, we may come to a stop. You say you saw a small animal, possibly a red shrimp about one inch long? Wonderful, wonderful! Three fathoms—you can see the bottom through the port? Good—we've made it!"

The bottom appeared light and clear, a waste of snuff-colored ooze. We were landing on a nice, flat bottom of firm diatomaceous ooze. Indifferent to the nearly 200,000 tons of pressure clamped on her metal sphere, the *Trieste* balanced herself delicately on the few pounds of guide rope that lay on the bottom, making token claim, in the name of science and humanity, to the ultimate depths in all our oceans—the Challenger Deep.

The depth gauge read 6,300 fathoms—37,800 feet. The time—1306 hours.

The depth gauge was originally calibrated in Switzerland for pressures in fresh water, considering water a non-compressible fluid as is usual in these cases. After the dive, the gauge was recalibrated by the Naval Weapons Plant in Washington, D.C. Then several oceanographers (especially Dr. John Knauss of Scripps Institution of Oceanography and Dr. John Lyman of the National Science Foundation) applied corrections for salinity, compressibility, temperature, and gravity. Agreement was reached that the depth attained was 35,800 feet—or 5,966 fathoms. This computed depth agrees well with the deepest sonic soundings obtained by American, British and Russian oceanographic ships, all of which had reported the round-trip sounding time in the Challenger Deep at almost precisely fourteen seconds. The corrected figure confirmed that the *Trieste* had indeed attained the *deepest* hole in the trench.

And as we were settling this final fathom, I saw a wonderful thing. Lying on the bottom just beneath us was some type of flatfish, resembling a sole, about 1 foot long and 6 inches across. Even as I saw him, his two round eyes on top of his head spied us—a monster of steel—invading his silent realm. Eyes? Why should he have eyes? Merely to see phosphorescence? The floodlight that bathed him was the first real light ever to enter this hadal realm. Here, in an instant, was the answer that biologists had asked for decades. Could life exist in the greatest depths of the ocean? It could! And not only that, here apparently, was a true, bony teleost fish, not a primitive ray or elasmobranch. Yes, a highly evolved vertebrate, in time's arrow very close to man himself.

Slowly, extremely slowly, this flatfish swam away. Moving along the bottom, partly in the ooze and partly in the water, he disappeared into his night. Slowly too—perhaps everything is slow at the bottom of the sea—Walsh and I shook hands.

Walsh keyed the UQC four times, the prearranged signal for "on the bottom." We assumed that we were far beyond the range of voice communication. Simply as a matter of routine and perhaps to enjoy the companionship of his own voice, Walsh called on the voice circuit. "*Wandank, Wandank.* This is the *Trieste.* We are at the bottom of the Challenger Deep at sixty-three hundred fathoms. Over."

To our complete astonishment a voice from nowhere drifted through to us. "*Trieste, Trieste,* this is *Wandank.* I hear you faint but clear. Will you repeat your depth? Over."

Walsh repeated the depth and added, "Our ETA on the surface is seventeen hundred hours. Over."

The voice came back to us charged with excitement, "*Trieste,* this is *Wandank.* Understand. Six three zero zero fathoms. Roger. Out."

Heartened by our voice link with friends above, we set about quickly to make our scientific observations. The temperature of the water was an icy 2.4° C. (36.5° F.). It had warmed gradually and continuously from the lowest reading of 1.4° C. at about 2,000 fathoms.

Next, I peered intently through the port. I sought visual evidence of bottom current. The depths of the Mediterranean are usually still, but I had become accustomed to strong bottom currents off San Diego. The water here appeared still; at least, below any threshold I could detect. Of course I knew it could not be completely stagnant. This life on the bottom was ample proof of some interchange of water. Water interchange is a prerequisite for oxygen replenishment, which, in turn, is a prerequisite for life.

As the turbidity that we had stirred up in landing began to clear, I saw a beautiful red shrimp. The ivory ooze was almost flat. There were none of the small mounds and burrows such as those so common in the Mediterranean. Nor was there the usual churning of the sea floor by the bottom-living animals. No animal tracks could be seen anywhere. The bottom was not perfectly smooth, however. I noted some minor undulations suggestive of animal plowings.

For twenty minutes we made our scientific observations. The vertical current meter had been destroyed during the tow; the horizontal meter was undamaged but it gave a reading of zero. I exposed some film to check on radiation from radioactive sources. (When later developed, the results were negative.) Then I switched on the aft searchlight for Walsh who peered out. Suddenly, he called out, "I see now what caused the shock at 30,000 feet!"

He pointed to the large plastic window that

permits us to see through the antechamber to the sea beyond. The plastic window was fissured with small open horizontal cracks. It had failed because of the differential contraction between the metal of the sas★ and the plastic. Fortunately, it was still in place. It presented no immediate threat to our safety. But I realized at once that it could mean some real trouble later on.

Once we surface, that antechamber is our only escape way. Before we could get out of the sphere all of the water must be blown from the sas by compressed air. If the plastic window didn't remain watertight we would be in a desperate predicament. True, we always carried a metal cover plate on the tug to seal off the window against this eventuality. But under the present sea conditions, it would be next to impossible to send divers down to attempt to install it. The numerous white-tipped sharks in the area would be an added obstacle. If the sas could not be cleared the alternative wasn't happy: We would have to be towed back to Guam, sealed in our cell, to be finally extracted upon arrival. In the event of head seas, this could be a five-day ordeal. Flooding the sphere or escaping by aqualung was another possibility but this would be very difficult. (Let me say here that a new mode of construction, very simple indeed, will make it possible to definitely avoid a similar accident in the future.)

Our original plan was to remain on the sea floor for thirty minutes. This plan we hastily revised. As it stood, our ETA of 1700 hours gave us only a scant daylight margin of ninety minutes to effect any necessary emergency measures. There was no time to lose. At 1326, I cut the current on the electromagnet for thirty-six seconds, releasing 800 pounds of ballast. Slowly, the *Trieste* lifted her massive 150-ton hulk off the bottom. The long seven-mile return trip to the world of man began. . . .

The sphere rose ever up and up, reversing the diurnal rhythm of her descent. We were borne upward from night into gray predawn. Faster and faster we rushed towards the light of day. I noted at 1617 hours, that the temperature gauge for the gasoline read 32° F. When we arrived on the surface, the gasoline temperature was 10° F. below freezing. The piping had not caused us any trouble.

As we approached the surface, our rocketing speed attained five feet per second. But the *Trieste* was still stable with no sign of oscillation or fluttering. Once we had pierced the thermocline, the warmer surface water decelerated us slightly. The lighter water increased the apparent weight of the bathyscaph by about one ton.

At 1656—almost exactly on our ETA—the *Trieste* broke the surface. The rocking of the sphere told us we had returned to the heaving breast of the sea. Our seven-mile elevator ride was ended.

In a matter of moments we would know the worst. We had conserved our meager supply of chocolate bars against the real possibility that we would be trapped in our sphere for several days. I decided to bleed air into the sas extremely slowly so that as little pressure as possible would press against the cracked window. In the daylight I could see that the plastic window had expanded again, closing up any fissures. Walsh slowly, very slowly, fed three bottles of compressed air into the sas. I watched, tensely, through the port controlling the bleeding of the air. Finally, the water level dropped below the level of the window in the door. A dense fog appeared in the antechamber produced by the sudden release of pressure. The sas was cleared!

We wasted little time pushing open the door to our vault. We clambered up the sas, through the hatch, and finally topside to the sunshine and steaming tropic heat.

★*Ed. note:* The entrance tube through which the divers descended into the cabin.

The Floor of the Sea

William Wertenbaker

The crude jigsaw fit of the continents first became apparent when the New World was mapped during the age of discovery in the sixteenth century. In the last two decades, during what may well be regarded as a second age of discovery, scientists have found new evidence to support a theory that the continents have drifted apart. The excitement of the scientists as they discover the Mid-Ocean Ridge, the sources of earthquakes, and reversal of magnetic fields is transmitted in the following selection by William Wertenbaker, who writes frequently on scientific subjects. His account, excerpted from an article that was originally published in The New Yorker *and subsequently, in somewhat altered form, in his book* The Floor of the Sea *(1974), explains the concepts of sea-floor spreading and plate tectonics—that the continents and oceans rest on a series of huge, rigid plates of the earth's crust that are constantly in motion.*

Geology is a coinage of the eighteenth century, when the past became a subject for inquiry instead of a dogma of the church. One of the first geological thinkers, Dr. James Hutton, a Scot and a physician who spent all his time diagnosing rocks, wrote of the earth, "No powers are to be employed that are not natural to the globe, no action admitted of except that of which we know the principle, and no extraordinary events to be alleged in order to explain a common appearance. . . . Chaos and confusion are not to be introduced into the order of Nature, because certain things appear to our practical views as being in some disorder." Hutton's was a new view of the earth (though men in every age have believed their ideas to be natural). The urge to disable a fact with revealed truths has been strong in geology, but geologists believe that, unlike earlier doctrines, Hutton's, which came to be called uniformitarianism, is true—and probably they are right, since they often have been tempted not to abide by it. . . .

Hutton's ideas, known to only a handful of friends in his lifetime, became after his death the kernel of Sir Charles Lyell's uniformitarianism. It was a great many years before Lyell's "Principles of Geology," published in 1830, could be improved by anything but detail. "This earth, like the body of an animal," Dr. Hutton had written, "is wasted at the same time that it is repaired." Geologists could see that rains and rivers wear down the earth, and that, indeed, the very substance of mountains is sediments worn off in a yet earlier time, crumpled, turned to stone, and lifted above the plains. Anyone who drives along a highway cut through the shoulder of a mountain can stop and see as much. The earth has lived four and a half times a thousand million years, through cycle after cycle of growth and decay, whose remnants are recent Alps, old Appalachians, and mountains so ancient they are only stumps. But if the earth is like the body of an animal, what quickens it? A hundred years ago, the Irish geologist and seismologist Robert Mallet suggested that the globe was shrinking. He supposed that the earth was getting cooler, and he thought that the solid crust, or surface layer, would have to crumple like the skin of a drying apple, creating mountain ranges. Mallet computed a shrinkage of six hundred miles in the earth's circumference since its formation. Many geologists rejected the theory immediately, but they had nothing as satisfying to propose in its stead. The theory was, according to some authorities, still at the heart of geological thinking eight or nine years ago, though in 1889 the geologist Clarence Dutton announced implacably, "It is quantitatively insufficient and qualitatively inapplicable; it is an explanation which explains nothing which we wish to explain." Great chains of mountains are still rising today, while others have long grown cold. Why? And why are the ranges long and narrow, instead of being, perhaps, round? Why are there mountains at all? Why are there earthquakes? What is tearing the earth apart, and why in California when not in Connecticut? None of this could geologists explain with any assurance, or agreement, even a few years ago—though they could tell the history of the earth in some detail—and, curiously, it has been the exploration of the floors of the oceans which, beginning about 1966, started supplying the answers.

That the ocean floors, which constitute two-thirds of the earth's surface, are no longer terra incognita is due in great part to the relentless curiosity of Maurice Ewing, who died [in] May [1974], at the age of sixty-seven. Ewing began staking out the bottom of the sea forty years ago. He always wanted very badly to understand the earth, and he spent most of his life exploring it. In 1949, he founded Lamont (now Lamont-Doherty) Geological Observatory, an offshoot of Columbia University, which has become the leading institution in the world for exploring the earth at sea. Ewing had an insatiable appetite for information about the earth (year after year he sent two ships around the world and kept them at sea eleven months in twelve), and he institutionalized it in Lamont, which has gathered a large portion of the important information about the earth under the sea. His methods revolutionized investigation of that part of the world. As director of Lamont, he provided the means for others to work, and as a teacher he often formulated the problems they worked on. His own discoveries were numerous, and superlative enough to establish a reputation in half a dozen fields of science. Two, in particular, underlie the entire fabric of global geology as it has come to be understood in the last few years: the nature of the earth's crust under the oceans, where it is both different from and thinner than the continental crust, and the extent

of the Mid-Ocean Ridge, the most important geological feature in the world. . . .

In 1956, Ewing and Bruce Heezen, who was then a student at Lamont and is now a researcher there, made a laconic and elegant association between depth soundings and earthquake data, showing that a narrow belt of earthquakes coincided with a rift valley in the center of the Mid-Atlantic Ridge—a wild, ragged range several miles high and about a thousand miles wide, whose peaks lie nearly a mile beneath the surface of the ocean—and, further, that the earthquake belt continued out of the Atlantic into unsounded areas and was, in fact, worldwide. The earthquake belt is forty thousand miles long. It runs through every ocean. Wherever ships have crossed it, they have confirmed Ewing and Heezen's prediction that a ridge—the Mid-Ocean Ridge, as it came to be called—would be there as well. The Ridge rises above water at Iceland, Tristan da Cunha, the Galápagos, and other islands. It comes ashore into the rift valleys of East Africa, out of the Gulf of Sinai into the Dead Sea, and into the Imperial Valley of California; the earthquake belt goes northwest from the Gulf of California along the San Andreas Fault. One geologist said during the sixties, without overstatement, that the Mid-Ocean Ridge was the most exciting geological discovery in several decades. As it was mapped, it became obvious that the Ridge is the largest single feature on earth, equal in area to all the dry land. The frequent occurrence of earthquakes along its entire length shows that it is neither dormant nor mature but still active. For more than a decade, the nature of the Ridge and its effect on the remainder of the world were the leading problems before marine geologists and geophysicists.

Exploration during the fifties showed that the ocean bottom was not the uniform, motionless, monotonous, almost featureless accumulation of sediments it once was thought to be. On the contrary, something momentous was going on there, and a handful of scientists . . . groped for what it was. A . . . paper on sediments in the oceans observed that they were unbelievably thin, compared with what had been expected in a world several billion years old. The sediment in the Atlantic is only fifteen hundred yards thick, and the sediment in the Pacific is half that. And no one, anywhere, had come up with a piece of sediment that was really old. For years, it had been assumed that if you could only find the right slope there would be unimaginably old sediments waiting to be sampled, but a few pieces between seventy and eighty million years old—an eighth the age of trilobites, a sixtieth the age of the earth—were the oldest anyone had been able to find. "It dawned on us near the end of the fifties," says David Ericson, a micropaleontologist who wrote about the sediments with Ewing, Heezen, and others, "that in order for there to be so little sediment in the oceans it could not have been collecting for more than a hundred million years, and that something must have happened about a hundred million years ago to renew or drastically reorganize the earth's crust under the oceans." It was a startling, even brazen, suggestion to make about two-thirds of the planet. What was the reorganization? The answer was in the Mid-Ocean Ridge. Maurice Ewing and his brother John, a geophysicist, who is now an associate director of Lamont, wrote in the *Geological Society of America Bulletin*, in 1959, "The formation of the [Mid-Ocean] ridge requires the addition of great quantities of basalt magma [molten rock material] and raises the question of its source. . . . We suggest that a convection current system [a form of circulation according to the principle that hot fluids rise and cold ones sink] has contributed the basalt magma and applied the extensional forces to the crust to produce the axial rift. We assume that all [continental] crustal material was collected into one hemisphere by [an] initial current system. The second current, [which] is assumed to persist to the present . . . broke the continental mass into fragments which moved [into] the present pattern." Here was a renewal of the sort the sediment samples implied—the moving

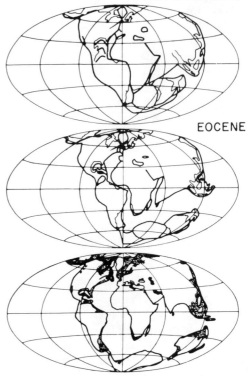

UPPER CARBONIFEROUS

EOCENE

LOWER QUATERNARY

CONTINENTAL DRIFT AS ENVISIONED BY ALFRED WEGENER

In his book Origin of Continents and Ocean Basins *(1929), Wegener suggested that the land masses of the world emerged from a single continent. Modern theorists place the initial breakup of the continent at about 200 million years ago, as opposed to Wegener's 300 million years. Stippled areas in the lowest diagram are modern continental shelves.*

continents wiped out an old ocean floor and left a fresh one in their wakes.

Wandering continents—continental drift—were the widely scorned idea of Alfred Wegener. Wegener was not a geologist but a meteorologist—a fact that later seemed significant to critics—and others before him had thought of mutable continents. Yet it was he who formulated the idea in acceptable geological terms—uniformitarian rather than diluvial, that is—and who attracted professional discussion. In "The

Origin of Continents and Oceans," published in 1915, he wrote, "The first concept of continental drift came to me as far back as 1910, when considering the map of the world, under the direct impression produced by the congruence of the coastlines on either side of the Atlantic. At first I did not pay attention to the idea because I regarded it as improbable." In 1911, however, he saw evidence for a land connection between Brazil and Africa and was sufficiently impressed to begin looking for more. In his book, he assembled a vast quantity of evidence, of varying quality, and constructed from it a coherent hypothesis. . . .

Wegener concluded that there was only one, jumbo continent in the Paleozoic era, and, from the dates of his evidence, that it began to break up during the Mesozoic era into the present continents. Antarctica, Australia, India, Africa, and South America, like the wedges of a pie, had had a common center near South Africa, which was at the South Pole. The separating continents barged their way through the ocean floor like ships in pack ice; at their bows were earthquakes and young, growing mountains, and in their wakes new ocean floor was formed. In fact, the Pacific, on which the drifting continents hypothetically encroached, is today ringed by earthquakes, volcanoes, and new mountain ranges.

Wegener candidly admitted that his evidence was circumstantial; he did not know what force moved the continents, and his few suggestions were flimsy. And though he believed he had proved that the act had been, and was still, going on, there were experts who contested most of his evidence. . . .

Many geologists found Wegener annoying—not worth discussing but necessary to discuss, nonetheless, because he developed a following among laymen and Southern Hemisphere geologists. . . . During the forties and fifties, Bruce Heezen has said, a young geologist who wanted to get or to keep an academic job in this country could not support continental drift cautiously enough. The late crystallographer Sir Lawrence Bragg once said that his showing of a paper of Wegener's to a well-known geologist afforded the only occasion on which he had ever seen a man actually foam at the mouth. A wise man might have found the reaction to Wegener a reason for believing him, though Wegener really was wrong in many respects. "Altogether, the polemics on continental drift are a sobering antidote to human self-confidence," a geologist reflected recently. "Practically all the arguments against it and many for it now appear to have been fallacious." Ideas are currency in science, but they are paper money, worthless without good coin to back them up; even quacks may have the right idea for the wrong reason, but making an idea seem indispensable instead of expendable requires some genius.

When an international oceanographic congress convened in New York City in 1959, a number of the best minds in marine geology were there, including J. Tuzo Wilson, who was then at the University of Toronto and is now director of the Ontario Science Center, one of the most creative men in geology; Sir Edward Bullard, of Cambridge University, who had done pioneering work at sea and on the theory of the earth's magnetism; the late Harry H. Hess, of Princeton, one of geology's global thinkers; H. William Menard, of the Scripps Institution of Oceanography, who almost singlehandedly was discovering and explaining the seafloor geography of the Pacific; Robert Dietz, who was then a Navy oceanographer and is now with the National Oceanic and Atmospheric Administration, an inventive student of both land and ocean geology; and Ewing and Heezen. Most of them believed that the Mid-Ocean Ridge must be the source of some sort of wholesale motion of the earth's crust, like continental drift. What form of motion was quite unknown—except that it was not Wegener's—and was the subject of the liveliest curiosity. So tentative that it got no more than passing mention in the papers delivered, the question was discussed in the halls. Too much was known about the ocean floor—this, indeed, was one reason for having a congress—for geologists still to be able to imagine almost anything they wanted on the bottom, as they once had. The Mid-Atlantic Ridge, for example, was not a heap of detritus that the continents left behind when they parted, as Wegener had asserted. And the continents were not plowing through the seafloor like icebreakers, with brand-new seafloor forming in their wakes. The Mid-Ocean Ridge was emphatically the youngest part of the ocean floor, and was still active. Besides, Ewing remarked, "they can't move their continents through our ocean floor without leaving marks." Bruce Heezen and J. Tuzo Wilson had spoken of an expanding earth, in which the growing Ridge and the separating continents were results of the growth of the earth's surface as a whole. Ewing, Sir Edward Bullard, and Harry Hess thought that thermal-convection cells in the earth's mantle (the layer under the crust) were making the ocean widen at the Mid-Ocean Ridge—specifically, at the rift valley—while an equal amount of crustal shortening and destruction occurred elsewhere. H. William Menard preferred a similar but more limited process, which did not displace continents. In 1956, in their first paper describing a worldwide ridge, Ewing and Heezen had written, "It must be borne in mind that the rift zone may be the primary feature of the combination, and the ridge simply a consequence of the rift." . . .

After the oceanographic congress, Harry Hess and Robert Dietz wrote papers describing a variation of continental drift that was in accord with everyone's new information about the seafloor. It was a daring variation at a time when moving continents were considered improbable, because instead of modifying Wegener they upped his ante. Hess called his paper an "essay in geopoetry," suggesting a good deal of uncertainty; Dietz named the process "the spreading seafloor." Hess and Dietz's continents, unlike Wegener's, moved passively with the seafloor instead of through it—the entire seafloor was conceived as being in motion—and

it was convection in the mantle that moved the crust about. According to them, the seafloor spread away from both sides of the rift valley, where new oceanic crust was formed from the rising magma of the convection currents; the convection currents descended at mountain ranges and at oceanic trenches (which are the deepest places in the oceans), and the oceanic crust riding atop them was destroyed, making room for what was manufactured at the Ridge. This mutation of the theory was generally accepted as the form in which continental drift was to be proved or disproved. It differed from Wegener's not only conceptually but in its incorporation of the oceanic crust and the Mid-Ocean Ridge. Curiously, the Ridge itself became one of the chief causes of skepticism. It was easy to see how the Atlantic could widen. But with a worldwide ridge it was difficult to see how there could be widening everywhere—east-west here, north-south there, other directions in between. . . .

"There was a kind of marvellous tension for several years," says a scientist at Lamont, "knowing the ocean must be very young, knowing the Ridge was there cooking. You had a feeling every time you went to sea that there were secrets there and we were very close to the answers, and that this cruise might sail right over one." It was so simple to say that magma rises out of the earth at the rift valley and the seafloor spreads away—carrying the continents along—and later disappears into the trenches. But the ocean floor did not look simple—not even, and perhaps especially not, the Ridge itself. The closer researchers . . . looked, the more complicated the entire ocean floor began to appear; and in a few years the simple pattern of convection and seafloor spreading had come to seem hopelessly naïve. . . .

In the nineteen-thirties and forties, Ewing had explored the ocean floor with seismic-reflection measurements—exploding TNT at the surface and timing the return of the echoes—but the measurements were difficult to record and process and were made only intermittently.

In 1960, John Ewing developed the seismic-reflection profiler, which automatically converts echoes into a continuous tracing of strata beneath the seafloor. With it, Lamont began to see for the first time the disposition of ancient layers within the ocean-bottom sediments—ocean floors of past ages—and the shape of the crust beneath all the sediments. The layers and the crust could be followed and mapped with the profiler just as the present ocean floors are followed and mapped with echo sounders. The layers in the sediment have now been traced through both the Atlantic and the Pacific. They look like buried plains, completely flat and of enormous extent. To Ewing, the flatness of the old layers was a strong argument that they had not been disturbed since they were formed—and how could the thin ocean crust have been shoved hundreds of miles by seafloor spreading without the sediments' being in the least rumpled in the process? (This remains one of the mysteries, or miracles, of seafloor spreading.) The profiler also shows that there are even more buried layers, just as flat, at the bottom of trenches; and here, according to the advocates of continental drift, an irresistible force was meeting an immovable object. Such findings did not win new supporters for seafloor spreading.

The Ridge, too, became more complex. In 1962, ships from Lamont and the Woods Hole Oceanographic Institution found the Mid-Atlantic Ridge to be offset near the equator by as much as several hundred miles along fracture zones, or faults. Huge fracture zones had been mapped in the Pacific through the late fifties by H. William Menard, at Scripps. But no one had known that they were a characteristic of the Mid-Ocean Ridge. . . . It was soon clear that the Ridge did not wind its way through the oceans sinuously like a snake but zigzagged like stairs; between the fracture zones that offset it sidewise, the Ridge was straight. It was soon clear also that, as with a stairway, the direction of the whole was not always the direction of the parts. There was confusion in many geological camps. What if the segments of the Ridge were so oriented, say, between Australia and Antarctica that they could never have separated those two continents? With vast lengths of Ridge virtually unseen, and offset by unknown numbers of fractures, exploration turned briefly, as it had several times before, to the study of earthquakes. And, by coincidence, there now existed new computers and a new worldwide network of seismograph stations, built to spot nuclear tests (and modelled after one established by Ewing and his colleague Frank Press in the late fifties to study earthquake waves), that could locate earthquakes precisely enough to reveal a few of the steps and offsets in the Ridge.

The magnetic field of the earth has only recently been extensively used to study it. . . . Ewing began towing magnetometers behind ships in 1948, getting continuous measurements of the field and finding local variations, or anomalies; since then, every Lamont cruise has towed one, because Ewing insisted on it, even through years of apparently fruitless effort. In 1952, Ronald G. Mason, of Scripps, borrowed Lamont's magnetometer and towed it in the Pacific, and later he built one of his own. The Mid-Ocean Ridge is volcanic and therefore highly magnetic, and Lamont found an obvious magnetic high over the rift valley; but, except for that one, it was unclear for years what was causing the local variations that the magnetometer was responding to. Enthusiasm for making magnetic measurements at sea waned so much at Scripps that in 1955 the opportunity of letting a government ship tow their magnetometer was nearly ignored. The ship was making a detailed topographical survey—steaming courses five miles apart—of a two-hundred-by-fourteen-hundred-mile area between San Diego and British Columbia. When Mason plotted the data that the magnetometer had registered, he had a startling pattern, unlike anything seen before—stripes of high magnetic intensity alternating with stripes of low intensity. As the ship continued on its monotonous way, the pattern filled the entire quarter-million square miles it surveyed. The pattern stimulated the liveliest interest and speculation. . . . No physical connection between the anomalies and the Mid-Ocean Ridge (between Oregon and Baja California, the Ridge appeared to have vanished underneath the continent) was seen for several years. "If they hadn't been messed up with the continent," Ewing later said of the magnetic anomalies off California, "they would have been understood immediately." . . .

For several centuries, it was assumed that the earth is a permanent magnet, but it is not. Rocks and metals that are permanent magnets lose their magnetism when they are heated beyond a certain temperature, called a Curie point. The earth is both hot and liquid inside—an embarrassing discovery, made around 1900, that left the magnetic field apparently existing for no reason. The best bet today is that the field is produced by some sort of fluid dynamics in the core—which the magnetohydrodynamicists will hit upon eventually. Mars has no core and no field; neither has the moon. When lava cools below its Curie point, the ferrous particles in it align themselves with any magnetic field present—the earth's, in the absence of anything stronger—and thereafter are a sort of fossil of that field at that time. Measuring a fossil field is an exeedingly delicate operation, which was not often performed before the fifties, but when it was done it revealed that there had been significant changes in the field. . . . By 1955, a group of researchers at Cambridge University had developed, with a huge quantity of rocks from all over Europe and covering a span of hundreds of millions of years, a consistent magnetic history of the continent that was hard to argue against. S.K. Runcorn, the leader of the group, wrote, "Appreciable polar wandering seems indicated. There does not as yet seem to be a need to invoke appreciable amounts of continental drift to explain the paleomagnetic results so far obtained." Two years later, however, he did invoke continental drift, to explain why North American lavas record a different path of polar wandering from the European.

Even more bizarre than a planet with wandering poles is a planet with reversing poles. Seven hundred thousand years ago, and—with the exception of a few relatively brief intervals—for more than a million and a half years before that, the north magnetic pole was in Antarctica; it will eventually return there, creating navigational difficulties. In accordance with the general mystery of the magnetic field, why it makes the excursions it does has not been revealed to magneticians, though it leaves plentiful signs of its passage. Reversely magnetized rocks—that is, rocks whose magnetic polarity is the opposite of the polarity that the earth's field has—were discovered in India in 1855, in France in 1906, and in Japan in 1929; by the nineteen-fifties, it was becoming clear that at least a large minority of magnetized rocks were reversely magnetized. One way such rocks could be created is by the polarity of the earth's field itself having been reversed at the time they became magnetized. . . .

[I]n the early sixties three researchers, Allan Cox, Richard R. Doell, and G. Brent Dalrymple, of the United States Geological Survey, reported on large numbers of sample rocks from places all over the world which embodied the same sequence of nine reversals at the same times (from radioactive dating) during the last three and a half million years. At one time, all fresh lava flows had been reversely magnetized everywhere; at another time, all had been normally magnetized. Sediment heated by the lava had the same polarity as the lava. Such a regular, worldwide, repetitive pattern could not be coincidence. Today, most scientists agree that the field reverses from time to time, though an occasional rock may perform the service for itself. Cox and his associates wrote, "The idea that the earth's magnetic field reverses at first seems so preposterous that one immediately suspects a violation of some basic law of physics, and most investigators working on reversals have sometimes wondered if the reversals are really compatible with the physical theory of magnetism." Neil Opdyke, a paleomagnetician at Lamont, recalls that in the sixties he and his

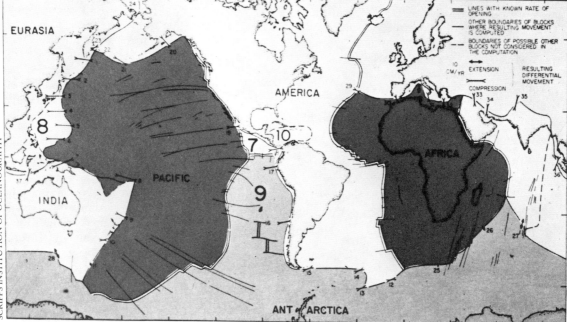

SEA-FLOOR SPREADING AND CONTINENTAL DRIFT

Map showing mosaic of plates that forms the earth's outer shell and the direction of plate movement. The ridge axes where plates are diverging and new oceanic crust is being formed are indicated with a double line.

colleagues often endured being called paleo-magicians. One gets a history of reversals by drilling down through accumulated lava beds. Chronologies have been constructed and are still being extended; detail keeps being added. The poles have switched position with some regularity for nearly a billion years—the present extent of the evidence. . . . Although the present state of affairs is understandably called normal, reversed intervals have accounted for an equal part of the world's history. By and large nowadays, the field stays the same for about a million years, an interval that the paleo-magneticians call an epoch; but the epochs may be punctuated by shorter intervals of opposite polarity, called events, which last only tens of thousands of years. . . . At present, we are in

an epoch that has lasted nearly seven hundred thousand years and is called the Brunhes, after an early finder of reversed rocks. . . .

In the fifties and early sixties, attempts . . . to show polar wandering or continental drift with lava samples from the ocean (from seamounts) proved inconclusive. But research ships each year produced tens of thousands of miles of magnetic profile of the ocean floor, and there clearly were large variations in magnetism across the Mid-Ocean Ridge in the North Atlantic and Indian Oceans. What caused the variations, however, was not clear. They reflected topography—any sort of volcanic feature meant a magnetic variation—but they reflected more: a seamount, for instance, produced a strong anomaly, but the anomaly might be positive and it might be negative. There was a whopping anomaly at the crest of the Ridge, from which the ups and downs gradually subsided toward the flanks.

In 1963, a few months after Cox and his colleagues published the first of their overwhelming evidence for field reversals (but before there was any clear calendar of them), F.J. Vine and D.H. Matthews, of Cambridge University, proposed that the magnetic anomalies of the Mid-Ocean Ridge were caused by a combination of reversals of the earth's field and seafloor spreading. If seafloor spreading occurred, strips of molten lava rising from the Ridge would cool at the crest, be magnetized according to the polarity of the earth's field at that time, and then be split and pushed to either side of the Ridge by fresh lava rising from below. If the field reversed itself, the lava that rose afterward would have the reversed polarity. As the process was repeated over the course of time, the result would be alternate strips of normal and reversely magnetized seafloor on either side of the Ridge. Vine and Matthews wrote, in *Nature,* "If spreading of the ocean floor occurs, blocks of alternately normal and reversely magnetized material would drift away from the centre of the ridge and parallel to the crest of it." Dietz had described "strips of juvenile seafloor" forming at the Ridge, and these, differently magnetized, would create a magnetic pattern like the magnetic stripes found in the Pacific off California. The seafloor would record reversals of the field as tree rings record wet years and dry years. This was a remarkable piece of imaginative thinking, but no one seemed to believe it might be true. Vine and Matthews had some magnetic profiles of the Ridge in the Indian Ocean, and by programming a computer with their theory they produced theoretical profiles that were fairly similar to the real ones. The demonstration was considered interesting but not good enough to be convincing. In the meantime, a United States Navy aeromagnetic survey, Project Magnet, made fifty crossings of the Reykjanes Ridge, the segment of the Mid-Ocean Ridge immediately south of Iceland, and Lamont obtained the data. When the crossings were correlated, they produced a pattern of parallel stripes, like those in the east Pacific; and one side of the Ridge was nearly the mirror image of the other, at least in the two-hundred-mile-square survey. This was in accord with Vine and Matthews' idea: if strips of magnetized seafloor were being created on the Ridge and then split and pushed to either side by new strips rising from below, the result should be that one side of the Ridge would be the mirror image, magnetically, of the other. No one had ever correlated profiles of the Ridge into a pattern before; if the east-Pacific pattern turned out to be characteristic of the Ridge, that would be a major discovery. Though the Navy had supplied better evidence for Vine and Matthews than they had supplied for themselves, detailed examinations in the next year or two made their idea seem less ravishing. The anomalies were symmetrical about the axis of the Ridge, but the one at the crest of the Ridge was apparently of quite a different breed from the ones farther away, though the anomalies of the Ridge flanks had been created at the crest if seafloor spreading occurred. No one was able to imagine a mechanism for turning crest anomalies into flank anomalies. There were also certain technical difficulties in producing any anomaly at all by seafloor spreading. Nonetheless, Vine and Matthews was the only hypothesis around, and other papers referred to it, if only in such (for scientists) savage terms as "dubious" and "untenable." A scientist at Lamont said recently, "The Vine and Matthews theory was interesting, but they hadn't the data to back it up." "Not many took it seriously here." Vine himself wrote, "At the time this concept was proposed, there was very little concrete evidence to support it, and in some ways it posed more problems than it solved." . . .

Seafloor spreading became an established fact, and not wishful thinking, in 1966, at Lamont. One ideal piece of evidence turned up there— not surprisingly, considering the hunger with which Ewing and Lamont gathered evidence— and once the pattern was seen in a single bit, it was quite visible in others, of which great quantities were soon brought out of Lamont's files. For a number of people, there was a period of almost unbearable excitement, lasting about two years and spreading to other geological establishments. Some people think it the most remarkable period in the history of geology. Today, young men speak like old war veterans, with the feeling of having lived through an era the like of which will not be seen again. By the end of 1968, a hypothesis called plate tectonics, which incorporated the ideas of seafloor spreading and continental drift with the mechanics of how they were accomplished, was accepted almost universally among those who had heard of it. By 1970, it had suffused and was reshaping traditional geology; the impetus for mountain building was accepted as coming from seafloor spreading, and geologists are now trying to read its history and consequences in the rocks of folded mountains of the West Coast, the East Coast, and the Mediterranean. Today, a paper on continental mountain building that took no note of plate tectonics would be surprising.

The mystery of the earth's surface began to unravel at Lamont during the winter of 1965–66. . . .

Ewing had been encouraging people to study magnetism, and late in 1965 some graduate students found magnetic-field reversals in sediment cores—long tubes of sediment from the ocean floor brought up in a device Ewing liked to call a cookie cutter. For many scientists, it was this discovery that established field reversal as fact, for sediments and lavas have almost nothing else in common. . . .

Neil Opdyke had worked in one of the first laboratories to pursue polar wandering, and he was an early believer in continental drift. Ewing, though he disagreed with all of Opdyke's driftist conclusions, had sought him out and hired him from the University College of Rhodesia. Opdyke is enthusiastic about the results his group at Lamont obtained. "It was one of the high points of my scientific life," he says. . . .

During the winter, spring, and summer of 1966, Opdyke and his graduate students did the tedious work of buttressing their findings. . . . They worked through a dozen cores—each from sixteen to forty feet long—that had already been painstakingly analyzed, dated, and correlated by fossils. Samples were taken every four inches and measured; when a reversal was found, continuous samples were taken. One by one, reversals appeared in each core, were dated by fossils, and were found to correlate not only from core to core but with the existing chronology established from lavas. The cores reached much farther into the past than the chronology from lavas did, and Opdyke and his group slowly approached a time when they would leave familiar ground.

Before that happened, there was excitement across the hall, where the magnetic pattern of the seafloor at last began to make sense, and the two magnetic investigations began to influence each other. In January, Walter Pitman, a graduate student (he has since become one of Lamont's senior scientists and adjunct professor of geology at Columbia) with an office opposite Opdyke's rooms, got three magnetic profiles of the Mid-Ocean Ridge from Lamont's computer, which had reduced them from a great many yards of paper to convenient size and had subtracted the normal effects of the earth's field, leaving only the interesting aberrations. The profiles had been made in the far South Pacific, aboard the National Science Foundation's ship Eltanin, which used to hover about Antarctica and did marine geophysics under Lamont's direction. Pitman had only recently returned from a voyage on her during which two of the profiles were made. Ships, even research ships, had rarely visited that remote part of the ocean, and information about the Mid-Ocean Ridge there was scarce. . . . That January, Pitman glanced over the two profiles made while he was on Eltanin, and passed on to the third, made on the previous voyage, Eltanin–19 (the name by which the profile came to be known, though some of Pitman's colleagues call it Pitman's magic profile). That one stunned him. Even to an untrained eye, the Eltanin–19 profile does not look haphazard or coincidental. To anyone who has tried unsuccessfully for years to stare meaning out of data, it has a special excitement.

It is too smooth, too regular, the variations of one side too faithfully reproduced on the other. Even to eyes glazed by the brassy interchangeabilities of mass production, its perfection is visible. Later on, when it was shown about, a man at Lamont said, "This is so perfect that I know seafloor spreading couldn't have done it." Norms are human obsessions; nature rarely bothers to construct one. The Eltanin–19 profile is probably the most remarkable piece of evidence to come along in any field in some years. It consists of vigorous peaks and valleys of magnetic intensity arranged in obvious symmetry, size for size, cluster for cluster, about the center of the Ridge. To an educated eye, it embodied,

almost too well, the Vine and Matthews hypothesis and seafloor spreading.

"It hit me like a hammer," Pitman said recently. "In retrospect, we were lucky to strike a place where there are no hindrances to seafloor spreading. We don't get profiles quite that perfect from any other place. There were no irregularities to distract or deceive us. That was good, because by then people had been shot down an awful lot over seafloor spreading. I had thought Vine and Matthews was a fairly dubious hypothesis at the time, and Fred Vine has told me he was not wholly convinced of his own theory until he saw Eltanin–19. It does grab you. It looks just the way a profile ought to look and never does. On the other hand, when another man here saw it his remark was 'Next thing, you'll be proving Vine and Matthews.' Actually, it was his remark that made me go back and read Vine and Matthews. We· began to examine Eltanin–19, and we realized that it looked very much like a profile that Vine and Tuzo Wilson published just before our data came out of the computer. Their profile was from the North Pacific—from the Juan de Fuca

MAP OF THE OCEAN FLOOR

The Mid-Ocean Ridge can be seen zigzagging like stairs through the mid-Atlantic, between Australia and Antarctica, and through the Pacific, disappearing beneath the continent between Baja California and Oregon.
A rift splits the crest along its length; the lines perpendicular to the ridge are fracture zones. White dots indicate the hundreds of holes drilled in the ocean floor by Glomar Challenger *between 1968 and 1975.*

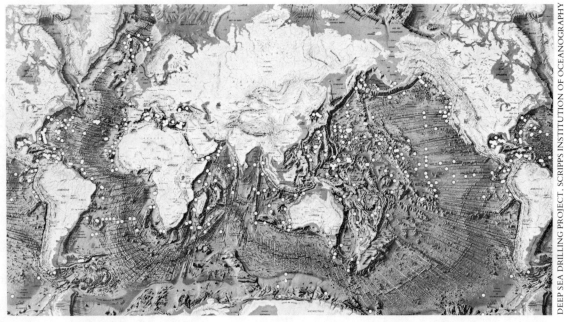

DEEP SEA DRILLING PROJECT. SCRIPPS INSTITUTION OF OCEANOGRAPHY

Ridge, a segment of the Mid-Ocean Ridge which lies off the Juan de Fuca Strait, south of Vancouver Island. We said, 'Isn't this interesting?' If the pattern was the same eight thousand miles away, there was the more reason for thinking that it was determined by something worldwide like field reversals—Vine and Matthews' idea. Vine and Wilson had tried to relate their magnetic profile to the known history of magnetic reversals, but they had failed—in part because, as it turned out, that history was still incomplete. One brief interval of magnetic polarity—an event, as they're called—hadn't been found yet. Without it, you couldn't get a fit between the history of reversals and a profile without false assumptions or distortions. But we were lucky. Just then, Opdyke began to put together the history of reversals from sediments, and pretty soon he found the missing event."

With a chronology of reversals of the earth's magnetic field, which was then complete, Pitman and Heirtzler attempted a demonstration that would show whether magnetic reversals and seafloor spreading really did have something to do with the Eltanin–19 profile, as they so abundantly appeared to. "Now we had a good time scale, and we used it to make a model of recent seafloor spreading in the South Pacific," says Pitman. "It was the crudest kind of model—just alternate blocks of normally and reversely magnetized seafloor. . . . But it fit. We programmed a computer to produce a profile from it, and the profile almost exactly matched Eltanin–19. That gave us the confidence to make a profile for one centimetre a year, at different latitude and longitude, and that profile matched the magnetic profiles of the Reykjanes Ridge, below Iceland. It was an important match—not only because our model was still good halfway around the world from where we started, in the South Pacific, but also because those Reykjanes Ridge profiles had been used against seafloor spreading. Lamont had published them with arguments on why they didn't fit the Vine and Matthews hypothesis, but everyone had remarked on their rough symmetry. I don't think Vine himself

had the nerve to claim the Reykjanes Ridge then. There's no rift valley there, and everyone thought rifting had to go on in a rift valley. There's no rift valley in the South Pacific, either. It's one of the ironies that the rift, which started everyone thinking about continental drift again, isn't found where seafloor spreading is working most smoothly—just where it's slow and difficult. Slow spreading makes for the big topography, but the beautiful, even magnetic pattern comes only with fast spreading—and puny topography. . . ."

In June of 1966 Pitman and Heirtzler were sufficiently satisfied with their findings and provings to call in a few people to look at what they had. Two, Jack Oliver and Lynn Sykes, were seismologists and taught in Columbia's Geology Department. "Sykes and Oliver," says Pitman, "went roaring back to look at their data." . . .

The seismologists were able to show, with their records of earthquakes, that one large segment of the earth's crust is being pushed, or drawn, under another at the submarine trenches, and also at certain mountain systems, like the Andes—the process that removes old oceanic crust from the surface of the earth as the Mid-Ocean Ridge generates new crust. "In the early sixties," says Sykes, "seismology here was on a vein that was about mined out, and we decided, Oliver, Bryan Isacks—another seismologist—and I, to set up a study of Tonga, because it has so many deep earthquakes. Though it wasn't clear what problem this would solve. Deep earthquakes were first observed in the nineteen-twenties, but they weren't believed possible until the thirties. To get any earthquake, you have to have something hard that can rupture. It was thought that down deep, in the mantle—the layer beneath the crust—high temperature and pressure should melt everything. The Japanese showed that, possible or not, deep earthquakes do occur. In 1964, when we set out our first instruments in Tonga, it was still a mystery why. Deep earthquakes occur only under trenches and island arcs. They lie in distinct zones dipping down from the trenches, under the island arcs—which are volcanoes—to a depth of six hundred miles. In our first papers, we assumed that the zones were about a hundred miles thick, but we found that they are really only about fifteen miles—very thin indeed. After we put in our instruments around Tonga, we noticed a few things right away, such as that—this was Bryan Isacks—the high seismic frequencies were travelling along the plane of the zone. This we hadn't expected at all. High efficiency in transmitting waves means hardness, coldness—all that the mantle was supposed not to be. So we knew we had to revise our ideas about what a deep-earthquake zone is like. It all came together in a model late in 1966. By then, all our thinking was influenced by what was going on in magnetics over seafloor spreading. Jack Oliver had been toying with the idea that this zone was a piece of something different hanging down there. Well, waves also propagated very easily along the seafloor fifteen hundred miles to our seismic station in Raratonga. All of a sudden, it became simple and obvious. We were led to the conclusion that the earthquake zone is the same as the seafloor, and then—Pitman had just given evidence for seafloor spreading—that the seafloor is being pushed, or pulled, down into the mantle to form the zone." With less than the usual professional diffidence, Isacks and Oliver wrote, "The evidence is largely qualitative, but it is not subtle and it is definitive. . . . The differences . . . are gross and evident even to the casual observer."

Meanwhile, Maurice Ewing and Dennis Hayes, a geophysicist, looked at profiler records of the trenches and found that they could actually see crumpled sediments scraped from pieces of seafloor that were being pushed or pulled down there. Accounting for how crust was disposed of had been the greatest vulnerability of the theory of seafloor spreading. Ewing said later, "We'd had from the start the idea that the crust was pulling apart. When we found the rift, I wrote Felix Vening-Meinesz, the Dutch geodesist, who made the first gravity

measurements at sea, who had thought there was convection in the earth, that this proved it. What we couldn't understand was how crust was gotten rid of. It's hard to see why. I repeatedly hammered the table and said, 'These trenches are empty.' Well, that's where the steel hits the rolling mill, and we can see where the sediments are scraped off against the inner wall as the crust dips under." . . . Lamont changed almost overnight from the most dangerous enemy of continental drift to its most fruitful advocate. And Lamont had over half the magnetic and profiler data on the ocean basins and eighty per cent of the cores waiting to be examined for evidence of seafloor spreading. Lamont had been gathering this data for years, in every part of the world, even when it was of no apparent interest, because of Ewing's belief—and insistence—that everything should be examined to find out what was interesting. After a visit, a geologist wrote Ewing, "I see why it is that your geophysicists were so largely responsible for the explosive development of plate tectonic data and concepts." At a meeting at Lamont, the geophysicist S. K. Runcorn, an advocate of drift for over a decade, remarked genially, "I feel like a Christian visiting Rome after the conversion of Constantine."

It took us a long time to work out the implications of what we had just done," says Walter Pitman. "To people who had essentially rejected an idea for years, as we had rejected seafloor spreading and continental drift, every ramification of it came hard. We had to get used to the concepts we were dealing with. But Opdyke knew what it all meant—he knew damn well. When I first spoke on Eltanin-19, at the American Geophysical Union meeting in the spring of 1966, he went around telling people—Runcorn, Cox—that they'd see something to blow their minds. Not many people came—the abstract was so dull—but those who did come were convinced. That was the catalyst. Then NASA sponsored a meeting in November at Columbia, and that broke down all the resistance. We all presented our evidence for the first time. One man who had been violently against continental drift just got up and walked out. But I remember that Menard, from Scripps, who had also opposed it, sat and looked at Eltanin–19, didn't say anything, looked and looked and looked. At the next A.G.U. meeting, in 1967, when we gave four papers on the complete magnetic pattern, it was a question of breathing room. There were three special symposiums on seafloor spreading, and thirty people spoke, where only Opdyke and I had spoken the year before. There was no opposition." One prominent geologist attributes "the wholesale, overnight conversion of American earth scientists to continental drift" to that meeting. By the following autumn, J. Tuzo Wilson was sounding nostalgic. "It is only ten years since Ewing and Heezen recognized that the mid-ocean ridges are continuous and of great size," he told a symposium in Zurich.

Several writers proposed the theory that the spreading seafloor moves as a solid plate rotating around some point on the earth's surface—a piece of basic solid geometry. Xavier Le Pichon showed that the earth's surface is composed of seven large plates moving in different directions. The Americas have recently joined into one plate, moving west. Most of the Pacific is moving northwest into the trenches between Alaska and the Philippines. Antarctica and much of the surrounding ocean floor appear stationary. Africa is moving northeast. Eurasia is moving east. The Indian Ocean, Australia, India, and Saudi Arabia are moving north. A map compiled at Lamont of the sites of all earthquakes from 1961 to 1967 seems designed to depict Le Pichon's plates. Again nature has been startlingly unsubtle. The earthquakes sharply define the seven large plates, plus a few small plates in the chinks; they run in spidery lines along the fracture zones and the crest of the Mid-Ocean Ridge. There are thick swarms of them at the trenches and mountain belts where plates come together and one plate is, in current parlance, subducted. The earthquakes are all at the plate borders; the plates are white space. By rotating the plates when he had found their poles, Le Pichon was able to translate the rate of seafloor spreading at the Mid-Ocean Ridge into the amount of destruction at the trenches for different plates (which often meet the trenches obliquely). Unencumbered seafloor moves fastest. At New Guinea, the Pacific and Indian plates come together at a rate of eleven centimetres a year, while South America overrides the Pacific at only six. Out of the pendant zones of earthquakes below trenches, Isacks, Oliver, and Sykes wrung evidence confirming Le Pichon's predictions. They summed up all the earthquake and other geophysical evidence in an influential paper they called "Seismology and the New Global Tectonics." The idea of plates is so central that "seafloor spreading" has become a department of "plate tectonics." The rotation of a rigid plate about an axis accounts for the different kinds as well as the different amounts of activity at the Ridge, at the fracture zones, at the trenches, and in the interior of the plate. All parts of the fabric are related and affect one another. Spreading on the East Pacific Rise is matched by activity along the San Andreas Fault and is balanced by consumption of the other end of the plate at Alaska, under which it descends, via the Aleutian Trench, with shattering earthquakes. Periodically, as the two plates disengage in an earthquake, a section of coastline pops up several feet. The meeting of North America and the Pacific floor is also tied to spreading at the Mid-Atlantic Ridge. As the Pacific plate moves northwest, it carries volcanoes away from a source of magma in the mantle; they become dormant and new ones form—for instance, those in the Hawaiian chain, which is oldest to the northwest and active only to the southeast. Cycles of mountain building correspond with cycles in seafloor spreading and can be matched by date. When all intervening seafloor is consumed, and two irresistible objects, like India and Asia, meet head on, the pattern readjusts. "It would be difficult to overstate the success of these ideas in bringing together the different disciplines that constitute earth sciences," a committee of geologists reported a few years ago. "[It is] a concept comparable to that of the Bohr atom in its simplicity, elegance, and ability to explain a wide range of diverse observations."

When Pitman and his colleagues had the magnetic pattern in the oceans explained, they also had the tale of continental drift—the one Wegener had tried to tell. The diaspora began two hundred million years ago. There was then a single continent, Pangaea, which had assembled out of smaller continents in a former convulsion of seafloor spreading. . . .

India forsook Australia and Antarctica and went tearing off north, and, after a mere seventy million years, ran into Asia. Following that collision, forty million years ago, there was a lurch and then a readjustment in the pattern of seafloor spreading. Australia split from Antarctica while India lost momentum against Asia. New Zealand separated from Antarctica at a rapid clip and, of late, has been catching up with Australia.

Spreading in the North Atlantic has been slow and hard, and Pitman and Manik Talwani, a geophysicist, who recently became director of Lamont, were three years deciphering the pattern. They had it complete two years ago. North America, Africa, and Europe did not leave each other at once. Africa went (or was left) first, a hundred and eighty million years ago. Dakar would then have been near Jacksonville, Las Palmas near Boston, Tangier around Cape Breton, France off Newfoundland, and England by Labrador and Greenland, there being no Labrador Sea yet. The Atlantic opened rapidly between North America and Africa for almost a hundred million years. Africa scraped past the southern edge of Europe, for a while dragging Spain ignominiously along and so opening the Bay of Biscay. Eighty-one million years ago, Europe and Greenland began to move, opening the Labrador Sea; but after twenty-five or thirty million years the center of spreading moved east of Greenland, which was left behind. Until about fifty million years ago, Europe travelled fast and was gaining on Africa. Since then, they have kept pace, but Africa has steadily crowded north against Europe. The motion has not been easy on the soft underbelly of Europe, which has been repeatedly squeezed. "It's one of the ironies of geology that the Alps, where classical geology began, have the most complicated sequence of orogenies in the world," Pitman said recently, after describing the relative motions between the two eastbound continents. Spain and other small fragments have been traded back and forth. Italy belongs to the African plate. The Alps are where the African plate is thrusting against the European—the plates are being, as it is called now, sutured. With a seismic profiler, the sediments beneath the Mediterranean can be seen, upheaved and crumpled in the embryonic stages of mountains. Africa will presently collide with Greece. . . .

While plate tectonics is a beautifully simple notion, the causes of it and the changes rung on it are not. Plate margins and sutures very likely will be the problem of the next five or ten years in geophysics and geology, and the mantle the problem of the next twenty. "Nothing is going to be as simple as plate tectonics now makes it seem," Ewing said in 1974, "just as no atom is exactly like the Bohr atom. But the Bohr atom was an idea capable of sufficient refinement, and so is plate tectonics. There are other things going on in the earth that are just as cunning as this magnetic pattern."

Minerals and Plate Tectonics: A Conceptual Revolution

Allen L. Hammond

The "need to know" has always been one of the strongest motives in scientific inquiry, but new knowledge often has very practical applications. Such is the case with the revolutionary ideas of plate tectonics, which promise to turn the search for precious mineral resources from a hit-or-miss operation into a rational program for exploration. In the following article, written in 1975, Science *magazine staff writer Allen L. Hammond discusses the implications of the theory of plate tectonics for mineralogy and, eventually, for the economy.*

Drastically higher prices for oil and declining U.S. production have drawn attention to supplies of other key industrial materials, especially minerals. Although immediate shortages do not appear likely, some authorities have expressed concern about the extent of U.S. dependence on other countries for supplies of chromium, manganese, and other metals. Moreover, depletion of high grade ores and environmental regulations affecting mining and ore processing are expected to increasingly constrain the availability of minerals.

Fortunately, renewed interest in minerals comes at a time of excitement and sweeping new ideas in the study of mineral deposits. The new ideas reflect the impact on economic geology of plate tectonic models for the evolution of the earth's crust. Many ore deposits, for example, are now known to occur at present or past boundaries of the huge crustal plates whose movements have shaped and reshaped the earth's surface. What ores are formed and where they are placed in the crust, it is proposed, depend principally on the tectonic history of a particular region; several models of the processes involved have been put forward. Similarly, it is proposed that the interaction of seawater with cooling volcanic rock is the principal means by which many metals are extracted and concentrated into economically valuable ore bodies. . . . These proposals and others have stimulated a host of more detailed investigations. Many geologists believe that these developments portend a fundamentally new understanding of the origin of minerals and are laying the scientific foundation for a new era in mineral exploration.

Not all mineral deposits fit the new conceptual framework, but many major classes of metal ores are explicable in its terms. The evolving theoretical models provide detailed if still controversial explanations for the chemistry, mineralogy, and stratigraphic location of these deposits and thus a host of clues with which to look for still undiscovered mineral deposits, some of which are finding tentative use in the mineral industry. They also have implications for the evolution of the earth's crust; similarities between recent and more ancient ore bodies are seen by some researchers as evidence that tectonic processes not unlike those of the present geologic era occurred throughout most of geologic history.

Many metallic ores are now widely recognized to be of volcanic origin in the sense that they occur in volcanic or igneous rocks and were formed at the same time as those rocks. According to plate tectonic theory, volcanism occurs in several circumstances: at diverging plate boundaries (mid-ocean ridges or other centers of sea-floor spreading), where mantle material rises to form new oceanic crust; at converging boundaries, where crustal plates descend into the mantle in a process known as subduction, leading to volcanism that forms chains of mountains

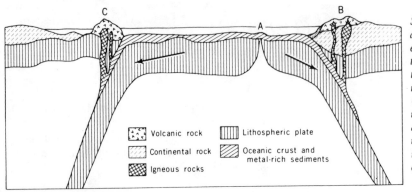

Schematic showing three different ore-forming environments and the plate-tectonic phenomena postulated to give rise to them: **(A)** *a mid-ocean ridge or rise;* **(B)** *a subduction zone underlying a continental margin; and* **(C)** *a subduction zone underlying an island arc. Arrows indicate direction of motion of the plates.*

Volcanic rock Lithospheric plate

Continental rock Oceanic crust and metal-rich sediments

Igneous rocks

or oceanic island arcs; and, less frequently, over hot spots caused by ascending plumes of mantle material (see illustration). Each of these processes, except possibly the last, is now thought to give rise to a characteristic type or types of ore deposits.

One of the clearest examples—and one which has had major impact on the thinking of economic geologists—is found on the Mediterranean island of Cyprus, long a rich source of copper. The copper sulfide ore occurs in the Troodos area of Cyprus in a distinctive sequence of rocks: on top, sediments of a type formed on the ocean floor; beneath the sediments, pillow lavas formed when molten volcanic material erupts into seawater; farther down, vertical sheets or dikes of basaltic rocks formed as rifts or cracks in the ocean floor are filled from below with volcanic material; and on the bottom, ultramafic rocks (rich in magnesium and iron) that are believed to be characteristic of the earth's mantle. This progression of rock types is known to geologists as an ophiolitic sequence. About the time that economic geologists recognized copper sulfide deposits as an integral part of these rocks on Cyprus, other geologists recognized the ophiolitic sequence as exactly that which should be formed at a mid-ocean ridge. Thus the Troodos area is now thought to be a largely unaltered piece of oceanic crust thrust up when Cyprus was formed, and the mineral deposits it contains are thought to be characteristic of those formed at mid-ocean ridges.

The minerals include sulfides of copper, iron,

and sometimes zinc embedded in the pillow lavas, small masses or "pods" of chromium ore near the top of the ultramafic layer, and asbestos deposits, also in the ultramafic rock. Although not present on Cyprus, lateritic nickel deposits are sometimes found in sections of oceanic crust where ultramafic rock (which is rich in nickel) has been exposed and weathered. Mineral deposits of the Troodos type are found in many parts of the world, including the northeastern United States and eastern Canada. They range in age from the geologically young deposits of Cyprus to older deposits that originated as many as 600 million years ago.

A second major type of mineral deposits—large bodies of low grade ores known as porphyry coppers—are commonly associated with converging plate boundaries. A prime example is the extensive copper deposits in the Andes, where the eastward-moving oceanic crust of the Pacific plunges under the lighter material of the westward-moving South American continent. Partial melting of the downward-moving oceanic plate is believed to generate magmas★ that rise through the overlying continental rocks, sometimes reaching the surface to form volcanoes. The upper portions of the pipelike stalks or cores of these magmatic intrusions into the surrounding continental rock often contain copper and molybdenum, and sometimes gold and silver as well. Several investigators [who] have studied this process . . . propose that for-

---

★*Ed. note:* Molten matter under the earth's crust.

126

mation of porphyry ore deposits is a normal facet of the processes that generate the igneous rocks in which they occur. Sillitoe, for example, suggests that the metals of the porphyry ores were initially incorporated in oceanic crust at the mid-ocean ridge, transported horizontally by the movement of the plate, and then released as the downward-moving plate is heated.

Porphyry copper deposits account for more than half the world's supply of that metal. In addition to the porphyry deposits in the Andes, there are deposits in western North America, in parts of the Alpine belt of Europe, and in Iran and Pakistan. Although normally associated with continental rocks, porphyry deposits are also found in some of the larger volcanic islands of the southwest Pacific. Most of these deposits are geologically youthful, less than 200 million years in age. Highly eroded remains of older deposits have been found, however, in northeast North America. Even in the richest porphyry ores, however, copper rarely exceeds 1 percent, and 0.5 percent is more common, so mining consequently involves extracting and processing large tonnages of ore. The association of porphyry deposits with the subduction of oceanic crust into the mantle and the concomitant magmatic activity is so strong that it has been the basis for exploration efforts. A major exception, however, may be the porphyry deposits in Arizona, which according to J. David Lowell of the University of Arizona do not show evidence of a subduction zone. In recent years new porphyry deposits have been found in Okinawa, Panama, and British Columbia.

Deposits of a third distinctive type, known as massive sulfides because they often occur as large, nearly pure lenses of high grade ore, are found in modern island arcs and some geologically older island arc materials that are now incorporated in continental margins. These deposits, like the porphyry coppers, are associated with the convergence of two crustal plates. They are typically polymetallic, containing copper, zinc, lead, gold, and silver.

The prototype deposits for investigators unraveling the origin of these massive sulfides have been those in northeast Japan. This black "Kuroko" ore is thought to have been formed by submarine volcanic processes and deposited in shallow, near-shore environments late in the evolutionary history of a volcanic island chain. The volcanic rocks associated with these deposits are correspondingly highly evolved and often include fragments from explosive eruptions. Marine sediments are also often found with such deposits.

A second variety of massive sulfide ores—those of the Besshi type—are also found in island arcs. Besshi copper and iron sulfide ores (named after a deposit on Shikoku Island, Japan) are, like the Kuroko ores, commonly thought to be submarine volcanic emissions, but deposited on the underwater slopes of volcanoes early in their evolution. Still other classes or subclasses of island arc mineral deposits, corresponding to additional stages in the evolution of these fragments of land can be distinguished. In fact, a model of the process proposed by [Andrew] Mitchell and J. D. Bell, also of Oxford University, describes seven such stages. They give the timing and accompanying rock types for the formation of Besshi and Kuroko massive sulfides, porphyry coppers, and exogenous mineral deposits (those not formed at the same time as the surrounding volcanic rocks) found in island arcs.

As the plates move, island arcs may be swept into and incorporated in continental masses. And because continents collide, it is not surprising that island arc fragments have been identified in what are now continental interiors. This is significant because a second and major source of massive sulfide and precious metal deposits is the so-called greenstone belts found in ancient Precambrian areas of continents. These belts have historically been the source of much of the world's mineral wealth, with rich deposits ranging from iron ores, important gold deposits, copper and zinc to lead and silver ores.

The Precambrian mineral ores, like younger massive sulfide deposits, are believed to result from submarine volcanic processes. The volcanic rocks associated with these ancient mineral deposits also show chemical and mineralogical similarities to those of island arcs. Hence some geologists believe that greenstone belts represent ancient island arcs. Since some of the Canadian belts date back at least 3 billion years, this would imply the existence of tectonic mechanisms similar to those that create modern island arcs throughout much of the earth's history, a conclusion that is still controversial. If crustal plates did exist in the Precambrian era, 600 million years ago and earlier, they were apparently much smaller but possibly more numerous; the greenstone belts tend to be hundreds of kilometers in length, not thousands of kilometers like modern island arcs. In any case, the similarities and differences between old and young ore types may be important for exploration—rocks, presumably ancient, that underlie the upper portions of the continents are largely unexplored.

A final class of mineral deposits, whose tectonic derivation is much more speculative, are those ores thought to be formed within a crustal plate, rather than at its boundary. Here the proposed mechanism is penetration of mantle material up through the crust to form a hot spot, possibly as a result of the mantle plumes which have been hypothesized as a driving force for the motion of the crustal plates. Hot spots, investigators are suggesting, may have heated the crustal rock, mobilizing metals from sedimentary or crustal materials and concentrating and depositing them nearer the surface. [P. W.] Guild, for example, proposes that the rich lead-zinc ores of the Mississippi Valley may have originated in this fashion. Similar proposals have been made for lead-zinc deposits in northwest Africa. In some instances, Guild believes, the minerals themselves have come from the mantle, propelled up through the crust by the heat of the plume. Diamonds, niobium, and some rare-earth deposits, for example, are associated with the explosive eruption of mantle materials to the surface and may be attributable to a plume mechanism. Heat from the mantle,

perhaps rising near subduction zones, may also provide the energy to mobilize metals present in lower crustal rocks and concentrate them into ore deposits in some circumstances. This mechanism has been proposed to explain the eastward shift from dominantly copper to dominantly lead-silver ores in western North America, and the repeated emplacement of tin ores in only a few areas of the earth. . . .

[A] pattern of mineral deposits with identifiable plate tectonic origins may well be common to the entire Appalachian-Caledonian chain. Limestones bearing lead and zinc are found from Norway to Alabama, always on the westernmost edge of the mountain chains. Also extending along the length of the chain are ophiolites with, in many places, Troodos-type copper and iron sulfides. Known occurrences of polymetallic island arc deposits are more scattered, according to [David F.] Strong, but appear to lie in the central and eastern portions of the Appalachians. Tin occurs in Alabama and Virginia still farther east. The mineral patterns constrain tectonic models for the Appalachians, especially by implying the existence of a southeastward-dipping subduction zone during the formation of the mountain belt, according to Strong. They also have implications for mineral exploration, he believes, since discoveries in Norway could lead to similar finds in Newfoundland and Tennessee or vice versa. Exploration of Newfoundland and eastern Canada has, in fact, accelerated in the past several years.

The new models of ore formation are far from complete and are still some distance from being completely accepted. Still newer ideas concerning the geochemical processes involved and the role of seawater are being proposed. . . . But perhaps the most significant aspect of the emerging synthesis between plate tectonic theory and metallogeny is the prospect that, in a resource-hungry world, mineral exploration can increasingly be guided by a detailed understanding of how, and perhaps where, ores are formed and deposited.

128

The Search for Atlantis

J. V. Luce

One of the most enduring legends of all time concerns the "lost continent" of Atlantis. The story, allegedly told to the Athenian poet and statesman Solon by Egyptian priests in the sixth century B.C., *was recorded by Plato 200 years later. Ever since then, countless theories have placed the lost continent in such varied places as the Azores, the Canary Islands, North Africa, and Bimini, off the coast of Florida. In the following selection from* The End of Atlantis *(1973), J. V. Luce, professor of classics at Trinity College, Dublin, presents a convincing argument for equating the civilization of the lost continent with that of Minoan Crete, which flourished in the second millennium* B.C.

The Short Account of Atlantis in Plato's Timaeus

It is now time to consider in some detail the primary source for the Atlantis legend. Plato outlines the story in his *Timaeus,* and describes Atlantis in more detail in the (unfinished) *Critias.* . . . The following quotation comprises the key portion of the statement of the Egyptian priest to Solon, as narrated by Critias in the *Timaeus:*

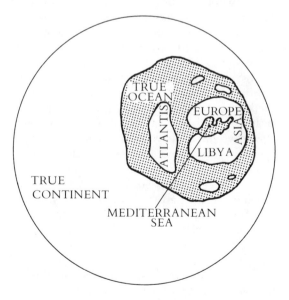

Plato's conception of the terrestrial sphere.

Many great and wonderful deeds are recorded of your state [*i.e.* Athens] in our histories. But one of them exceeds all the rest in greatness and valour. For these histories tell of a mighty power which unprovoked made an expedition against the whole of Europe and Asia, and to which your city put an end. This power came forth out of the Atlantic Ocean, for in those days the Atlantic was navigable; and there was an island situated in front of the straits which are by you called the pillars of Heracles; the island was larger than Libya and Asia put together, and was the way to other islands, and from these you might pass to the whole of the opposite continent which surrounded the true ocean; for this sea which is within the Straits of Heracles [*i.e.* the Mediterranean] is only a harbour, having a narrow entrance, but that other is a real sea, and the land surrounding it on every side may be most truly called a boundless continent (see illustration). Now in this island of Atlantis there was a great and wonderful empire which had rule over the whole island and several others, and over parts of the continent, and, furthermore, the men of Atlantis had subjected the parts of Libya within the columns of Heracles as far as Egypt, and of Europe as far as Tyrrhenia [*i.e.* Etruria in North Italy]. This vast power, gathered into one, endeavoured to subdue at a blow our country and yours and the whole of the region within the straits; and then, Solon, your country shone forth, in the excellence of her virtue and strength, among all mankind. She was pre-eminent in courage and military skill, and was the leader of the Hellenes. And when the rest fell off from her, being compelled to stand alone, after having undergone the very extremity of dangers, she defeated and triumphed over the invaders, and preserved from slavery those who were not yet subjugated, and generously liberated all the rest of us who dwelt within the pillars. But afterwards there occurred violent earthquakes and floods; and in a single day and night of misfortune all your warlike men in a body sank into the earth, and the island of Atlantis in like manner disappeared in the depths of the sea. For which reason the sea in those parts is impassable and impenetrable, because there is a shoal of mud in the way; and this was caused by the subsidence of the island.

★ ★ ★

Plato makes Atlantis "larger than Libya and Asia together." This is a very vague measure of size, and obviously relative to Plato's own conception of the size of "Libya" and "Asia" which we cannot know for certain, but it is likely that he is thinking of a land mass as big as North Africa and Asia Minor put together. He also dates the Atlantis invasion between 8000 and 9000 years before the time of Solon (c. 600 BC). I shall argue later that the large dimensions of Atlantis and the extreme antiquity of her aggression against Greece are distortions and exaggerations imported by Plato himself into a historical tradition which was garbled before it reached him, and which he failed to identify correctly. Here I only emphasize the hard core of the legend as I see it, namely the tradition of a great and highly civilized island empire which had once menaced the autonomy of Greece and Athens in particular, and which came to an end as the result of a natural catastrophe. Plato, I believe, did not invent this tradition. It came to him from his ancestor Solon, as he tells us, and Solon in turn derived it from Egyptian priests.

The Atlantic Location of Atlantis

The quest for lost Atlantis has inspired an immense literature. . . . There are said to be over two thousand works dealing with the lost continent. Indeed, the theme of Atlantis amounts to a licence to write on almost anything in pre-history. Evidence for the lost civilization has been sought from the floor of the Atlantic Ocean, and in the art of the Incas. Devotees of the occult have received special information of its whereabouts from the spirit voices of ancient Egypt. Berlioux thought he had found Atlantis in Morocco; Frobenius favoured Nigeria; Spannuth claimed to have seen the walls of its citadel on a submarine reef near Heligoland. Plato's description of the elaborate artificial harbour of the Atlantean metropolis has prompted identification with ancient Tartessus (not yet itself exactly located). And so on.

No review of the full sweep of Atlantis literature can be attempted here. My aim is to review *Plato's story* in the light of recent archaeological discoveries. But Plato does put Atlantis in the Atlantic Ocean west of Gibraltar, so I feel obliged to say something about modern opinion on this particular solution of the problem.

A book which has been very influential in forming modern opinion, particularly in the United States, is Ignatius Donnelly's *Atlantis: The Antediluvian World.* Donnelly's book, first published in New York in 1882, had run through eighteen editions by 1889, was translated into German in 1894, and was still thought worthy of re-issue in 1949. Donnelly was a man of parts—idealist, novelist, Congressman, and assiduous researcher, not merely in pre-history but also in the field of Shakespearean scholarship. . . .

Himself a very uncritical researcher, Donnelly bemuses his readers into a mood of infinite credulity. There was once a land connection between Europe and America: *ergo* Atlantis. Primitive and civilized peoples alike all over the world have Deluge legends: *ergo* Atlantis. Mexican and Peruvian civilizations were as advanced as anything in the Old World: *ergo* Atlantis. Having accepted all this one is in no mood to question such propositions as "*Genesis* contains a history of Atlantis," or "The Carians of Homer are the same as the Caribs of the West Indies."

Donnelly abounds in information, but is weak in argument. In its day his book was an interesting, though very uncritical, compilation. One must regard it as fundamentally unsatisfactory nowadays because its accounts of the origin and diffusion of higher civilization is completely at variance with what is now the accepted outline picture based on the archaeology of the Middle East. Donnelly's time-scale and location of Atlantis are totally incompatible with what we now know of the Mesolithic and Neolithic periods, and the emergence of the great cultures in the river valleys of the Nile, the Tigris-Euphrates, and the Indus. It is a curious reflection on human gullibility that a revised edition, with a puff for Hoerbiger's theory of the moon as a "captured" planet, should still be offered to the public as a reliable account of how civilization developed.

Donnelly's best-known disciple is probably Lewis Spence, who published three books on Atlantis in the 1920s. Spence tried to put Donnelly's views on a firm scientific footing in two main respects. First, he marshalled some geological support for the notion of a large land mass occupying most of the present North Atlantic Ocean in Late Tertiary times. This Atlantic continent, he argues, first broke down into two island-continents, Atlantis not far west of Spain, and Antillia near the present West Indies. These island-continents with associated smaller islands persisted until Late Pleistocene times. Atlantis was finally submerged about 10,000 BC, and Antillia still has a recognizable existence in the West Indian archipelago. . . .

Spence's views are more moderate and plausible than those of Donnelly. There is, undoubtedly, a mid-Atlantic Ridge running from Iceland to the South Atlantic. This great Ridge lies at an average depth of one mile below the surface, with ocean basins averaging three miles deep to the east and west of it. In a few places it actually comes to the surface, as in the Azores, Ascension Island, and Tristan da Cunha. But modern scientific opinion is against treating it as the remnant of a sunken continent. On the contrary, geologists and oceanographers now think that the Ridge has been raised from the floor of the Ocean, probably by volcanic activity.

To return to Plato's Atlantis: no matter what criticisms are made of Donnelly and his school, people who believe in Atlantis will continue, I know, to look for it in the Atlantic Ocean. That is where Plato put it, and that is what the name suggests. But the name "Atlantis" is a most deceptive guide. Atlantis is *not* derived from Atlantic. Linguistically both names are in the same generation, so to speak, like brother and sister, and both trace their parentage back to Atlas, the giant Titan who held the sky on his shoulders. In Greek they are adjectival forms of Atlas, meaning "(the island) of Atlas" and "(the sea) of Atlas" respectively. They differ in form because the nouns with which they agree differ in gender. So if you decide to use the name of Atlantis as a clue to its location, you must consider what was the original location of the mythical Atlas. Now Atlas may once have been located well inside the Mediterranean before the gradual extension of Greek geographical knowledge pushed him to the west and located him on the High Atlas range in Morocco. You must also consider the possibility that Solon invented the name Atlantis (see below, p. 135). . . .

Professor Andrews has recently made the ingenious suggestion that Plato misread Solon's notes on the location of Atlantis. Instead of the true reading "*midway between* Libya and Asia" he read "*larger than* Libya and Asia." In Greek it is the difference of only one letter—the difference between *mezon* and *meson*. But this clever suggestion is, I suggest, unnecessary. Given his geographical preconceptions, Plato would have put Atlantis outside the Mediterranean in any case. . . .

The Minoan Hypothesis

Minoan civilization was rediscovered only in this century. The ancient Greeks had almost completely forgotten it. They did remember that Minos once ruled the seas, and they could learn some interesting details about Bronze Age Crete from Homer: its ninety cities, the mixed population of the island, the badly sheltered harbour at Amnisos, the dancing floor in broad Knossos. But these were disjointed recollections and beyond that everything shaded off into the vagueness of myth: Theseus and the Minotaur, Ariadne and the Labyrinth, Europa and the Bull, Zeus born in the Dictaean cave, the craft of Daedalus. We can now see that there was historical substance even in the myths, but they would have meant little or nothing to Plato and his contemporaries. For them Crete was a quiet Dorian backwater completely out of the mainstream of history. They remembered nothing of the splendours of the palaces or the glories of Minoan art and technology.

Nothing could have induced Plato, or any other ancient writer, to equate Atlantis with Minoan Crete. Such an identification was impossible until the discoveries of Sir Arthur Evans which began at Knossos in 1900; and now, nearly seventy years later, we can survey the complex architecture of the great palaces at Knossos, Phaistos, Mallia and Kato Zakro. We can wander along the little streets of Gournia or through the villas at Tylissos. In the Herakleion museum we can study the graceful forms and bold swirling decoration of Kamares pottery. The artistry of the frescoes and the craftsmanship of the seal-stones compare most favourably with the ill-formed, ill-executed puerilities of much modern art. In short, as we begin to form an overall picture of Minoan culture, we have to admit that the "first civilization of Europe" was also one of the most accomplished and inventive that the world has ever known.

Did the memory of this civilization survive in the Atlantis legend? I believe it did. But of course the suggestion is no new one. So far as I can discover, the credit for first making it belongs to a certain K.T. Frost, who was for a time on the staff of Queen's University, Belfast, and who was killed in action in the First World War. Frost published his theory anonymously in an article entitled "The Lost Continent" in *The Times* of 19 February 1909. He later argued his case in greater detail in an article "The *Critias* and Minoan Crete."

Frost's main points are very well made in his 1909 article. In the following quotations I have emphasized his crucial suggestions by the use of italics.

The recent excavations in Crete have made it necessary to reconsider the whole scheme of Mediterranean history before the classical period. Although many questions are still undecided, it has been established beyond any doubt that, during the rule of the 18th Dynasty in Egypt, when Thebes was at the height of its glory, Crete was the centre of a great empire whose trade and influence extended from the North Adriatic to Tel el Amarna and from Sicily to Syria. The whole sea-borne trade between Europe, Asia, and Africa was in Cretan hands, and the legends of Theseus seem to show that the Minoans dominated the Greek islands and the coasts of Attica . . . The Minoan civilization was essentially Mediterranean, and is most sharply distinguished from any that arose in Egypt or the East. In some respects also it is strikingly modern. The many-storeyed palaces, some of the pottery, even the dresses of the ladies seem to belong to the modern rather than the ancient world. At the same time the number of Minoan sites and their extraordinary richness far exceed anything that Crete could be expected to produce, and must be due in part to that sea-power which the ancient legends attributed to Minos.

Thus, when the Minoan power was at its greatest, its rulers must have seemed to the other nations to be mighty indeed, and their prestige must have been increased by the mystery of the lands over which they ruled *(which seemed to Syrians and Egyptians to be the far West),* and by their mastery over that element which the ancient world always held in awe. Strange stories, too, must have floated round the Levant of vast bewildering palaces, of sports and dances, and above all of the bull-fight. The Minoan realm, therefore, was a vast and ancient power which was united by the same sea which divided it from other nations, so that *it seemed to be a separate continent with a genius of its own.*

Frost then describes briefly the sudden eclipse of Cretan power (which he attributes to a raid on Knossos) and continues:

132

DOLPHIN FRIEZE in the Queen's apartment in the palace of Knossos, Crete, circa 2000 B.C.

Recent evidence suggests that "Atlantis" might have been Minoan Crete.

As a political and commercial force, therefore, Knossos and its allied cities were swept away just when they seemed strongest and safest. *It was as if the whole kingdom had sunk in the sea, as if the tale of Atlantis were true.* The parallel is not fortuitous. If the account of Atlantis be compared with the history of Crete and her relationship with Greece and Egypt, it seems almost certain that here we have an echo of the Minoans . . . The whole description of Atlantis which is given in the *Timaeus* and the *Critias* has features so thoroughly Minoan that even Plato could not have invented so many unsuspected facts. He says of Atlantis: "The island was the way to other islands, and from these islands you might pass to the whole of the opposite continent which surrounded the true ocean." It is significant too, that the empire is not described as a single homogeneous Power like Plato's *Republic* and other States in fiction: on the contrary it is a combination of different elements dominated by one city. "In this island there was a great and wonderful empire which had rule over the whole island and several others, as well as over parts of the continent." *This sentence describes the political status of Knossos as concisely as the previous sentence describes the geographical position of Crete.*

The above extracts from Frost's original article seem to me to make their case with such conciseness and cogency that it is surprising the learned world did not take more notice of his views. Once the general equation of the Minoan power with Atlantis is established, many of the details fall neatly into place as Frost went on to point out:

The great harbour, for example, with its shipping and its merchants coming from all parts, the elaborate bath rooms, the stadium, and the solemn sacrifice of a bull are all thoroughly, though not exclusively, Minoan; but when we read how the bull is hunted "in the temple of Poseidon without weapons but with staves and nooses" we have an unmistakable description of the bull-ring at Knossos, the very thing which struck foreigners most and which gave rise to the legend of the Minotaur. Plato's words exactly describe the scenes on the famous Vapheio cups which certainly represent catching wild bulls for the Minoan bull-fight, which, as we know from the palace itself, differed from all others which the world has seen in exactly the point which Plato emphasises—namely that no weapons were used.

Frost's work laid down indispensable guide-lines for the source criticism of the Atlantis legend. . . . Frost started with Evans' discoveries behind him, and reemphasized the role of Egypt in the transmission of the story. He rightly pointed out that the Atlantis legend makes good historical sense *if its materials are viewed from the Egyptian point of view*. I have already argued that Solon really did acquire information from the Saïte priests. It therefore seems worth trying to visualize what the priests themselves knew about the period, how they would have presented their knowledge to an intelligent Athenian in the early sixth century BC, and what he would have made of it.

Ancient Egyptian inscriptions are full of praises of the Gods and of the Pharaoh, but manage to contain disappointingly little historical material in proportion to their bulk. On the whole the Egyptians of the Bronze Age knew little and cared less about foreign countries. They were not great travellers or seafarers, and their geographical horizons were quite restricted. Their world was bounded by Nubia and Punt (Eritrea?) to the south, by the Euphrates on the east, and by the Libyan desert tribes on the west. They knew something of Cyprus, the south coast of Turkey, and Crete from long-established trading connections. By the first half of the fifteenth century BC they were becoming aware of Mycenaen Greece. But Egypt remained always very much in the centre of their limited universe, and they took note of foreigners only as providers of desirable imports, or as hostile invaders.

Late Bronze Age Egypt left much the same sort of record of itself to the Saïte period as it has to us. I venture to suggest that New Kingdom Egyptian records still extant allow us to reconstruct the sort of information which the priests must have passed on to Solon. From material still available to us we can see how the Atlantis legend could have taken shape, though doubtless the Saïtes had some annalistic information on papyrus which has not survived.

What sort of records, then, would the priests of Saïs have expounded to Solon, knowing that he was interested in the early history of Greece, and in early contacts between Egypt and the Aegean world? These records must have included documents like those still extant which refer to "Keftiu" and "the isles which are in the midst of the Great Green [Sea]." Keftiu is almost certainly Minoan Crete. The identification has been challenged, but is now generally accepted. Keftiu is the same as the Akkadian Kap-ta-ra, a land "beyond the upper sea," *i.e.* the eastern Mediterranean, and the same as the Biblical Caphtor. Keftiu is first mentioned in an Egyptian document which may go back to the third millennium, and it disappears from reliable records before the end of the fifteenth century BC. The "isles of the Great Green" are thought to include the islands of the Aegean, or at least the southern part of it, and also some coastlands of the Greek mainland. The "Great Green" is certainly the Mediterranean. The designation "isles of the Great Green" first appears in Egyptian records about 1470 BC, and is not found after the middle of the twelfth century. Its use coincides very closely in time with the rise, zenith, and decline of Mycenaean Greece. . . . To Egyptian eyes in the fifteenth century BC Minoans and Mycenaeans must have seemed more alike than different. Mycenaean art and religion, as we now know, were at this time very much in debt to Minoan. Mycenaean traders must have used the same sort of ships and followed the same routes as the Minoans. The Minoan penetration into Greece in the sixteenth century was followed by a Mycenaean penetration of Crete in the fifteenth century. There is evidence that the Egyptian court in the decade 1460–1450 knew of an alteration in the political status of Knossos. Where diplomatic protocol required they could make a distinction between Greeks and Minoans, but for practical purposes such distinctions were perhaps not very firmly drawn. It certainly seems unlikely that Egyptian antiquarians of the Saïte Dynasty could have distinguished at all clearly between the various "peoples of the sea" who either traded or fought with their ancestors in the Late Bronze Age. We may suppose that Solon's informants gave him a composite picture of Aegean-Egyptian relations based on the sort of documents which are still extant, but supplemented by some detailed annals which we do not now possess. . . .

The above references are sufficient to cast Keftiu and the other islands in the role of a great power hostile to Egypt. If Solon had enquired more particularly about Keftiu he would have been told it was an island far away to the west. The Ipuwer papyrus uses the phrase "as far away as Keftiu." The priests could have expatiated on the rich "tribute" of costly vases which it was able to send, and may perhaps have known something of its influence over the other Aegean islands. They may even have had some record of the domination of Attica by the Keftiu, which would have been particularly interesting to Solon. The possibility of such detailed knowledge is inherent in the religious link between Neith of Saïs and Athena of Athens. If Solon pressed them on the ultimate fate of Keftiu the priests could certainly have told him that it disappeared from their records about the middle of the XVIII Dynasty. Whether they had any information about the Thera catastrophe is conjectural. . . .

In general, one can see how this, and similar, material, expounded from the Egyptian point of view without any clear chronological or geographical framework, could have been worked up by Solon into something like the Atlantis legend as Plato gives it.

We know that Egyptian schoolboys of the XVIII Dynasty, *c.* 1500 BC, were set the exercise of writing out Egyptian and Cretan names in parallel columns. There is a writing board of this date, now in the British Museum, which is headed: "How to make names of Keftiu." We may surely infer from this that Egyptian archives contained Minoan place and personal names together with what the Egyptians conceived to be their equivalents either in sound or sense. Plato tells us that Solon made Greek names by translating what he took to be the sense of the Egyptian names that he was given. If this is true he was only doing the same as Egyptian schoolboys 900 years before.

In the light of this we may put a question which has an important bearing on the origin of the Atlantis story. Is there any semantic connection between the name Atlantis and the name Keftiu? "Keftiu" means either "the island of Keft" or "the people of Keft," depending on what determinative is added in hieroglyphic. The root "keft" has been connected with *caput* and *capitul,* and it has been pointed out that in the Old Testament "kaphtor" is used for the capital of a pillar. The ancient Egyptians probably regarded remote and mountainous Crete as one of the four "pillars of heaven" which supported the sky at the four corners of the world they knew. . . . It is even possible that the Egyptian priests had some recorded material about the worship of sacred pillars which was so prevalent in Minoan Crete. Imagine Solon's reaction when confronted with this sort of information about ancient Keftiu. He could not have failed to associate it with the myth of Atlas, who, according to Homer, had a daughter in a remote western island and kept "the pillars which hold the sky round about." My suggestion is that Solon translated Keftiu by Atlantis, the island of Atlas, and, since the inhabitants bore the same name, he called them the descendants of Atlas. This must have seemed to him a very reasonable equivalence. The name would scan well in the epic poem he planned, and the "island of Atlas" would have the right sort of mysterious and western flavour that his plot required.

Frost did not use all the detailed facts and arguments in the preceding paragraphs, but they are all advanced from the point of view which he was the first to suggest as appropriate for source criticism of the Atlantis story. There is, I think, great plausibility in his crucial contention that "Solon really did hear a tale in Saïs which filled him with wonder and which was really the true but misunderstood Egyptian record of the Minoans."

Frost's views made little or no impact on learned opinion. His theory sank so much into oblivion that a German scholar in 1951 completed the final draft of a work arguing out a similar conclusion before learning of Frost's priority in the field.

The Volcanic Destruction of Minoan Crete

Frost's case has subsequently been strengthened in only one major respect. But the additional evidence is so striking and so important that it deserves the fullest consideration. Sole credit for advancing it belongs to Professor S. Marinatos, at present Director-General of the Greek Archaeological Service. His article "The Volcanic Destruction of Minoan Crete," published in *Antiquity* in 1939, makes a major advance towards the solution of the Atlantis problem. Frost noted the sudden downfall of the Minoan empire, but was brief and vague about the cause of the collapse. Marinatos strengthened the whole hypothesis at this crucial point. His theory had begun to take shape in his mind as early as 1932 when he was excavating at Amnisos. He found there a pit full of pumice stone, and noted great orthostats which had been tilted out of position as though by the powerful suction of a mass of water. And then, as he reflected on the sudden and simultaneous destruction and abandonment of so many Minoan palaces and villas, the conviction grew in his mind that the downfall of Crete was due, not to foreign invaders, but to a natural catastrophe of unparalleled violence and destructive power. The source and focus of this cataclysm was to be sought, he suggested, on the volcanic island of Thera, which lies only 120 km. due north of Knossos.

The editors of *Antiquity* published Marinatos' article with a note saying that they regarded the case as needing further evidence to support it, and expressed the hope that more excavation would be undertaken to solve the problem. Then the Second World War intervened, and no progress was possible for some years. But since the war further work of great relevance and interest has been done. . . . Here I shall remark only that the long north coast of Crete lay very exposed to the effects of any violent eruption on Thera. If this Bronze Age eruption was as great as is now supposed, it is hardly possible to exaggerate the sudden destruction and loss of life that must have been caused on Crete. Was this the Bull from the Sea that was sent to plague Minos? Have we here the grim historical reality behind Plato's words: "But afterwards there occurred violent earthquakes and floods; and in a single day and night of misfortune the island of Atlantis disappeared in the depths of the sea"?

Deep Water, Ancient Ships

Willard Bascom

At about the time that Sir Arthur Evans was making his archeological finds in Crete, scientists acquired a new tool—marine archeology—for unlocking some of the mysteries of other ancient civilizations. The Mediterranean Sea was the main highway not only for Crete but for the ancient Greek, Roman, Trojan, and Phoenician civilizations as well, and its depths contain the well-preserved wrecks of many ships that sailed some 2,000 years ago. Divers have recovered statues and other treasures from these wrecks as well as invaluable knowledge of the ancient world. They have also located sites of ancient cities below the sea off the coasts of North Africa, France, Italy, Sardinia, and Sicily.

Haphazard and crude at first, marine archeology became an increasingly exact science with the development of the aqualung, which allowed divers the necessary time to carefully map and recover their finds. Sonar is now also being used to locate wrecks. Yet even today, the chance nature of some of the most important finds remains, as indicated in the following article from Deep Water, Ancient Ships *by Willard Bascom, himself an oceanographer, diver, and treasure-hunter.*

Marine archaeology can be said to have begun in 1900, when a party of sponge divers accidentally found an ancient wreck and were then recruited by the Greek Government to salvage statues from it. The beginning was crude and destructive of much historic material, yet it aroused world-wide scholarly interest in antiquities under the sea.

The first fifty years was characterized by sporadic accidental discoveries, vandalism and looting of wrecks, archaeologists waiting topside for whatever the divers brought up, and little public interest in anything except statuary. The big change of recent years came with the invention of convenient self-contained underwater breathing apparatus which permits archaeologists to go below and do the work themselves. This . . . [is a review of] the development of modern methods by sketching the most important and interesting wreck excavations.

The purpose of retelling these familiar stories here is not only to give proper credit to the great men of underwater archaeology but to underline the differences between the work done in the past and that which is now proposed for deep water.

Here then are the accounts in chronological order of the wrecks that are mainly responsible for the present status of marine archaeology.

Antikythera

At the southern tip of the Peloponnesus is Cape Malea, around which all ships passed as they headed southwest into the Mediterranean. Not far off the cape is the large island of Kythera. Both these rocky promontories were surrounded by rough waters and tricky currents, so sailors tended to give them a wide berth. Midway between Kythera and the western tip of Crete, some fifty miles away, is a much smaller island: Antikythera, well situated to intercept ships running before the northeast wind at night.

The little island also could serve as a shelter from winds from the other direction, and so it was that in the fall of 1900 a Greek sponge boat returning from Africa took shelter from a southerly storm in the quiet lee of its steep cliffs. After a day or two of boredom, the captain, Dimitrios Kondos, decided to send a man down to see if there were sponges in the area. The undersea slope was steep, and although the boat was close in to the cliff, the diver first touched bottom at thirty fathoms (sixty meters). He had been on the bottom only a few minutes when he blew himself to the surface in a fright and told a wild tale of a city below populated by men, women, horses—their features eaten away as though by syphilis.

Captain Kondos scoffed and made the next dive himself; when he surfaced again, he carried the arm of a bronze statue. He had found a wreck that lay on the narrow ledge parallel to the cliff face; below that, the bottom sloped steeply into deep water. The divers recognized that the ancient ship might be valuable, so they stayed at the site for several days, each making several short dives a day and recovering such artifacts as they could easily break off or dislodge. When the weather eased, they sailed for home. The stories vary on whether they tried to sell some of the salvaged pieces and were turned in to the police or whether in a burst of patriotism they immediately reported the find to the Greek Government. Anyway, the story of the find was soon out and it led to great public excitement.

By the end of November, Captain Kondos and his crew were back at the site, accompanied by Professor A. Economou of the university at Athens and Minister of Education Spiridon Stais, who had arranged funds and a navy support ship. The divers had agreed to raise the rest of the statues if they were paid properly and given the use of a winch. Unknowingly, they had offered to do the deepest salvage job in history up to that time. In the first few days, they raised the philosopher's head, a bronze sword, and parts of various statues.

There were headlines about archaeological treasures, for as Peter Throckmorton points out, the country was naturally proud that Greek divers were raising Greek antiquities that had been taken long before by a Roman conquerer. Later the godlike "athlete," a very lifelike majestic bronze statue of a naked man with porcelain eyes and calm demeanor, was recovered.

The divers poked in the bottom with rods to find objects, which were then attached to the winch cable. Not surprisingly, many statues broke as the winch dragged them free of the mud, but the pile of marble statues on deck grew, though most of them were badly eaten by borers. After six months of exhausting work, all the small, loose objects had been recovered. These included blue glass bowls, a gold brooch of Eros, tiles, pottery, a bronze bed, human bones, and, of course, amphorae. When they left, the divers reported that they thought there were other statues still at the site.

The age and origin of the wreck was a subject of long debate, because the bronzes were from the fourth century B.C. while some of the pottery seemed to be first century. A radiocarbon date on some of the wood gave 120 B.C. ± forty years. But the ultimate evidence was the setting on the clockwork "computer," which precisely put the date at 82 B.C. and suggested the statues were part of General Sulla's★ booty en route to Rome.

Mahdia

In 1907 Alfred Merlin, then a director of antiquities for the Tunisian Government, was browsing in a bazaar when he noticed a small, bronze, lime-encrusted figurine in one of the stalls. It was much like some he had seen in the Louvre, but it seemed to have been recently found beneath the sea. He acquired the piece for the museum and questioned the apprehensive shopkeeper about its origin. He learned it had been found by Greek sponge divers near the coastal town of Mahdia and set off at once on camelback to find out more. After some initial reluctance, Merlin persuaded a diver to show him the site of the wreck. He then obtained the assistance of a French Navy tug, raised the money to pay the divers, and in a few days was ready to go.

After eight days of searching, the wreck was found—sixty marble columns, weighing over two hundred tons, in water forty meters deep. No statues were to be seen, but there was wood

---

★*Ed. note:* Roman general, later dictator, who defeated Mithradates, king of Pontus in Asia Minor.

beneath the columns. The divers then lifted and moved the columns to enter what seemed to be the muddy interior of the ship. Even though the mud was only a few centimeters thick, it had helped to protect the treasures beneath, and traces of paint still remained on the wood of the hull. Marble statues like those that had fared badly on the Antikythera wreck shone like new when they were washed down on deck. There were also busts in bronze, candelabra, kraters for mixing wine, and pieces of many statues.

As the divers worked, Merlin pondered the origin of the ship and how and when it came to be wrecked at that site. One clue to the date of the wreck came from a small terra cotta lamp of a design that had been popular for a short time at the end of the second century B.C. Based on the markings on the statuary and a stele from Piraeus, he decided the origin of the voyage was probably Athens. As for the owner of the ship, his first guess was that it was Roman loot, possibly from Sulla's campaign of 80 B.C. Later, however, Merlin decided that the sculptures were not originals but copies made for the extensive art trade of that time. His opinion was that thousands of shiploads of such statues, columns, decorated furniture, and jewelry had been exported from Greece to Italy.

The wreck was more than three miles from shore. Since its columns occupied a space thirty meters by ten, the corbita was a fair-sized ship— or it was very much overloaded. Apparently, it had been driven many days before a northeast storm to have gotten so far off course—or perhaps its intended port was on the African coast. Disabled, possibly by losing rudder or mast, the ship had been overwhelmed by the sea. It went down quickly and landed right side up on the bottom, with the deck cargo still in place. The wooden parts that projected above the seabed were rapidly destroyed, but the mud preserved the rest. Probably, the Mahdia wreck behaved almost exactly as a ship would have that went down in deeper water; it is a good model for one to think about.

The excavation lasted five seasons, until 1913, by which time the art works filled six rooms of the El Alaoui Museum in Tunis.

The statuary includes a bronze statue of "the spirit of the games" and a "Herm" of Dionysus with a long beard, both signed by Boethus of Chalcedon. Other statues included Aphrodite, Artemis, Athena, and a winged Eros. There are bedsteads, inlays, and two-handled wine kraters as tall as a man, decorated with pictures of Bacchus. The columns turned out to be a complete prefabricated temple. This great find caused Salomon Reinach, an archaeologist-philosopher, to speak of the Mediterranean as "the treasure vault of the ancient world." . . .

Cape Gelidonya

In order fully to appreciate the finding and excavation in 1960 of this bronze-age wreck on the south coast of Turkey, one should read the writings of Peter Throckmorton, who was largely responsible for its beginnings. His books best capture the spirit of the Turkish sponge divers who first told him of the wreck at Cape Gelidonya; for generations, these brave and rugged men of the sponge trade have passed down the secret locations of the best sponge beds and of the places where amphora mounds can be found that supply water jugs for the diving boats.

Throckmorton's method for finding old wrecks was straightforward. He would visit the waterfront bars of coastal towns, pleasantly engage divers in shop talk over a powerful drink called raki, and eventually turn the conversation to "old pots in the sea." Every diver knew places where one could find the old pots—amphorae— and other things too. Such things as lead from ancient anchors or bronze from old tools; these were, until recently, sold to junk dealers for scrap value. With patience, much time at sea on the diving boats, and a lot of raki, Throckmorton began mapping the dozens of old wrecks the divers showed him along that deserted coast. Always he sought some special one that would make archaeologists really sit up and take notice.

Eventually during a long raki session in 1958,

his old friend and sponge-boat skipper, Captain Kemal, asked how one would go about dynamiting some flat bronze ingots he knew about to get them free of the sea bottom. The ingots, he said, were "rotten." Previously, he had salvaged some knives, a spear point, and a hatchet of bronze but they were not strong and broke when the children played with them.

Clearly this was a very old wreck (bronze knives meant before 1000 B.C.), and so, in a reckless moment, the chronically broke Throckmorton offered Kemal double the scrap value if he would hold off dynamiting the wreck until it had been seen by archaeologists.

By Christmas that year, he had returned to New York and enlisted the help of George Bass, then a research assistant at the University of Pennsylvania Museum. They agreed to do a precise underwater excavation together, uncompromisingly following land standards, and set about raising funds and locating equipment. By May of 1959 they had an expedition on the site at Cape Gelidonya; George Bass was the director, Throckmorton the technical adviser.

The main problem was to plot the location of each piece exactly, so that when the excavation was finished it would be possible to reconstruct the position each piece had come from. A whole series of ingenious devices for measuring, drawing, photographing, tagging, lifting, and surveying were employed under difficult conditions. The party camped on a narrow shingle beach at the base of a steep cliff from which rocks fell at random moments, and so they were literally between a landslide and the deep blue sea.

The material they found showed that the ship was about contemporaneous with the Trojan War (1200 B.C.) or with the time of the mysterious Sea People.

There were important finds of tools (hammers, chisels, axes, adzes, awls, knives, and plowshares) that were not unlike the ones used in Turkey today except that they were made of bronze instead of steel. Some of these had incised letters in the still-undeciphered Cypro-Minoan script used on Cyprus in the late Bronze Age. The sea diggers also found bronze blanks for making more tools, and these led to the theory that the ship was in the business of trading new tools for old, which were then repaired and sold at the next stop. The oxhide-shaped bronze ingots were the important find. The oxhide shape is sort of a webbed X, about eighty centimeters in the longer direction, three centimeters thick, and weighing some twenty-five kilograms.

Forty of them, twenty-seven with founding marks, almost certainly had come from the copper mines of Cyprus. There were also bits of treasure such as a black cylinder seal five hundred years older than the wreck, a Syrian scarab, a mass of blue and white Phoenician beads, and three almost complete sets of scale weights. A fragment of a basket made of matting and rope was found between two copper ingots, indicating that very delicate fibers can survive beneath the sea if the conditions are right. Enough of the wood remained to give a cross section of the ship, whose hull planks had been held down by ballast stones above which were brushwood padding and a mass of ingots.

The Gelidonya wreck artifacts that survived three thousand years of difficult, shallow-water conditions hint at the possibilities that may exist in the deep. . . .

A Little Deeper

In the early 1960s, local sponge draggers working out of Bodrum, Turkey, retrieved a bronze statue of a Negro boy from three hundred feet of water, and shortly afterward a statue of the Roman goddess of fortune was recovered in the same area. Not far away to the southeast, near Marmaris, a statue of Demeter, Greek goddess of agriculture, had been picked up about 1950.

Dr. George Bass, while directing the diving excavation of the wrecks at Yassi Ada, learned of the nearby finds and was eager to see if their source could be located. The area to be searched was several square miles in extent, because the

DIVERS SEARCH A WRECK FOR ARTIFACTS

draggers pull their nets for miles before retrieving them; so there is no way to tell whether an object found in the net was picked up at the beginning or near the end of the trawl.

Dr. Bass first made a brief search with underwater television, but no wrecks were found, so he decided to enlist the help of the Marine Physical Laboratory of the Scripps Institution of Oceanography. The following is taken from the account, by Maurice McGehee, Bruce Luyendyk, and Dwight Boegeman, of the first use of side-looking sonar to find an ancient wreck.

"The sonar had to be completely self-contained and portable enough to be flown to Izmir, trucked to Bodrum and installed in a small Turkish fishing boat."

The boat used was the sixty-five-foot-long *Kardeshler*. It was "less than ideal for this work as it had a minimum speed of three knots and a very destructive vibration when underway." They first searched the area where the Demeter had been found, knowing only that the sponger had dragged its net parallel to the coast about a mile offshore when it brought up the statue. In order to be sure that the dragger's path was covered, they systematically searched from the coast outward. Eventually the area yielded about fifteen targets of various sizes, but these were not examined at the time.

The search then moved to the northern area, where two out of six days of operations yielded useful data. One problem was that of deciding if an indicated target was man-made or of geological origin. The criterion used was that if a given target was associated with several others of various sizes in close proximity and if its height above the bottom was large, it would be considered geological. A target was "probable" if it was relatively isolated and had a low profile.

Eventually a prime "probable" target was selected for the sixteen-foot-long submarine *Asherah* to dive on. The searchers hoped it would be the wreck from which the statue of the Negro boy had been recovered. Guided by transit operators ashore and a target confirmation by the side-looking sonar, the *Kardeshler* dropped a buoy at the target location where the water depth was two hundred and eighty feet. The submarine descended, landing directly on top of a large shipwreck, and photographed many amphorae and some unidentified debris, which were later classified as Roman. No statuary was seen and no further work was done.

This find was deeper than divers could go, and apparently the sub was not a very effective or practical tool for archaeology under such conditions. However, this test proved that a side-looking sonar can find old wrecks, whatever the depth, and that cameras can confirm them. As Bass and Katzev later said, "There is now every reason to believe that such ships as those that yielded the Marathon Boy in the Athens Museum and the Piombio Kouros in the Louvre will be found by similar methods in the future."

Geology and the Flood
Paul C. Tychsen

The Biblical story of Noah and the Flood, one of the best-known stories about the waters of the world, has its counterpart in the traditions of almost every culture. Are these stories all myths or legends? Or do they reflect an actual world-wide, catastrophic flood? People have been searching for evidence of the Ark on the ice-capped summit of Mt. Ararat for more than a century and a half; now geologists are presenting other evidence to confirm or deny the existence of such a flood. In the following article, part of a symposium sponsored by the Lutheran Church, Missouri Synod, from Rock Strata and the Bible Record, *Paul C. Tychsen analyzes some of this geologic evidence.*

The traditional Biblical date for the Flood, as computed in the Ussher* chronology, is approximately 2350 B.C., some 4,300 years ago. This date, however, is subject to the criticisms . . . [that the genealogies in Genesis probably omitted many generations]. Regardless of the exact date, the Biblical account states that great amounts of water descended upon the earth without ceasing for 40 days and nights and remained on the land surface for some 371 days. The account in Gen. 7:19, 20 (KJV) states:

> And the waters prevailed exceedingly upon the earth, and all the high hills that were under the whole heaven were covered. Fifteen cubits upwards did the waters prevail, and the mountains were covered.

The duration of the Pleistocene epoch is said to have been anywhere from three-quarter million to two million years. If one attempts now to pinpoint the occurrence of the Deluge by correlating it with commonly accepted conclusions of geology, Noah's flood would necessarily be confined to the waning stages of this epoch or the beginnings of the Recent (or Holocene) epoch.

---

* *Ed. note:* James Ussher (1581–1656), Irish Archbishop of Armagh, propounded a chronology for events of the Bible. He based his dates on the genealogical tables contained in the Bible, particularly in Genesis.

**Ed. note:* Rhodes W. Fairbridge, professor of geology at Columbia University.

Fairbridge** has presented a graph (see illustration) showing the constant rise of sea level from a time approximately 17,000 years ago to the present. The graph indicates a most marked rise in sea level to a time approximately 6,000 years ago. Fairbridge states that this marked rise indicates the most rapid upsurge of the sea identified in the geologic record. He states:

> The greatest and fastest rise yet discovered in the geological record reached its crest about 6,000 years ago. The cumulative incursion of the sea flooded low-lying coastal lands in every part of the world. This was the deluge that drowned the homes and troubled the legends of the ancients. The flood of the sea was joined by floodwaters brought down from the highlands by rivers. In the worldwide climatic shift that brought on and reinforced the melting of the glacier, climatic belts shifted everywhere on earth except near the equator. Regions of the temperate zone where the recorded history of mankind was just beginning became noticeably milder and wetter. Archaeological evidence in the valley of the Tigris and the Euphrates and geological evidence in the valley of the Mississippi testify to calamitous and spectacular inundations.

At this point it may be well to recall that the Deluge is a commonplace phenomenon in the folklore of many ancient groups. Greek legend, Babylonian scripture, and Hindu literature all attest to the presence of a flood. These varied legends cannot be accepted by geologists as unchallenged proof of an inundation. However,

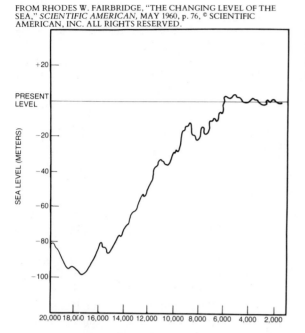

Graph indicating marked rise in sea level from about 17,000 years ago to an unusually high level approximately 6,000 years ago. This marked rise is one of the most rapid upsurges of sea level recorded in geologic history.

it can be argued that geological science may possibly produce evidence of the very phenomena which inspired the ancients to record the flood in their literature. Such evidence, at least in part, may be apparent on the graph. . . . The reader will note that the sea level curve for the past 6,000 years has been quite stable, indicating but slight recession of the water. Fairbridge notes that the variation of sea level during this span of time has been of the magnitude of ten or twelve feet above or below the present sea level. Certainly the lands throughout the world would have witnessed the effects of such a marked rise in sea level.

Centers of trade and civilization some 6,000 years ago were located along river valleys and at seaports where transportation junction points offered man an opportunity for survival. Rise in sea level at this time would be particularly noticeable along the shorelines characterized by shallow depths immediately offshore and where the coastline was relatively straight. Additional quantities of water were being contributed by rivers such as the Tigris and Euphrates whose valleys witnessed frequent floods. Considerable land about the Caspian is below the general level of the Black Sea and the Mediterranean, and one of the earthquake zones in the world extends east and north of the Persian Gulf and continues across Mesopotamia. Subsidence, however slow, of parts of this structurally weak region may well have initiated further flooding of extensive areas.

To an observer in Biblical times, the world was primarily confined to an area bounded in the north by the Black Sea, in the east by the region immediately beyond the Tigris, in the south by the Persian Gulf, and in the west by islands of the eastern Mediterranean. With such limited geographic information, communications, and scientific equipment, attempts to accurately assess the extent of the Flood would have been virtually impossible. This has led some to suggest that Noah's flood may merely have been a local affair and not world wide. However, Holy Scriptures are quite explicit about it having covered the entire earth.

One might conclude furthermore that if

144

water had remained on the earth's surface for merely a year or so, as recorded in Scripture, little evidence, if any, would be left. Strand lines demand considerable time before the landscape bears evidence of their existence, and the amount of sedimentation taking place during a single year on shallow sea floors would scarcely be noticeable. . . .

The thick limestone and shale units characteristic of the widespread and long-continued seas of the Ordovician and Cretaceous periods★ . . . would be entirely lacking, and even if a thin veneer of sediment had been deposited, subsequent erosion would have undoubtedly removed all evidence. Little or no sedimentary trace of a short-lived inundation would remain.

This line of reasoning is from the viewpoint of uniformitarianism and does not take into account the fantastic effects which would result from an inundation of the entire globe in a period of 40 days. Again, see Kramer's discussion on this point. However, it is entirely possible that the evidence has been obliterated by the passing millennia.

In summary it may be said that geologic research has discovered records of many great floods. It has also shown an unusual rise in the world's water level at a time approximately 6,000 years ago. But it is not possible at this point in time to identify any of these phenomena from the past with Noah's flood. Geology simply does not provide us with an answer of this type. Nor does the Bible enable us to establish the date of Noah's flood.

It would therefore seem best to allow Noah's flood to be a concern of theologians. Certainly geologists can neither prove nor disprove it. The believer does not doubt that God in a miraculous act could have inundated the entire globe. From this point he moves on to ponder the great lesson of sin and grace which the story of Noah and his world provides for modern man.

★*Ed. note:* The Ordovician period, from about 500 million to 425 million years ago, and the Cretaceous period, from 135 to 63 million years ago, were characterized by floods so extensive that they covered large portions of entire continents.

Installing Oil Production Platform off Aberdeen, Scotland

Unit Four

We part from the visions of painters and the hypotheses of scientists and plunge into a discussion of our ever-expanding need for resources. This unit is but the beginning of our selections related to the complexities of using the ocean. We float only halfway between the law and the deep blue sea, between what the sea has to offer and our ability not to take it but to share it.

"Full many a gem of purest ray serene, / The dark unfathom'd caves of ocean bear," wrote Thomas Gray in "Elegy Written in a Country Churchyard," and never was a poet more a prophet. From the continental shelf we already extract a vast treasure of oil. The hope for wealth from the deep sea colors the deliberations of the United Nations' conferences on the law of the sea.

What is this wealth? Are the hopes realistic? A selection from *Newsweek* magazine sketches the major marine mineral resources and speedily confronts the need for new laws to control development.

The exploitation of oil and gas on the continental shelf began off the Gulf Coast of the United States in the 1930s. American oil companies have since dominated the global search for oil at sea. American shipyards and industries have constructed the platforms and fabricated the pipelines.

Marine Resources

Oil exploration began off the shores of the United States but has not continued. A world map of offshore exploration shows prospecting in every likely place except off our own coast. Why is this? Do we not have a desperate need for oil?

Robert Bendiner, in a selection from the *New York Times,* describes how environmental and social concerns of Americans have caused this anomalous situation. People express concern about the destruction of our marine environment. In part this masks a greater concern about the disruption of the social structure of coastal communities. There are also concerns about "Alice-in-Wonderland" leasing procedures and double-blind auctions of public resources.

These concerns have led to elaborate legal steps that are required before leasing. Luther J. Carter indicates that more steps may be added.

Deep-sea manganese nodules are discussed by Allen L. Hammond. Their origin is uncertain, but there are enormous quantities of them and they contain valuable metals. Various ways of mining the nodules are being developed, and the environmental consequences of mining are under investigation.

Frank L. LaQue, when he was vice president of the International Nickel Company, was one of the first industrialists to take deep-sea mining seriously. He points out why such mining in the next few decades has little prospect of changing the world's mining industry. The potential tax revenues, of such concern in the United Nations deliberations on law of the sea, appear very unpromising.

What about the food resources of the sea? Can the sea feed the land? Hopes are widespread.

The sea is vaster and yet yields far less than the land, so prospects, at first glance, look good. However, we eat the plants of the land but not the tiny, indigestible plants that the poet Shelley called the "sapless foliage of the oceans." From the sea we eat the equivalents not of grass nor even of sheep, but of wolves.

A brief description of the marine food chain is extracted from the prize-winning book *The Sea Around Us* by Rachel Carson. Published in 1950, it triggered much of America's present interest in the sea.

In "Food From the Sea" C. P. Idyll and Hiroshi Kasahara describe the character of the existing fishing industry and the factors that control it. There are very large possible fisheries for species that people do not eat but that may have promise as hunger grows. Even so, they argue, the contribution of fishing to the global food problem will not change the final solution very much.

The extract from *The Drama of the Oceans* by Elisabeth Mann Borgese describes the history of the extraction of food from the sea and what happens to the catch. The greatest fishery is off Peru for tiny anchoveta. Does it feed starving millions? Yes, millions of American chickens! Borgese also discusses the prospects for farming the sea. Coastal waters have great promise for a Blue Revolution in food production. The Green Revolution, too, held out great promise of feeding the world, but an expanding population has eaten it up.

Not only do we take food out of the sea, we put poison into it. Our selections on pollution begin with a comparison of a common theme as expressed by a poet, an explorer, and a scientist. Anyway you look at it, the surface of the sea is becoming ugly. However, the sea is deep and voluminous, and until recently men have polluted it without concern. Bostwick Ketchum, associate director of Woods Hole Oceanographic Institution, describes this diverse pollution and how the ocean and the food chain respond. He proposes that the only ultimate solution is recycling the materials that our civilization produces.

The pollution of most topical interest is oil, which is discussed by Roger Revelle and three other experts. Oil enters the ocean in many ways, and the relative importance of various sources may be surprising. The oft-protested danger of oil spills from drilling platforms is not really the problem. There are many kinds of crude and refined oil, and their effects on the environment are very complex. The conclusions deal with the problems of national and international control of oil pollution.

Is the sea dying? Judge for yourself. Don't be an alarmist but accept that there is a limit to what the sea can purify. Remember Macbeth after he murdered Duncan, king of Scotland:

Will all great Neptune's oceans wash this blood clean from my hands?
Or will these, my hands, the multitudinous seas incarnadine?

The Riches in Davy Jones's Locker

For thousands of years, people have thought of the harvest of the sea in terms of food resources. Recently, however, attention has been turned increasingly toward harvesting the mineral resources that lie beneath the sea. Increasing industrialization and the rapid rise in world population—from less than three and one-half billion a decade ago to more than four billion today—has given new urgency to the need to find and exploit these resources. The following article, reprinted from Newsweek, gives an overview of the ocean's mineral resources and of the integrally related problem of who shall have the rights to those resources.

*W*hen Icelandic gunboats sent a few rounds across the bows of British fishing vessels last year for allegedly violating Iceland's territorial waters, the world tittered as though it were watching one of those charming British movie comedies. But the Great Cod War may actually have been the opening shot in a vast conflict developing as the nations of the world rush increasingly to sea in search of food and minerals. The scramble has been intensified by the oil embargo and threatened cartels by other raw-material-producing countries. The drive to exploit the sea poses entirely new technological problems—and legal issues that may be even tougher to resolve. . . .

There's no question that the oceans' resources are tempting enough to touch off bitter disputes. Fishing has traditionally been the main commercial activity on the seas. But in a few years, the biggest ocean business is certain to be oil. Nearly 20 per cent of the world's oil and 10 percent of its natural gas already come from offshore wells, and offshore production is likely to increase as other consuming countries try to match the luck of the North Sea nations in pumping petroleum from the depths. The potential rewards are well worth the effort; for instance, Great Britain, which once was entirely dependent on imports for its oil, now expects to be a net exporter by 1980.

Oil and Mineral Reserves

An estimated 26 per cent of the world's known oil reserves are submerged in the seas, and some experts think further exploration will raise that percentage enormously. . . . Fortunately for the oil-consuming countries, the underwater reserves aren't concentrated in the Persian Gulf; according to the experts, they are spread rather equally around the globe. By one count, oil exploration is being conducted off the shores of 80 countries.

To tap more underwater pools of fuel, the oil companies will have to drill in deeper water. The technology to do this is at hand, but the costs will be great—and they are already huge. It costs seven times more to drill for oil offshore than on land. As oilmen try to drill in deeper water, the costs rise geometrically.

The bottoms of the oceans hold equally tantalizing prospects for solid minerals. Experts say that the crumbling, coal-like manganese nodules lying on the ocean bed at depths of 9,000 to 18,000 feet contain enough copper and nickel at least to double known reserves—as well as increase the supply of manganese, cobalt and other minerals. One mining engineer claims the nodules hold ten times the amount of nickel on land.

Mining these nodules could do much to restore America's now wavering confidence over

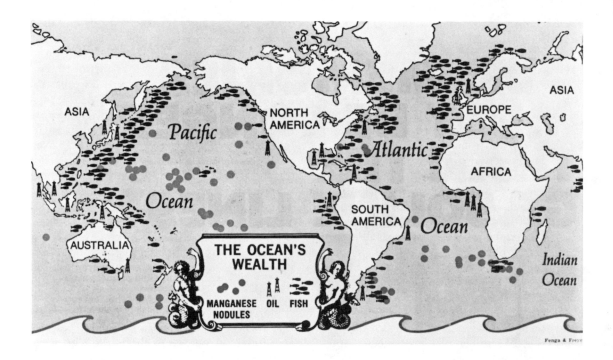

THE OCEAN'S WEALTH

MANGANESE NODULES   OIL   FISH

Fenga & Freye

its future resource supply. The nation, for instance, has imported as much as 20 per cent of its copper in recent years, and the trend is up. There is only one nickel-producing plant in the U.S., supplying less than 10 per cent of the nation's demand. The country has no reserves of manganese, the alloy used to toughen steel. And the nation—and the rest of the world for that matter—depends on just one country, Zaïre, for three-quarters of its cobalt, which is needed in certain electronic equipment and space-age alloys. According to a recent Interior Department report, just three sea-bed mines could change the picture dramatically; they could cut U.S. dependence on manganese imports by 12 per cent, on copper imports by 41 per cent, on nickel imports by 54 per cent and also produce three times the cobalt now used.

While oil and the nodules are the big game, there are a myriad of other minerals on the globe's continental margins—that is, the underwater areas that slope gradually away from the shore. Sand and gravel are now mined in the Atlantic, for example; limestone is being recovered off the Bahamas and tin off Thailand and Malaysia. Furthermore, the margins are known to contain other resources from diamonds to phosphorite rock to bauxite and from zinc to silver to gold. . . .

Law of the Sea

But while industry is excited by the potential riches, much of its efforts are being conducted in a no man's sea. The laws intended to cover exploitation of the oceans' resources are hopelessly outdated. . . .

[A] central issue at . . . [the Law of the Sea Conference*], however, is how the deep-sea bed, or that area beyond the 200-mile limit, will be mined. The United Nations declares that the riches of the ocean depths are the "common heritage" of all countries, not just those that have the capital and technology to exploit them.

Ed. note: The Third Law of the Sea Conference began in 1973 and is still underway in 1976 after months of negotiations.

The U.S. doesn't argue against sharing revenues from deep-sea mining with less-developed countries, but it insists that its corporations receive a relatively free hand to operate once they have obtained leases from a central sea-bed authority.

But less-developed countries fear that if multinational companies are given a free rein they will cheat on their revenue-sharing obligations by disguising their production activity; these poorer countries also worry that the multinationals will pollute the ocean as they mine it. "Without firmer control," says Frank X. Njenga, Kenya's chief representative to the conference, " 'common heritage' is just a phrase." The less-developed lands propose that all exploitation be controlled by a single international concern, which would in turn be controlled by all the nations. But the U.S. fears that such a multigovernmental corporation—which the less-developed refer to as The Enterprise—might stifle ocean production to protect, say, the copper markets of Zambia and Chile. Njenga, however, insists that it would reflect the desires of all countries and therefore resist cartelization efforts by a few. "It is possible to live with The Enterprise," says Njenga. "And it is the one issue the less-developed countries are united on."

Outwardly, U.S. officials are hopeful that the problems can be settled by next year. Chairman Russell Peterson of the Council on Environmental Quality says: "There is reason to be optimistic over the fact that the world is even having a conference. I predict success." But there is strong Treasury Department sentiment against any sort of controls over American business, and Sen. Lee Metcalf says, "At best, I am realistically pessimistic."

If the conference does fail, the U.S. is likely to pass legislation of its own to guard the investments of U.S. companies that go to sea without the protection of international law. But other nations would act unilaterally, too; the result could be a hodgepodge of arbitrary and retaliatory projects that would only inhibit peaceful attempts to mine one of the world's great remaining resoures.

Taking Oil Off the Shelf

Robert Bendiner

In November 1973 the Arab nations of the Middle East and North Africa, which controlled about 68 percent of the world's oil production, plunged the nations of the world into an energy crisis by announcing an embargo on oil shipments to pro-Israeli countries and a 25 percent cutback in oil production. The United States, which received approximately 6 percent of its daily consumption needs directly from the Arab countries and another 6 percent after processing in other countries, faced its severest energy shortage since World War II. President Nixon responded with Project Independence, a program designed to decrease United States dependence on oil imports. Among the key provisions was a plan to lease some 10 million acres of the Outer Continental Shelf—four times the usual yearly acreage—for offshore oil development. The plan aroused a storm of opposition from environmentalists, the coastal states, and critics—within and without government—of the present procedures for leasing. In the following article from The New York Times Magazine, *Robert Bendiner, a member of the* Times *editorial board, assesses the environmental impact of offshore drilling and the wisdom of our current offshore policies.*

It should be noted that, largely because of the kind of opposition cited by Bendiner, the government in the fall of 1975 abandoned its goal of leasing 10 million acres annually in favor of a plan to conduct six lease sales per year in "frontier" (that is, previously unexplored) areas in the Gulf of Alaska, the Atlantic seaboard, and the California coast. By year's end, 1975, the Department of the Interior had offered for sale the first such lease, 1.25 million acres off the southern California coast; bids were accepted on only 413,000 acres for a total of $438 million, a remarkable $1.5 billion short of the original estimate. Meanwhile, the percentage of domestic demand for crude and refined product that was met by imports actually rose, according to figures from the American Petroleum Institute, from an estimated 36.1 percent in 1973 to close to 40 percent by the end of 1975— a far cry from independence!

*L*anding on the helicopter pad of an Exxon oil rig 27 miles out in the Gulf of Mexico on a recent gusty morning, I could not escape the thought that the platform and its operations looked in truth like those depicted in the company's colorful television commercials. The production platform itself was spotless. Accommodations for the 30-man crew, who work alternate seven-day weeks in a complicated welter of pumps, pipes, compressors, separators and sensing devices, were maintained with military neatness. Personnel were forbidden to throw so much as a paper cup over the side. Even rainwater falling on the platform was filtered before it was allowed to reach the surface of the Gulf.

Looking down some five stories through the open grill of a catwalk, I could see scores of blue-fish darting about the structure's tubular steel legs, which extended downward 170 feet below the green surface. The fish are neither attracted nor repelled by oil for the simple reason that no oil is anywhere to be seen—either on the rig or in the adjacent water. It is the structure itself that is responsible for their presence. Like any topographical irregularity in the sea, it acts as a reef, attracting a food chain that starts with barnacles and works up through sea urchins and ever-larger marine creatures to the 400-pound shark that the crew had pulled in a few days before.

At least as significant as what could be observed in this casual way were data I had previously been shown by an official of the Louisiana Wildlife and Fisheries Department with a long record of vigilance against the incursions of the oil industry. His facts and figures proved that, after a history of drilling on and off the Mississippi Delta going back to 1927, the state's shrimp, oyster and fin fish were in as healthy a condition as ever.

Under the circumstances it seemed at least reasonable to take another look at the Federal Government's plan to lease 10 million acres of the Outer Continental Shelf this year for the possible extraction of oil and natural gas—and at the determined opposition it has encountered.

One of the reasons the environmental movement remains vigorous, in spite of economic pressures and dire talk of energy shortages, is that people who care little about protecting the environment at a distance are vehement about protecting it close to home. Since seven of the ten most populous states in the country are on the coasts, the Government's plan to lease immediately five times the offshore acreage ever leased before in a single year has aroused a volume of opposition surpassing the reaction to more localized threats of desecration, however serious. What is more, it is an opposition that comes as much from Governors, legislators and hardheaded men of business as from nature lovers.

Is all the uproar warranted? At the very least it is understandable, but curiously not for the reason usually cited—the fear of oil spills like those that marred a long stretch of coast on the Santa Barbara Channel in 1969 and threatened the Louisiana shoreline in 1970. The imposition of drastic liabilities for damage, the financial loss incurred in the waste of the oil itself, at fabulous current prices, and the dawning sensitivity of the companies to a sharp decline in public esteem—all have combined to hasten the development of a greatly improved technology in the extraction of offshore petroleum.

Automatic storm chokes, installed in the well itself below the seabed, react instantly to sudden rises in heat, pressure and rates of flow. These together with thicker pipe casing, electronic monitoring and a complicated group of blowout preventers on top, known as "Christmas trees," are designed to cut off and contain the oil in cases of emergency and reduce to a minimum the chance of blowouts. Practically all the big spills of the past occurred when companies deliberately neglected to use even the crudest preventers they had in order to cut costs.

Should even the new safety systems fail, as all things will from time to time, improved booms and skimmers are at hand to contain spills before they are very likely to get to shore. The ideal boom has been described as a vertical curtain or barrier so constructed and placed in the water that it can follow the motion of the waves, with its top never going beneath the top of the slick and its bottom never rising above it. As a result of considerable research and experimenting, much of it by the United States Navy, the newest booms come much closer to this ideal than those used in the past. But it should be said that what is effective in the usually calm waters of the Gulf may not do at all in the stormy Atlantic, much less in the coastal waters of Alaska.

The visitor to one of these impressive production platforms in the Gulf cannot overlook the possibility that he is being shown a specially tidied-up exhibit, but the likelihood is small. Too many observers have visited too many similar rigs in these waters and off the California coast for such trickery to work. A delegation of Federal and New York State environmental officials returned from one such inspection on the Pacific coast last fall with the conviction that "no casual spill was likely and that the chance of any substantial spill was fairly remote."

As long as pipes are used to carry the oil ashore, as they are in the Gulf, environmental danger at the source appears to be a matter for constant watchfulness rather than a decisive obstacle. Transfer to ships would be another matter, entailing a certain amount of chronic and unavoidable leakage. Among sources of oceanic pollution, spills from offshore drilling rank very far down on the list. Tankers and even normal shipping are far greater offenders, accounting, according to the Coast Guard, for 19 times as much oil on the waters as offshore production. River deposits and sewer drain-offs, including the discarded crankcase oil of thousands of gas stations, are much more significant sources. So, for that matter, is the natural seepage that is emitted from time to time from the oil seabeds of the world, that might indeed have been a sheen on the waters viewed by Columbus, Leif the Lucky or Ulysses.

Visual pollution by 2,700 rigs and wells is real enough in the Gulf, where the awkward-looking structures along with drilling ships, service boats and other attendant activity give the entire seascape a depressingly industrial look. But off the Atlantic coast the intention is to drill from 25 to 75 miles off the New Jersey, Delaware and Maryland shores. At 25 miles, the uppermost tip of the rig could just be seen from the shore on a clear day. Any distance beyond that would preclude all possibility of offending man's eye.

It is the shore itself where the great doubt lies. What happens on the Outer Continental Shelf is only the beginning of the threat. What drilling can do to the coast, unless it is carefully planned and controlled, is the real issue that has to be faced. This is the aspect of the question that received the least detailed attention in the Interior Department's environmental-impact statement on the great leasing program—not surprisingly in view of the position taken by Rogers C. B. Morton, who was Secretary of the Interior when it was promulgated. Responding to a flood of protests from Governors and Senators last year, Mr. Morton overrode their insistence on the need for preliminary coastal planning on the ground that both foreign policy and inflation forbade the country's continued dependence on foreign crude oil, leaving him no alternative. To let the state governments hold things up, he suggested, could "give an extremely important national decision to a very limited number of people."

Putting aside the question of how much of the Secretary's remarks were rhetorical, one may well consider at this point the possible consequences to the coast of what 20 United States Senators have jointly described as a "hasty and ill-conceived" proposal. Growing out of former President Nixon's Project Independence, the

155

plan is to lease as much of the shelf in 1975 as has been leased in all the years since the Federal Government started the practice in 1953. As with the sea, these consequences do not appear to include much of a threat from the oil itself. Pipe carrying the offshore fuel could be brought in even under a beach and invisibly carried any number of miles inland. Indeed, if that were all that was involved, the environmental risk would probably be worth taking. The states affected would have only to stipulate that facilities for handling the oil in most areas, except on such narrow strips as Long Island or Cape Cod, be located in some already industrialized region. Oil found off the New Jersey coast, for example, could be brought, partly underground, to storage tanks or refineries in the Wilmington area or in the vicinity of Newark Bay. Distance is hardly a technical problem, though it could be a cost factor in a country that is already traversed by more than 200,000 miles of oil and gas pipe. New lines especially are heavily protected against corrosion and present little danger of rupturing.

The real difficulty, social as well as environmental, comes before any oil is produced, as the current boom in the North Sea has been demonstrating. After the first serious look at what is happening in those waters, a team sent out by the Conservation Foundation concluded that "while the risk of oil spills warrants public attention, the Scottish situation suggests that the United States would be unwise to continue focusing disproportionate attention on spill hazards to the neglect of onshore development impacts."

If the companies are required to locate storage tanks, refineries and petrochemical works well inland, and if laying the pipe itself can involve at most a very narrow corridor, what onshore impact is left? In good part, the answer lies with the oil rigs themselves. Gigantic structures, they must be assembled at points as near to their intended operating sites as possible. At these shore points they are placed side down on great barges and towed to sea. The leg section of the largest platform is by itself some 23 stories tall. Once in place, a series of decks is built on top of it, rising seven stories above the surface of the water to accommodate landing pad, derrick, pumps and machinery, monitoring equipment, repair shops, offices and living quarters. Concrete platforms, sometimes preferred for deep and turbulent water, are even more massive and difficult to assemble and put in place.

The building and siting of these towering "islands" inevitably requires an onshore task force to operate the essential fleet of boats and barges, to supply the daily needs of the crews but, above all, to assemble the platforms in the first place. Along with their families, this working force, swooping by the thousands on a small coastal community, creates monumental problems—both social and environmental, both immediate and long-range. These problems are at least being given serious consideration in Scotland, though conclusions are not always allowed to determine the course of events. But if the Department of the Interior has its way, they will hardly be considered at all before the oil companies are given a green light here.

While the production of offshore oil does not in itself employ great numbers of people, this onshore activity does. Without the most deliberate planning, a small seaside resort can suddenly find itself host to a regiment of platform construction workers, supplemented by other builders needed to put up housing for them and their families. Next come additional schools, stores, restaurants, professional office and entertainment establishments, with roads and sewers to serve them—in short, all the bustling expansion of a boomtown, complete with the labor shortages, inflated land prices, soaring wages and social strains between natives and outsiders that everywhere characterize boomtown development.

But there are in this case several major departures from normal boom conditions that promise greatly to aggravate the difficulty. A small shore community is likely to be scenic and geared to the kind of pleasure activities that are especially vulnerable to a sudden upsurge of industry and construction. Even more, the fragile nature of coastal terrain leaves it open to irreparable damage from the heavy equipment that the building and hauling of oil rigs and heavy supplies require. The Mississippi Delta below New Orleans has suffered visibly from this activity—to the extent of actual subsidence of the marshland in some areas. Oil executives concede the point and indicate that they would not do today, along the Atlantic coast, what they did without objection on the Gulf coast before there was much consciousness of the environment and its vulnerability. But without detailed cooperation between states and companies, in advance, they cannot avoid doing it.

Above all, the boom can end as abruptly as it began. The leased acreage may turn out to have nothing like as much oil as expected. A field's potential is revealed only by actual drilling, and the odds are only one in seven that a given well will yield any oil or gas at all. Even if most of the wells on a given tract have been productive, moreover, the time comes—say, in 20 years—when the oil is gone, an inevitability with a depleting resource. Then the offshore activities cease: No more platforms are built; no more pipe is laid; no more crews and supplies have to be ferried out day after day. Workers drift away, perhaps to move on to some other spot on the coast. Stores close their doors, schools disappear, houses are boarded up as boomtown gives way to ghost town—all the worse because its preboom resort quality can never be recovered.

It is fear of just such a future along the North Sea that caused a tough official of the Shetland Islands to take a firm line with representatives of Shell Oil. When their exasperation at the Shetlanders' strict planning regulations took the form of a threatened withdrawal from the proposed site, the official's instant response was, "That's the best news we've heard since you arrived." It is the same fear that motivates some coastal-state Governors to demand more time for planning and laying down conditions such as the Shetlanders successfully forced on Shell. Maryland is even working out a plan to confine

North Sea Giant.   *The three 310-foot-high concrete supporting shafts, or pylons, of "Beryl A," the world's first concrete oil drilling and production platform, tower above Gants Fjord near Stavenger, Norway.*

onshore oil facilities to certain designated tracts, which the state would then buy up for eventual sale to the companies.

At the same time, there is no doubt that many critics of outer-shelf development are motivated as much by a failure to share in the return as they are by fears for the environment. In the case of state and local officials, the attitude is not unreasonable, since the states and localities themselves, not the Federal Government or the oil companies, will sooner or later pay the price in whatever damage is done to resort business, fisheries and tourist trade.

While the Supreme Court has decided that the states have no jurisdiction over waters beyond the three-mile limit, what happens to their coasts is another matter. No doubt the Federal Government has the right, for example, to lease drilling sites 50 miles from Long Island if it wishes, but it will fall to the state of New York to protect Montauk or the Hamptons or the Rockaways from the ravages of attendant onshore activities. It is on the basis of this wholly legitimate interest that the coastal states and their representatives in Congress are demanding not only a share of the income from what is, after all, a public resource but also a voice on where exploratory drilling is to be done—and even who is to do it.

The present arrangement for leasing the Outer Continental Shelf has a backward, almost Alice-in-Wonderland, character. The Interior Department's Bureau of Land Management leases the 5,760-acre tracts on the basis of sketchy estimates of their potential yield in oil and gas. The estimates are necessarily sketchy for the simple reason that the bureau is obliged to depend on information from the oil companies themselves in the form of seismic and geophysical data made prior to any exploratory drilling. This raw information is sold to the Government, but not the oil companies' interpretations of it. For competitive reasons these are claimed by the companies as "proprietary" and remain their secret.

On the basis of its own largely uneducated guesswork, the Interior Department is then free to consider financial bids from companies operating on at least somewhat educated guesswork. It is a kind of two-handed poker game in which only one of the players is permitted to see his own hand. Only after a lease has been awarded—and the Government has thereby lost substantial control of the tract—is the purchaser of its mineral rights given a permit to do exploratory drilling, the only real way to find out what the tract contains.

The result of this curious procedure is that the Government, eager for the revenue and inclined therefore to be modest in its evaluations, can innocently dole out a publicly owned treasure for a fraction of its worth. When competition among the companies for available tracts was high, the danger of this happening to any appreciable degree was not serious; even on the basis of their own fragmentary preliminary data, the companies would raise the prices they were prepared to offer the Government in the simple process of outbidding each other. But with the Interior Department now flooding the market with more than five times the normal annual acreage, competition is bound to be less.

It is this topsy-turvy system that is now the target of considerable legislative attention. Among others, Senators Ernest F. Hollings of South Carolina, Henry M. Jackson of Washington, Clifford P. Case of New Jersey and Alan Cranston of California want to separate exploration for oil and gas from their development and production, making the former either wholly or partly a function of the Federal Government.

Embodying recommendations of the Senate Commerce Committee's National Ocean Policy Study, proposed bills would have the Geological Survey conduct seismic, geomagnetic, geophysical, geochemical and exploratory deep-drilling activities in "frontier areas" of the shelf, where development has not yet occurred. In addition, the Interior Department would be required to prepare a 10-year leasing program based on the nation's projected energy needs and the coastal-zone management plans of the individual states, none of which will have time to do an adequate planning job if the present program goes into effect this year. Without any idea of where oil is to be drilled, they could do no more than hypothetical planning in any case.

The department is considering better methods of bidding, including a decreased reliance on the initial bonus payment and more on royalties after production has started, as well as the elimination of joint bidding by the big companies. But so far it is as firmly opposed to Government explorations as the oil companies themselves. The argument is that private industry can do the job more efficiently than Government and that the public ought not to pay—and pay dearly—for the risk taking involved in exploration. Though the efficiency argument is not overwhelming, because the Government would be hiring the same commercial survey experts who now serve the companies, there is some cogency in the contention that taxpayers would be forced to gamble. It can be just as convincingly argued, however, that the companies now pass their exploration costs along to the consumer anyway, so there would be nothing really new about what some company men criticize as "socialized" risk taking.

Overshadowing all such objections to the program, however, is the contention of Government and oilmen alike that the country needs the offshore oil now in order to avert dependence on foreign sources—needs it too desperately, in fact, to afford the luxury of delay. It is this contention which, in view of the damage that a hasty and poorly planned program can do, calls for serious consideration. The consideration given it so far—for example by the Federal Energy Administration, the National Ocean Policy Study and the General Accounting Office—has indicated such a lack of realism in the expansionist plans for the continental shelf that the Department of the Interior is already pulling in its horns and hinting, unofficially, of a somewhat scaled-down leasing program. The reasons for grave doubt are these:

Although nobody knows how sizable are the petroleum deposits to be found on the Atlantic Continental Shelf, the Geological Survey has drastically reduced its estimates from those it made in 1968. Instead of 48 billion barrels of oil and 220 trillion cubic feet of natural gas, it now figures 10 billion to 20 billion barrels and somewhere between 55 trillion and 110 trillion cubic feet. These figures are still significant, of course, but for a country that consumed 6.3 billion barrels of refined products in 1973 alone (2.2 billion of them imported), the hoped-for supply could not in any case offset present imports for more than four to nine years unless per capita use were, at the same time, quickly and drastically curtailed. To satisfy the assumptions of Project Independence, the General Accounting Office roughly estimates that 15 million to 28 million acres of the Outer Continental Shelf would have to be leased *and drilled* by 1985. For reasons to be considered next, that is clearly impossible.

Whatever reserves are discovered offshore, none of them will add significantly to the country's usable supply for something like a decade. Between the department's call for tract nominations by the companies, issued a few months ago, and the beginning of exploratory drilling, three years can be expected to be consumed in the bidding process, public hearings, impact statements and sale. After that, four or five years are likely to pass under the most favorable conditions before such drilling is completed and production platforms are erected at the appropriate sites. Beyond that, it would be two years more before production could be brought to a peak.

What threatens to attenuate even this timetable considerably—aside from possible lawsuits—is acute shortages of mobile rigs for exploratory drilling, of equipment materials, highly skilled manpower and capital. According to the Government's Project Independence Report, drilling facilities will fall short of the requirements of an accelerated program by some 38 per cent. Drill pipe, casing and tubing are all in short supply, and experts in both government and industry concede the serious shortage

Oil from the British North Sea fields being taken from the tanker Theogenitor. *American opponents of offshore drilling fear boom-town conditions that have marked North Sea exploitation.*

of engineers, geophysicists, and other professionals—all needed in force for projects in the unfamiliar deep water and turbulent environment of the Atlantic shelf.

Not least among the factors to be considered in the current offshore plan is that the dumping of so vast a seabed area on the market at one time—with similar acreages planned successively in the years immediately to come—calls for a capital outlay that has even the oil companies staggered. Frank Ikard, president of the American Petroleum Institute, has been quoted to the point: "On one side, I say, 'Great; let's get on with offshore development.' On the other side, I say, 'Where are we going to get the money to bid?'"

Add to these formidable obstacles the fact that, under the terms of the leases, exploration must take place within five years after sale and you have some idea of why the Comptroller General's report to Congress found the program "hastily conceived by Interior under pressures exerted by the presence of the energy crisis and fears that the newly formed F.E.A. [Federal Energy Administration] would assume responsibility for the shelf leasing program."

If that is the case—and the department's current hesitation is almost evidence enough that it is—the first order of business would seem to be a revision of the 1975 program before any tracts are parceled out. An interim schedule could then be worked out which could have several highly important advantages. The delay that will be caused by shortages in any case could be put to excellent use, to begin with, if the department were to take the advice of Russell E. Train, Ad-

160

ministrator of the Environmental Protection Agency, and shift its focus to areas "where the resource potential is high and where the adverse environmental effects would be low."

While the reduced and selective program was being worked out—with the advice of the most skilled technicians the Geologic Survey could command—a broader kind of planning could be proceeding on both state and national levels. The states would have the time to complete their own coastal protection plans in accordance with the Coastal Zone Management Act of 1972. Once these are drawn up and approved by the Commerce Department, any future offshore leasing programs will have to conform as far as their onshore impacts are concerned. Now in the process of development, these plans should begin coming in sometime next year. They should assure a genuine role for the coastal states down to monitoring and surveillance where the standards they have set up for their own shore areas are potentially affected.

The same time could be used by the Federal Government to evolve a sensible method for joint exploration of the Outer Continental Shelf—if Congress fails to impose one—and to do the one great and obvious thing it has so far signally failed to do—that is, to frame an over-all energy policy for the nation. It is impossible to know how immediately essential offshore drilling really is (it could be, instead, rash waste of a valuable resource we will desperately want at a later date) unless we know how much energy we will need in the next few decades, how much can be conserved and what other potential sources exist for supplying it.

All energy taken from the earth exacts a price in environmental damage. Until there is an imminent prospect of harnessing the sun or fusing the atom, the crying need is for a panel of the nation's wisest heads—a Presidential commission of its leading scientists, industrialists, economists, environmentalists and, yes, philosophers—to decide, all things considered, which is the best course to follow, whether that course is nuclear energy, coal or the last drop of oil from the Outer Continental Shelf. Only then will full concentration on any given source make sense. Only then will Congress be likely to feel both the pressure and the confidence to move ahead. Only then will the nation, including even its most ardent environmentalists, feel free to accept proposals perhaps questionable in themselves as the price demanded by a coherent energy policy—a policy that becomes more urgent with every passing month.

The Outer Continental Shelf: Energy Needs vs. Environment

Luther J. Carter

The difficulties encountered by the accelerated offshore leasing program are in part attributable to the existing complicated leasing procedures, outlined in the following selection from Science *by Luther J. Carter, a staff writer for the magazine. The uncertainty about the pending legislation to change these procedures, which, by the spring of 1976, had been passed in the Senate but not in the House of Representatives, has further deterred the oil companies from active bidding on these leases.*

The significance of . . . [proposed legislation regarding offshore oil] can only be understood in light of the current OCS leasing procedures. As these procedures now stand—and they have undergone major changes in response to the requirements of the National Environmental Policy Act of 1969 (NEPA)—several distinct procedural steps are involved in opening up the frontier areas.

First, an environmental impact statement has been prepared on the overall accelerated leasing program and public hearings have been held on that statement. Then, in the case of each frontier province where leases are to be offered, there is the usual "call for nominations" whereby the Department of Interior asks industry and the public to recommend tracts to be included in the sale (or excluded, as the case may be).

Interior then selects the tracts for the proposed sale, and prepares a draft environmental impact statement for circulation among interested federal and state agencies and private groups. After a public hearing, the statement is revised and reissued in final form. No sooner than 30 days after issuance of this final statement, the Secretary of Interior decides whether to proceed with the sale as tentatively scheduled, and whether terms of the sale should include any unusual stipulations or restrictions to protect the environment. If he decides to proceed, and the presumption seems to be that he will usually so decide, the sale follows a month later. The successful bidders can then begin exploratory drilling on the basis of geophysical and other data collected prior to leasing.

If there are no court injunctions or other special difficulties, a bit over a year is taken up with the preliminaries to a sale. After the sale, another two or three years normally elapse before oil or natural gas are discovered (many leases are, of course, unproductive) and a production plan is approved by Interior. Then, a production platform must be ordered and put in place, and this also requires a year or two. Altogether, allowing for the completion of production wells, as many as eight years may elapse from the call for nomination to the time oil actually begins to flow.

By its amendments to the OCS Lands Act, the Senate would modify the procedures described above by separating exploration and development into more distinct phases. At the time of sale, bidders would be given to understand that even if oil is found, they will not be allowed to produce it if the environmental consequences or other adverse impacts of such production are found to outweigh the benefits.

The assessment of risks versus benefits would come once the leaseholder completes exploratory drilling and submits his production plan to Interior, to the governors of potentially affected coastal states, and to a regional OCS advisory

162

board if the governors have formed one (with themselves the voting members). The Secretary of the Interior could then tentatively approve those parts of the plan pertaining to activities confined to the OCS and, on the strength of such approval, the leaseholder could decide to order the production platform despite the at least outside chance that final approval would eventually be denied. Depending on whether leaseholders would indeed elect to run this risk, the new procedural steps might delay production either not at all or by a year or longer.

The preparation and circulation of an environmental impact statement on the production plan would almost certainly be necessary. Also, state and local authorities would be evaluating the plan in terms of its compatibility with their own plans developed under the Coastal Zone Management Act (CZMA), which provides that once a CZMA plan has received Washington approval, all federal actions and programs ordinarily must be consistent with it.

Furthermore, under the OCS bill, if a governor or regional advisory board asked that a production plan be disapproved, the Secretary would have to accept this recommendation unless he determined that to do so would be contrary to the "national security or overriding national interests." Even earlier, at the preleasing stage, the governors and the regional board would have this qualified right of disapproval with respect to the size, timing, or location of a lease sale.

Mining the Sea: Manganese Nodules

Allen L. Hammond

Oil has been recovered from the continental shelf in substantial quantities for more than thirty years, but now both industrialists and oceanographers are intrigued with the prospects of recovering solid minerals from the bed of the deep sea. Manganese nodules, containing significant quantities of copper and nickel as well as manganese and cobalt, have been found at depths of up to 18,000 feet or more. Recovering them poses new problems of technology and of international law. In the following selection, Allen L. Hammond, staff writer for Science *magazine, discusses the various experimental techniques for deep-ocean mining and for extraction of the valuable resources from the nodules. He also assesses the possible environmental impact of this new form of exploitation of the ocean's wealth. (The problem of ownership of these resources is discussed further in Unit Five.)*

*F*ishing and offshore oil drilling are still the most significant of man's efforts to tap the resources of the 70 percent of the earth covered by oceans, although minerals such as salt, tin, and limestone are extracted from the sea or dredged from the continental shelf in growing quantities. The most glamorous and perhaps the most mysterious marine resource, however, is the billions of tons of metals contained in manganese nodules that lie scattered over the seabed in the once inaccessible depths of the open ocean.

The nodules contain iron, nickel, copper, cobalt (and traces of two dozen other metals) in addition to manganese, often in concentrations comparable to those in land ores. This wealth of metals and the prospect of deep sea mining operations to recover them have stirred up interest among oceanographers and industrialists alike. Indeed, to judge from newspaper stories about the inexhaustible riches of the deep sea floor and from the intensity of the international debate over the ownership of these resources, a new gold rush might be beginning. But despite investments estimated at $100 million in the development of mining and processing technology, the economic viability of exploiting the nodules is still uncertain. . . . Nor is the resource unlimited. Although nodules occur on the seabed in many parts of the world, they are unevenly distributed. Only a few areas have nodules with a high metallic content in sufficient abundance to make mining feasible. Efforts to explain this distribution pattern and to account for the formation of the nodules are leading, however, to a better understanding of deep ocean geochemistry. . . .

Distribution and Structure

Manganese deposits on the sea floor occur as thick slabs or crusts, as thin coatings on basalt rocks, and as nodules, some of which were recovered as early as the 1872 expedition of the *Challenger*. Nodules, the commercially significant type of deposit, are black or brown agglomerations of manganese oxide and iron oxide minerals. The nodules range in size from about 1 to 15 cm across. They are extremely porous, and differ widely in shape, internal structure, and composition from one deposit to another—even within the space of a few hundred meters. Nodules are found predominantly in areas of oxygen-rich waters and low sedimentation.

The nodule beds most favorable for mining appear to lie in the Pacific Ocean, southwest of Hawaii and just north of the equatorial zone of high biological productivity. . . . The richest deposits occur in a narrow band, perhaps 200 km across and 1500 km long, running roughly east-west around 9° latitude. Not only are the nodules in this band high in copper and nickel content, but they are also very abundant. . . . The seabed in this area is about 5000 m deep and the sediments include silicon-rich remains of plankton (Radiolaria). High concentrations of nodules are also found along the southern edge of the equatorial belt, but the deposits, like those of the Atlantic and Indian oceans, are relatively poor in copper and nickel. (Deposits high in cobalt, however, have been discovered in the South Pacific.) As a result, both scientific and commercial interest have been focused on the North Pacific beds.

The structure of nodules from different regions varies widely, but most appear to be layered, often in the form of concentric rings around a small nucleus. . . .

Radiometric dating of nodules indicates that they grow very slowly. Growth rates between 1 and 100 mm per million years have been reported. By comparison, sediment accumulates on the ocean floor at a rate exceeding 1 m per million years even far from sources of continental debris and outside the equatorial zone of high biological productivity (where the rate of accumulation is much higher). Nodules are mostly found on the surface of the sediments or only partially submerged, however, and there is little agreement as to what keeps them from being buried. Some investigators believe that deep sea organisms that feed on protozoans nudge the nodules, while others believe that the protozoans themselves buoy up the nodules. . . .

Prospects for Mining

The history of mining is replete with examples of recovery and extraction techniques whose development was stimulated by new ores. The discovery in some parts of the Pacific Ocean of manganese nodules that also contain nickel and copper in higher concentrations than most ores being mined today is having a similar effect. Whether the methods now being developed to dredge nodules from the deep seabed and to extract pure metals from the nodule ore will become the basis of a new mining industry is still uncertain—economic, environmental, and legal questions about the feasibility of deep sea mining remain to be answered. Nonetheless, development of what may prove to be a major new marine technology appears well under way.

The participants include both industrial concerns and governments. Leading the field appear to be . . . [two] U.S. firms— . . . Deepsea Ventures, Inc. (a Tenneco subsidiary), and Kennecott Copper Company, both of which are engaged in exploration and the development of mining and processing systems. Others include Ocean Resources, Inc. and the Canadian-based International Nickel Company (INCO). In Japan and Western Europe the national governments have taken an active role in partnership with industrial groups such as the Sumitomo group in Japan, an association headed by Metallgesellschaft AG and Preussag AG in West Germany, and the French Société le Nickel. The Japanese, in combination with Ocean Resources Inc., have actively worked on the development of a mining system, while INCO, the Germans, and the French have so far emphasized exploration and process development.

The potential profits from deep sea mining operations will depend heavily on the metal content of the nodules, their size and abundance on the ocean floor, and the characteristics of the underlying sediments. Consequently, those interested in mining have spent considerable effort in exploring for rich deposits. The principal technique has been to scan the sea floor with an underwater television camera and to collect samples at periodic intervals for later analysis. Typical high grade deposits are reported to assay 27 to 30 percent manganese, 1.1 to 1.4 percent nickel, 1.0 to 1.3 percent copper, and 0.2 to 0.4 percent cobalt (although nodules with higher assays have been found) and to have about 10 kg of nodules per square meter.

Recovery Methods

The dredging methods now in use are limited to about 300 m, and to recover nodules from

Manganese Nodules on the Ocean Floor

Each nodule is several inches in diameter.

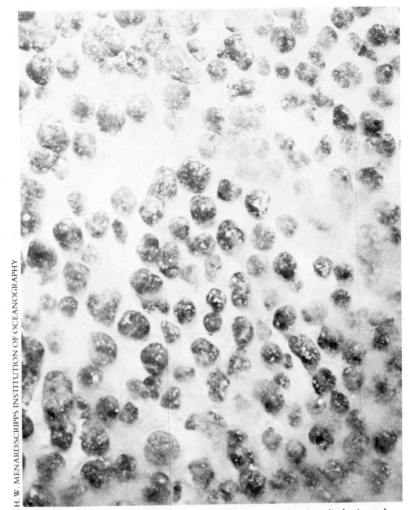

deposits at depths of 5000 m or more requires new techniques. Two principal types of deep mining systems are being developed. One, a mechanical system known as the continuous line bucket (CLB), consists of a long loop of cable to which specially designed dredge buckets are attached at intervals of 25 to 50 m. A traction drive (on the mining ship) moves the cable so that buckets descend into the ocean, are dragged across the seabed to scoop up nodules, and return to the surface to deposit their load.

A second, hydraulic recovery method consists of a length of pipe suspended from a mining ship; a sea floor device (dredge head) designed to collect nodules, screen out those larger than a certain size, and feed the rest to the bottom of the pipeline; and some means of pumping water up the pipeline with sufficient velocity (about 4 m/sec) to transport the nodules as well. The bottom device is either self-propelled or dragged across the bottom by the pipe string, depending on the design. Both conventional centrifugal pumps and compressed air injected into the pipe (air-lift pumping, in which the air bubbles provide enough buoyancy to raise the entire column of water and nodules) are being considered.

Mining Nodules Will Not Be Easy

Either system must contend with the difficulties of operating a mining ship in all kinds of weather, of controlling an extremely long length of pipe or cable, and of collecting nodules from a seabed of uncertain and variable physical properties. According to John Mero of Ocean Resources, Inc., the CLB system has the advantage of simplicity and flexibility, since it has no underwater machinery, can recover nodules of any size, and does not need to be designed for a specific depth and type of sediment as do hydraulic systems. Moreover, he claims that the system will cost much less than the more complex hydraulic equipment, although most other marine engineers dispute his estimates.

Mero admits, however, that such measures as shark-proofing the polypropylene cables and using two surface ships instead of one to keep the bucket loop from becoming entangled when the system is shut down may be necessary. A more serious difficulty is that the buckets must actually pick up a good load of nodules but little sediment for the system to be economic, and there is no way to control how the buckets interact with the bottom. Perhaps the most fundamental limitation of the CLB system, according to critics, is that cables are not strong enough and cannot be moved rapidly enough to mine 3 or 4 million tons of nodules per year with a single unit—a capacity that hydraulic systems can achieve and that many companies believe to be the economic minimum.

Hydraulic systems, on the other hand, require more elaborate gear, and a ship specially equipped to handle a long string of pipe. Centrifugal pump systems must be electrically or hydraulically powered underwater, a tricky and trouble-prone arrangement in the marine environment. Air-lift systems, however, are less efficient and must cope with the complexities and potential instabilities of three-phase flow (the pipe will simultaneously contain water, air, and nodules moving together). The bottom device, the key to a successful mining system, must traverse or sweep the loosely compacted sediments of the sea floor without becoming bogged down. Hydraulic systems are limited to recovering only part of the available resource, since larger nodules that might block flow in the pipe must be left behind. And the complete system may be expensive.

The advantages and disadvantages of each system depend in part on who is doing the talking, of course, and in a competitive arena such as deep sea mining, no company will disclose precise details of its own system, so that much of the available information comes from competitors. Claims and counterclaims are therefore to be regarded with some caution.

Experience with both systems is limited. The CLB system was tested in 1970 by an agency of the Japanese government and again in 1972 by a consortium headed by Ocean Resources, Inc. Deepsea Ventures has successfully tested an air-lift hydraulic system. . . .

The hydraulic mining system being developed by Deepsea Ventures, Inc. . . . includes a . . . small collecting device that is essentially a passive dredge head dragged across the ocean floor on the end of a pipeline. Sets of tines much like those on a rake collect the nodules, screen out those that are oversize, and help to separate the nodules from the sediment. An air-lift system is used to bring the nodules to the surface. Because of its simplicity, the system (it involves no underwater pumps or other machinery) has the advantage of low cost and reliability of operation, according to [company spokesman, Raymond] Kaufman. Because of the relatively crude collecting device, however, it must be carefully designed to match sea floor conditions at a particular mine site. And because the dredge must be towed across the bottom without exerting undue stress on the pipeline, the mining area must be very carefully surveyed before mining can begin. . . .

Deepsea Ventures intends to recover about 1 million tons of nodules per year. . . . [Other companies are] known to be aiming for 3 to 4 million tons a year or more. The difference is

167

related to the metals which are the ultimate aim of the mining efforts. . . . Kennecott, INCO, and apparently many others interested in deep sea mining, believe that copper and nickel will be the prime metals obtained from the nodules. To obtain enough of these metals for profitable operation, large tonnages of nodules are needed. Deepsea Ventures, on the other hand, is planning to extract and sell manganese and cobalt as well, rather than consign them to the scrap heap. Because there is so much manganese in the nodules, fewer are needed to yield a marketable amount of metals. The decision to include manganese influences not only the design of Deepsea Ventures' mining and processing system but also the company's business strategy, since they must find new markets (in the steel industry, for example) for manganese metal.

Most other companies are waiting . . . before committing themselves to mining systems. Kennecott, for example, is known to be designing a hydraulic system, but is moving very slowly. The Germans have published research on hydraulic mining designs, but are not far along in developing a system. The syndicate of companies which owns the rights to the CLB system will probably raise the money for a final test of that technique in the next couple of years, according to Mero, but no definite plans exist. INCO is maintaining an interest in both types of systems but has made no commitment to either.

Processing the Ore

Work on methods of processing the nodule ore is much further advanced. Kennecott has operated several pilot plants and is proceeding with the design of a full-scale system. Deepsea Ventures is nearly as far along. European companies, especially in Germany, have large research efforts on processing under way. Since a processing plant is expected to account for more than half of the cost of a complete nodule-mining operation, the stakes in choosing the best extractive method are high and four or five processes are being investigated.

Straightforward methods of extracting metals do not work because of the nature of the nodules. The manganese oxide and iron oxide minerals that are the main metal constituents of the nodules are extremely fine-grained. Consequently, physical means of separating the metals have not proved successful. Smelting the nodule ore has been tried, also without great success. Although it is possible to reduce the oxides by heating in a furnace to a temperature of about 1500°C, the result is an alloy of various metals (including iron) that is difficult to separate further. In consequence, chemical (hydrometallurgical) separation methods seem to be of most interest to the industry.

In the chemical approach, the nodules are partially or completely dissolved and the metals separated from solution. The task is made easier for those who seek to recover only nickel and copper, since these metals are predominantly found in association with the manganese minerals (adsorbed onto them or embedded in the crystal lattice). Cobalt is found mostly in the iron. Hence several different reagents, selective leaching techniques, and reducing treatments can be utilized, depending on the desired product. . . .

Environmental Effects a Concern

Nodules are found in areas of the ocean that are in many respects biological deserts, but that does not mean mining will have no effect on the ocean environment. The potential problems include local disruption of the sea floor ecosystem, and distribution of sediment particles—some of which may remain suspended for a year—throughout the water column. The sediment might alter chemical balances both in the more populated surface waters and below, and as it settles out may bury organisms that live on the sea floor. Many oceanographers do not believe that mining itself will do much permanent damage to the oceans as a whole, however, because of the relatively small areas to be affected, but they point out that it may destroy sediment records of scientific interest. More serious environmental effects may come from processing the nodules, especially if, as some have suggested, second-generation plants are built to operate at sea. Even on land, operations that involve discarding all but the copper and nickel must dispose of millions of tons of residues that contain manganese and other toxic metals in oxide form.

The impact of the new source of metals on world markets is also a controversial subject. At one extreme, deep sea mining optimists like Mero forecast a drop in metal prices within 10 years after production starts and the closing of many land mines within 20. Others, such as A.J. Rothstein and Kaufman of Deepsea Ventures, maintain that copper prices will not be affected at all in the foreseeable future and that only for cobalt are the nodules likely to cut into prices or land-based production significantly. An analysis prepared for the World Bank points out that mining of enough nodules (about 6.5 million tons) to meet the world demand for cobalt in 1967 would have provided 22 percent, 0.9 percent, and 13 percent, respectively, of the world demand for manganese, copper, and nickel. The analysis also assesses the economics of deep sea mining as uncertain, largely because of the still unknown costs of the mining itself (estimates range from about $3 to $30 per ton

of dry nodules at dockside). It concludes that exploitation of the nodules could reduce the dependence of industrialized countries on imports of metals and could seriously affect a few of the developing countries that depend on exports for revenue.

Commercial mining operations are not likely to begin before 1977 (it takes 3 years to build a processing plant and none are now under construction), if then. The prospects probably depend in large part on what happens to the international political and legal argument over the law of the sea, although several companies have hinted that they might force the issue by mining even in the absence of an international convention. At present almost all of the major companies are actively negotiating for partners to share the estimated $250 million investment (and the risks) of a deep sea mining venture. Kennecott, along with four other companies, . . . [in February 1974] announced a 5-year, $50-million research program to determine mining feasibility.

In any case, there appears to be growing conviction in the industry that deep sea mining is gathering momentum, that manganese nodules are indeed on the verge of commercial exploitation.

Deep-Ocean Mining

Frank L. LaQue

Both industrialists who are risking capital in developing techniques of deep-ocean mining and diplomats who are bargaining to secure a "fair" share of the surplus income from the ocean's mineral resources assume that mining the oceans will prove economically as well as technically feasible. But not everyone is optimistic about a taxable surplus. In the following excerpt from his 1971 article "Deep-Ocean Mining: Prospects and Anticipated Short-Term Benefits," Frank L. LaQue, retired vice president of International Nickel Company and former president of the International Organization for Standardization, warns that revenues from deep-sea mining will have little effect on the income gap between developed and developing nations. This article was originally published in Pacem in Maribus, *edited by Elisabeth Mann Borgese.*

The expectation of considerable revenue from the exploitation of ocean mineral resources has generated a great deal of discussion during the past few years. Until now these discussions have concentrated heavily on how exploration and exploitation of the anticipated resources should be made subject to some form of international regulation and on how the revenues from such exploitation should be applied for the benefit of mankind. Underlying this discourse is the concept that deep-ocean mineral resources represent "a common heritage" and that, therefore, the wealth derived from their exploitation should be held in trust by the international community and applied for the common good. There are some, also, who feel that the anticipated riches from ocean exploitation should go toward redressing the imbalance between the developed and developing nations of the world.

Those holding these views envision two prospects for the future. One suggested possibility is that a wild international scramble will take place among the highly developed nations to dominate the exploitation of undersea resources. International tensions would consequently be aggravated and the advanced nations would become even more prosperous in relation to the developing ones. The other suggestion is that an enlightened international social conscience will result in a general recognition that the substantial (often called tremendous) new resources in the ocean can provide mankind with a splendid opportunity. Imbued with a generous new spirit, men may seize the chance to organize exploitation so as to eliminate any possibility of increased international tension and may then distribute the derived wealth for the maximum benefit of mankind, with special concern for developing nations.

The purpose of this paper is to evaluate the prospects for exploiting deep-ocean metals and to examine the possibility of achieving the proposed international goals as a result of this exploitation. This study will deal only with metals—those that are sometimes called "hard minerals," as distinguished from such other minerals as petroleum, natural gas, sulfur, and phosphorites. Sand, gravel, diamonds, precious coral, and the like, will be excluded. . . .

The Anticipated Benefits From Deep-Ocean Mining

In the light of the uncertain future of deep-ocean mineral exploitation and its yet-to-be-established commercial value, the prospects for using revenue from this source to help developing nations seem poor.

In terms of prosperity, the nations of the

170

world range from affluence to poverty on a sliding scale. Because the variations are gradual, an agency charged with distributing tax revenue from deep-ocean mining, even if it were substantial, would have difficulty deciding which developing nations were entitled to a share and how the total should be allocated among them.

The gross national product of a country is a reasonable measure of its prosperity. For purposes of this discussion, it is assumed that developing countries, as candidates for revenues from deep-ocean mining, would be found in Latin America, South Asia, the Near East, the Far East (except Japan), Africa (except South Africa), and Oceania (except Australia and New Zealand). The total gross national product for these areas in 1967 amounted to 12.6 percent of the world G.N.P., or $291,254,000,000. . . .

It can be calculated that the value of world production of manganese, copper, nickel, and cobalt in 1967 represented only .28 percent of the total gross national product. The distributable revenue from taxation, 10 percent of the total value, would be about .028 percent of the world G.N.P. It may be noted that the total world production of these metals in 1967 had only about one-half the value of the world catch of fish in that year.

In addition, a substantial portion of the world's production of manganese (23.1 percent), copper (41.7 percent), and cobalt (89 percent) comes from developing countries. While most of the world's nickel now comes from Canada, New Caledonia stands second in nickel production, and new nickel projects are in various stages of exploration and development in New Caledonia, Guatemala, the Dominican Republic, Indonesia, the Philippines, and the Solomon Islands. . . .

If, as would be the case, only the revenue from taxes on the profits deriving from the exploitation of deep-ocean metals is available for adjusting the relative prosperity of developed and developing nations, this amount would be about 10 percent of the total market value of the metals and would represent only a little more than .025

percent of the world gross national product and only about .2 percent of the G.N.P. in 1967 of the developing nations. On a per-capita basis, this would come to forty-one cents a head if it were divided equally among the 1,594.9 million people in the developing countries.

It should be evident, therefore, that even in the unlikely event that the deep-ocean bottom replaced all land sources of manganese, copper, nickel, and cobalt, the assignable revenue from the exploitation of these deep-ocean metals could have little impact on efforts to close the current gap between developed and developing nations. Furthermore, substituting ocean for land sources of these metals would tend to detract from, rather than to advance, the prosperity of those developing nations with large deposits of metal-bearing ores.

Conclusions

Most of the current activity in the recovery of metals from deep-ocean nodules can be characterized as an examination of the technical and economic feasibility of various conceptual approaches. Some of these may lead to preliminary or pilot-scale projects that will precede full-scale commercial operations. No such operations are taking place at present, and the aim now is to provide a basis for future decision whenever new sources of ore may be needed. Such an eventuality may occur when per-capita consumption of metals in developing countries approaches the present level in the advanced nations.

While the future of deep-ocean mining cannot be predicted with precision, it is safe to draw a few general conclusions:

1. There will be no commercial-scale exploitation of deep-ocean nodules for several years—probably not before 1980 to 1985.
2. There is a need for an international program of ocean exploration that could be part of the International Decade of Ocean Exploration proposed in 1968 by the former American President Lyndon B.

Johnson to confirm the extent, the distribution, and the possible value of metals in deep-ocean nodules.

3. Since exploitation operations in the foreseeable future will probably be few in number and conducted on a small scale, any international control agency or mechanism should place emphasis on providing a regulatory environment, either national or international, that will provide incentives for risky exploitation and will not place the operations under undue restraint. Unnecessary restrictions can result from efforts to deal with unknown situations and circumstances that may never be encountered.

4. While appropriate international regulations will be needed in the future, details should not be worked out before the facts are in hand. International laws or regulations aimed at a codification of practice should logically await reasonably precise knowledge of the practice that is to be codified.

5. Since we currently do not know how much revenue for "the benefit of mankind" can be expected from the exploitation of nodules and since the amount will probably be small for the foreseeable future, the prime international emphasis should be on encouraging exploration and preliminary exploitation rather than on the disposition of revenue. Whatever revenue does accrue from deep-ocean mining will probably have no significant effect on the absolute or relative prosperity of the recipients and may well have a greater effect on the distribution of prosperity among developing nations than on the comparative position of the developing and developed countries.

6. Developing nations should not be encouraged to expect that the exploitation of deep-ocean metals will provide a major component of the funds they need for future development.

The Living Chain
Rachel L. Carson

Most scientists believe that life evolved in the sea billions of years ago, and it is to the sea that many people are now turning in hopes of sustaining life for an ever-growing world population. All hopes for using the sea to feed the land must depend ultimately on the food chain within the sea. In the following selection from The Sea Around Us, *winner of the 1951 National Book Award, marine biologist Rachel Carson describes the interlocking links in this chain.*

*W*ith the surface waters, through a series of delicately adjusted, interlocking relationships, the life of all parts of the sea is linked. What happens to a diatom in the upper, sunlit strata of the sea may well determine what happens to a cod lying on a ledge of some rocky canyon a hundred fathoms below, or to a bed of multicolored, gorgeously plumed seaworms carpeting an underlying shoal, or to a prawn creeping over the soft oozes of the sea floor in the blackness of mile-deep water.

The activities of the microscopic vegetables of the sea, of which the diatoms are most important, make the mineral wealth of the water available to the animals. Feeding directly on the diatoms and other groups of minute unicellular algae are the marine protozoa, many crustaceans, the young of crabs, barnacles, sea worms, and fishes. Hordes of the small carnivores, the first link in the chain of flesh eaters, move among these peaceful grazers. There are fierce little dragons half an inch long, the sharp-jawed arrowworms. There are gooseberrylike comb jellies, armed with grasping tentacles, and there are the shrimplike euphausiids that strain food from the water with their bristly appendages. Since they drift where the currents carry them, with no power or will to oppose that of the sea, this strange community of creatures and the marine plants that sustain them are called "plankton," a word derived from the Greek, meaning "wandering."

From the plankton the food chains lead on, to the schools of plankton-feeding fishes like the herring, menhaden, and mackerel; to the fish-eating fishes like the bluefish and tuna and sharks; to the pelagic squids that prey on fishes; to the great whales who, according to their species but not according to their size, may live on fishes, on shrimps, or on some of the smallest of the plankton creatures.

Unmarked and trackless though it may seem to us, the surface of the ocean is divided into definite zones, and the pattern of the surface water controls the distribution of its life. Fishes and plankton, whales and squids, birds and sea turtles, all are linked by unbreakable ties to certain kinds of water—to warm water or cold water, to clear or turbid water, to water rich in phosphates or in silicates. For the animals higher in the food chains the ties are less direct; they are bound to water where their food is plentiful, and the food animals are there because the water conditions are right.

The change from zone to zone may be abrupt. It may come upon us unseen, as our ship at night crosses an invisible boundary line. So Charles Darwin on H.M.S. *Beagle* one dark night off the coast of South America crossed from tropical water into that of the cool south. Instantly the vessel was surrounded by numerous seals and penguins, which made such a bedlam of strange noises that the officer on watch was deceived into thinking the ship had, by some miscalculation, run close inshore, and that the sounds he heard were the bellowing of cattle.

To the human senses, the most obvious patterning of the surface waters is indicated by color. The deep blue water of the open sea far from land is the color of emptiness and barrenness; the green water of the coastal areas, with all its varying hues, is the color of life. The sea is blue because the sunlight is reflected back to our eyes from the water molecules or from very minute particles suspended in the sea. In the journey of the light rays into deep water all the red rays and most of the yellow rays of the spectrum have been absorbed, so when the light returns to our eyes it is chiefly the cool blue rays that we see. Where the water is rich in plankton, it loses the glassy transparency that permits this deep penetration of light rays. The yellow and brown and green hues of the coastal waters are derived from the minute algae and other microorganisms so abundant there. Seasonal abundance of certain forms containing reddish or brown pigments may cause the "red water" known from ancient times in many parts of the world, and so common in this condition in some enclosed seas that they owe their names to it—the Red Sea and the Vermilion Sea are examples.

Food From the Sea

C. P. Idyll and Hiroshi Kasahara

What is the actual potential of the ocean for feeding the land? There is a wide range of opinion among experts in part, as C. P. Idyll and Hiroshi Kasahara point out in the following excerpt, because of the different methods used for estimating potential yield. The authors suggest that we can maximize our harvest of the sea by conservation measures, increased use of such stocks as krill, red crab, squid, and lantern fish, and the development of new fishing techniques. Yet they remain pessimistic that the sea alone can solve the problems of feeding a hungry world. Idyll is a marine biologist at the National Oceanic and Atmospheric Administration and a specialist in fish populations and marine ecology; Kasahara was in charge of fisheries for the United Nations Development Program. Their article was originally published in Exploring the Ocean World, *edited by Idyll.*

The sea has provided food for man as long as he has lived on its borders. . . .

Millions of years after man started to fish he is still taking most of his catch from coastal waters. This is partly because the nearshore areas are easiest to reach. In addition, despite the common impression that the sea is monotonously the same over its whole vast expanse, there are important differences in productivity among parts of the ocean. Depending upon whether the water is deep or shallow, cold or warm, salty or brackish, the quantities and kinds of animals to be found in it vary greatly. The result is that some areas of the oceans are deserts while others are extremely rich in life. The richest are the shallow, well-lit waters along the coasts.

The sea contains an extraordinary variety of creatures, comprising millions of species of plants and animals. There are, for example, more than twenty thousand species of fishes. A great many marine animals and plants have been used by mankind for food, but a surprisingly small number of them are consumed in large quantities. Only about a dozen kinds form three-quarters of the amount used. About 40 percent of the world's catch consists of herrings, anchovies, and sardines; a little less than a sixth is cod, haddock, and hake; the rest, in rapidly declining proportions, consists of horse mackerel, tuna, flatfish, salmon, and shark, followed by hundreds of other varieties.

The overwhelming proportion of food from the sea—approaching 90 percent—consists of fish. Shellfish supplies 7 or 8 percent and whales about 1 percent. Plants from the sea provide much less than 1 percent of man's food, compared to a figure of 80 percent for plants from the land in some parts of the world.

The scarcest and most valuable of all human food materials is animal protein, needed to build muscle and other body tissue. Plant protein does a good job, but some of the amino acids required by the human body are in short supply in many plant proteins. Fish is one of the best of all sources of these, and one of the cheapest. The productivity of ocean water is determined in much the same way as the richness of earth; the greater the amount of fertilizing minerals—chiefly nitrates and phosphates—in the water, the greater its productivity. And just as the productivity of land is increased by returning minerals to the surface layers through plowing, in the sea a kind of plowing serves this purpose, too.

The similarity between production of living material on land and in the sea is not mere coincidence; it is a consequence of the fact that in both regions the creation of edible animals depends on the growth of plants. These are eaten by animals; man either consumes the plants directly or he eats the animals that have eaten plants or other animals.

In the sea the plants are mostly very different from those on land, being microscopic in size so that they are rarely seen (unless they bloom in prodigious numbers in phenomena called "red tides") and often being encased in skeletons of hard calcium carbonate or silica. Their small size makes them hard to collect and their skeletons render them unpalatable to man. The relatively scarce large marine plants we call seaweed are eaten to a minor extent, but generally speaking, the quantity of marine plants providing food for man is insignificant.

But the role of the plants in the economy of the sea is enormous, and in a particular region the abundance of the sea creatures consumed by man—the fishes, the crustaceans (crabs, shrimps, lobsters), the mollusks (oysters, clams, scallops)—is first determined by the ability of that region to grow plants. The nutrient minerals are carried to the depths of the sea in the dead bodies of animals and plants. Once they sink below the level where light is strong enough to carry on photosynthesis, plant production ceases. This depth varies from about one hundred to three hundred feet according to the clarity of the water. The rich areas of the ocean are those where some mechanical process overturns the water layers and brings mineral-laden water to the sunlit surface again. In the temperate zones this takes place when the surface cools in the autumn, increasing the density so that the water sinks, which forces deep layers to the top. Productivity is also renewed when prevailing offshore winds skim off the surface waters; these are replaced by water welling up from the depths, bringing with it fertilizing mineral salts. . . .

While it is certain that the food resources of the sea are not being fully used, no one knows exactly how large these resources are. In spite of recent progress in oceanography and fisheries research, forecasting of future production is still a matter of educated guesswork. Much of

the necessary scientific data is missing, and a number of unknown technological and economic factors are involved. The recent development of the world fishing industry has clearly shown that most of the estimates made by scientists in the 1940's and 1950's were much too low. The question is being reexamined in the light of more recent information.

There are different methods of estimating the potential yield of food from the sea. One way is to start by calculating the amount of plant material produced in the whole ocean during a year. This difficult job must then be followed by estimations, based on uncertain data, of how much of this plant material is passed along to the herbivorous animals, how much of the substance and energy locked up in these is passed along to the first level of carnivores, and so on up through several more levels of the food pyramid. The results have been estimates of greatly different magnitude, ranging from twice to ten times (or more) the present harvest. Such estimates have little practical significance, for they do not indicate under what circumstances, from what areas, or in what kinds of animals the estimated potential yields might be realized.

The second method of estimating the potential yield is to extrapolate from known production of existing fisheries and from presumed possible exploitation of unused or under-utilized stocks. For example, it is believed that the Arabian Sea (the northwestern part of the Indian Ocean) contains virtually untapped fishery resources of large magnitude; fully exploited, these fish stocks would support several million metric tons of fish landings annually. Only recently has the food potential of the rich waters off the west coast of Africa begun to be appreciated. . . . Up to three or four million tons of additional production may be possible. . . . Many other areas of the world ocean, including various parts of the Indian Ocean, also appear to be greatly underexploited.

Even some parts of the sea already under heavy exploitation may be capable of greatly increased fish production. California was for many years the leading producer of fish in the United States, and even after the collapse of her big sardine fishery in the mid-1950's—caused by altered oceanographic conditions and over-fishing—she still ranks high, with about 250,000 tons a year. But despite the vigorous exploitation of California waters, additional millions of tons of fish could be caught. These are hake, anchovy, jack mackerel, and deep-sea lantern fish—all species not familiar to, or sought by, buyers in the United States, and this, of course, is why they are underexploited. But they are species that can make fish meal and fish-protein concentrate.

Indeed, according to fishery scientists, the waters adjacent to the United States (including the Pacific, Atlantic, and Gulf of Mexico) are potentially capable of supporting annual yields of fish and shellfish totaling some 20 million tons, compared to 2.5 million tons caught by United States flag vessels and about 1.5 million taken by foreign fishermen in 1966.

This does not by any means exhaust the list of underexploited fish stocks of the world. . . . But there is a finite limit to these, and the day will come when there will be no new stocks to be harvested by conventional methods. Unless spectacular new ways of fishing are invented, it seems likely that production of the kinds of fish now being used will about double in the next fifteen to twenty years. The rapid rate of increase of the last two decades will probably slow down as readily exploitable new resources become more and more difficult to find. At that point the hope of substantial increases in food from the sea will depend on man's ability to find uses for the truly enormous variety of marine organisms now wholly or largely unused, and to develop methods for harvesting them profitably.

Stocks for the Future

The greatest of these untapped resources is plant plankton, of which perhaps some 400 billion tons in wet weight are produced in the sea in a year. Because it is unpalatable and very expensive to catch, it seems unlikely that we can use plant plankton soon, if ever. Animal plankton also exists, in such large quantities that every person in the world could have several tons every year if it could be caught, processed, and delivered to him—and if he would eat it. Unfortunately, despite its total bulk, animal plankton is scattered thinly throughout the enormous mass of the ocean—some 330 million cubic miles—and most attempts to harvest it have failed to pass the test of economics. Furthermore, much of this material is unpalatable or inedible (for example, stinging jellyfishes).

Nonetheless there are exceptions, and in some areas of the world there exist concentrations of animal plankton large enough to make economic harvest seem possible. The most likely of such plankton is the krill of the Antarctic, *Euphausia superba.* Krill are small crustaceans, relatives of the shrimp. Those found in the Antarctic are about one and a half or two inches long, and they occur in enormous numbers up to the edge of the ice pack. Observations by British whale biologists and more recently by Soviet fishery scientists put estimates of their production at somewhere between 40 million and a fantastic 1.3 billion tons a year. It is probable that a conservative 50 million tons a year could be caught without endangering the stocks—about as much as the present total world seafood production. . . .

[Meal made from krill] . . . has a strange orange color derived from the bodies of the little shrimplike animals. It has a high protein concentration—more than sixty percent. The high proportion of unusable chiton (the shell of the krill) and the very rapid spoilage rate (even in the cold southern ocean) are problems that so far have prevented an industry from developing. But technology has solved problems more difficult than these, and it seems likely that krill, in northern seas as well as in the Antarctic, will be an important crop in the future.

The "red crab" *(Pleuroncodes planipes),* in its young stages, may be another animal plankton creature that will eventually supply food for man. This animal exists in incredibly large numbers off the coasts of California and northern Mexico. One scientist describes how his vessel "crunched through" immense swarms of red crabs for miles. Hundreds of thousands of tons may be available for harvest after efficient methods of catching them and converting them into some saleable product have been devised.

Once the proper techniques have been developed, many animals may be successfully harvested. Herein lies the greatest hope for truly substantial increases in the sea harvest, since the abundant marine herbivores would be tapped—the level of the food pyramid in the sea equivalent to the cows, pigs, and sheep of the land. At present most of the catch consists of carnivores—salmon, cod, tuna—which are the equivalent of tigers, foxes, and weasels. The significance of the comparison is not so much the unattractive-

Women Fishing in Lagoon, Truk

ness of tiger meat compared to beef (maybe the public would prefer tiger if they had been raised on it) but the far smaller total quantity of the carnivores compared to the animals that feed on grass. The enormous differences in size and in other characteristics between land and ocean herbivores have prevented wide use of the latter in the past; perhaps this can be changed, and if so the world will be less hungry.

Squid are not herbivores, but they are another potential food resource of enormous size. They are distributed over most oceans of the world, and may be one of the most abundant of the sea's larger animals. They are not nearly as well known as the fishes, largely because little use is made of them. . . . In proportion to their abundance they are an enormously under-utilized resource. If more people accept them and if efficient methods of harvesting them can be developed, another great source of human food will be available.

It is also known that huge numbers of small, odd-looking bristlemouths and lantern fishes (so called because of their luminous organs) occur in deep layers of vast oceanic areas. They can be considered an important potential resource for fish meal. So far no techniques have been developed to catch them in large quantities at low cost.

Realizing the Sea's Potential

If mankind is to realize what is clearly a great potential for food from the ocean, efforts should be made in three directions. First, in order to maintain catches at the highest possible level, conservation measures based on scientific research will continue to be important. . . .

The second requirement is a greatly increased level of search for under-utilized stocks throughout the oceans of the world. Even after years of active investigation there still remain large areas that are apparently productive, judging from oceanographic data, but for which knowledge of potential resources is limited. . . .

Thirdly, if full use is to be made of the ocean's protein material, there is a need for research to develop new fishing techniques. . . . In order for food production from the ocean to keep growing at a high rate, spectacular technological developments will have to take place to utilize its enormous potential resources, which cannot be harvested economically by present methods.

Man has by no means reached the limit of the harvest of the sea. In the years ahead he will continue to increase the catch of familiar kinds of seafood as his needs expand and his technology improves. He will learn to use new and unfamiliar kinds of animals from the sea—squid, krill, lantern fish. He will get some additional fish and shrimp from the new shallow-water farms that are now being established.

But the importance and potential of food production from the ocean must be assessed realistically. There is no evidence to support the view that the problem of feeding the growing world population can be solved by exploiting sea resources. An overwhelmingly large proportion of the food requirement has been and will continue to be met from agricultural products, and the main contribution from the ocean will be to increase the supply of animal protein. The amount of animal protein available may well increase greatly in the future if man is wise enough to make full use of the sea's rich resources.

Farming the Seas: The Blue Revolution

Elisabeth Mann Borgese

Prehistoric man began to fish almost 10,000 years ago, but the technological changes that have accompanied the evolution of the fishing industry in the last century today threaten the very survival of some of the living resources of the ocean. In the following article, excerpted for Oceans *magazine from* The Drama of the Oceans, *Elisabeth Mann Borgese traces these developments and discusses new methods of farming the seas. She urges the application of scientific methods to aquaculture, which could lead to a Blue Revolution, similar to the Green Revolution in agriculture. Mrs. Borgese is a founder of the International Ocean Institute, which organizes the annual* Pacem in Maribus *conferences aimed at producing a peaceful international regime of the oceans.*

There is an old story from the Frisian Islands, retold by Grimm in "The Fisherman and His Wife." Its theme, which appears in the folklore of many peoples, sums up the story of all fishermen through the ages, from their simple beginnings up to the present point of man's technological clash with nature.

A poor fisherman, so the story goes, is sitting on a rock one day when he catches a large flounder on his line. The fish, amazingly enough, begins to speak with human voice. Claiming to be an enchanted prince, the fish begs the old man to spare his life and let him return to the depths. The fisherman agrees and releases him unharmed. The fisherman's wife, when told the story, is angered that her husband did not ask the enchanted fish to grant them a boon. She plagues the fisherman to find the flounder again and to demand a neat new cottage to replace the hovel in which they live.

Next day the fisherman calls out and the fish swims up. The old man makes his wish, and when he returns home he finds a neat new cottage where the hovel had stood. But the wife is soon dissatisfied. She urges her husband to ask the flounder for more: first for a castle and vast estates, then to become king, then emperor, then pope. The flounder fulfills every wish. But finally she demands power over the sun and moon, to control the universe. This is too much.

Amid tremendous waves and crashing thunder, the enchanted flounder destroys all he has bestowed. Empire and kingdom disappear; the fisherman and his wife end as they began, sitting in their hovel in poverty.

The history of mankind's fishing experience parallels the folk story. The poor fisherman—and most real ones have been as poor as the one in the fable—was probably well content with the fulfillment of his first modest wish for better snares. He appreciated his new netting and line, his improved trawls, traps, and seines. But as time went on, the technological imperative pulled him inexorably on to expand his fisheries to distant waters all over the world. He became king and emperor of motorized fleets spanning the world oceans. His aresenal included planes and earth satellites with infrared scanners to map the water temperatures. There were cameras to spot schools of fish, so that he could hunt not only on the surface and the seabed, as before, but could also sweep the midwater with teleguided nets. There were great factories that stayed at sea for months on end processing the fish.

And soon no effort was involved, nor much luck needed, as he wished for and received catches heretofore unimaginable, fishing indiscriminately and ruthlessly for anchovy and whale alike.

But, like the poor fisherman in the fairy tale,

man was not destined to control sun and moon or to become Lord of the Universe. That final wish was hubris, begotten of good fortune and begetting misfortune. Man had by this time so damaged and depleted life in the oceans by over-fishing and polluting that the fish grew fewer and fewer, and were foul when caught.

The fisherman and his wife were, in truth, poor once again.

In his first role in the ocean, Homo sapiens was probably a fisher. Swimming in the warm sea with the playful dolphins, he was fairly high up in the food chain, eating and being eaten. A spearfisher, probably, like the swordfish and the narwhal. Or a diver for abalone and sea urchin. He may have used obsidian flints to spear the fish or to pry shells open. Like the otter, he may have swum on his back while cracking sea urchins open against a stone. The marine environment, it seems, encourages the using of tools.

All the other technologies—hook and line, net, harpoon, sonar—that the creatures of the deep had perfected over millions of years, man had to reinvent from his land base.

Recent research by nautical archaeologists shows that the earliest cave dwellers in the Mediterranean region in Mesolithic times were hunters feeding on red deer. By the seventh millennium they had become fishermen, as indicated by the large number of fish bones found in their caves. Three thousand years before our own time, fishing had developed into a highly organized craft. Miniatures discovered in Minoan houses destroyed in an earthquake about 1500 B.C. show boats full of fishing tackle, rods, and hooks, and divers plunging into the sea with their bags, one of them carrying what looks like a large sponge.

Bronze Age Mediterraneans probably fished with gorge hooks—that is, bones attached in the middle to a line and sharpened at both ends. The bait was wrapped around the bone, and when a fish swallowed it, the hook was jerked into position in the fish's gullet. Hooks of this type have been found in Mesolithic and Neolithic caves. They were used by American Indians until modern times.

The Chumash and Kwakiutl Indians of the Pacific coast of North America, whom European explorers came upon in the sixteenth century, were extremely primitive fisher folk not unlike those who lived around the Mediterranean 10,000 to 20,000 years ago. The sea was bountiful. Six species of abalone crowded the kelp-enshrouded rocks, and when the tide went out, the Chumash could rise from a shady resting spot and pry a few dozen of the tasty creatures from their precarious moorings. They also made hooks, which they baited with squid and weighted with stones to sink them to where the halibut moved, a few feet off the bottom of the sea. Perhaps they brought the craft of line and hook making with them when they crossed the Bering Strait some 20,000 years ago or sailed the Pacific on vessels like Thor Heyerdahl's *Kon-Tiki*.

The Nordic people were fishermen in the Stone Age, 10,000 years ago, venturing far out in their skin boats after deepwater fish such as cod, pollack, and ling.

Through the ages, Eskimos and the Polynesians and Melanesians used hooks and lines, and they were sophisticated enough to use artificial baits. The Eskimos carved and painted colorful ones that resembled beetles. The Polynesians used shining mother-of-pearl; did they know of the anglerfish's luring light at the end of its inborn hooked fishing line?

Net fishing probably evolved similarly in the East and the West, with an observant spear fisher noting that the quick fish that darted away at the moment his spear was released could in the end be trapped if he could seal off some area with woven reeds or grasses. The craft of weaving grasses into traps later grew into true net-making when man the cultivator developed various materials for making line, such as cotton and wool. With the aid of fibrous plants, especially the one from which common hemp is made, he became a more efficient fisher.

Did he know about the larvacean and its net?

It was a big step from spear or line fishing to net fishing. The fisherman now took more fish than he and his immediate family could consume. Net fishing brought him onto the market and made him a trader. But although traditionally the trader is wealthy, the fisherman remained poor—he is poor in the lore of Nordic ballads, in Biblical parables and Indian tales, in the Odyssey and the Thousand and One Nights.

Net fishing, furthermore, is a communal or community-orienting activity. Tool making and mending, fish hauling and processing are tasks in which the whole village, young and old, male and female, can share. Those going to sea must act as a well-disciplined unit in braving the challenges and dangers of the elements. Those staying behind are united in their anxiety for their loved ones out in wind and wave, united in their grief when disaster strikes.

The apparently inexhaustible nature of the fish supply, together with the necessarily collective methods of the fishing enterprise, discouraged the formation of private or individual property concepts. But they did not discourage the technological imperative that drove our fisherman in search of better materials and implements—trawls, traps, lines, and seines.

Unhappily, technological change often does more harm than good. This applies even to very simple changes, such as the introduction of nylon nets. In Brazilian fishing villages, for example, this kind of innovation made the formerly free fishermen dependent on the importer and his middlemen, led to the depletion of the fish supply, introduced competition instead of cooperation, and disrupted communities. This is well documented in a study by John C. Cordell. "We should not delude ourselves," Cordell warns, "about the miracles technology can work in primitive settings. Since modernization is a socially selective process, the question remains: Who benefits from it, and at whose expense? In a highly stratified society like northeastern Brazil, the answer to this question is a foregone conclusion, since part of the cost of technological change is the maintenance of social inequality."

Regardless of its social impact, the new technology allowed fishermen in various countries to stop fishing indiscriminately and, instead, to specialize more systematically in certain varieties—cod, haddock, pollack, whiting, and hake; on flatfish such as sole, flounder, halibut, plaice; on salmon, seabass, and rockfish; on fish that travel in schools, such as mackerel, tuna, herring, sardine, anchovy, menhaden, and shad. And the greatest challenge of all, making fishermen drunk with visions of wealth, like dreams of a gold rush or an oil strike, was hunting the giant whale.

Whales had been hunted for many centuries before the great whaling ships were built late in the nineteenth century. In Europe, large-scale whaling began in the Bay of Biscay. The Basque whale fishery was well established in the twelfth century, and is probably at least two centuries older than that.

In 1150 King Sancho VI of Navarre, called "the Wise," made a grant to the city of San Sebastian which listed articles of commerce and the duties to be paid for warehousing them. This list includes whalebone (baleen) as a prominent item. The Basques were expert whale hunters, navigating their coastal waters in small boats and using hand harpoons; they towed the whales they killed to shore for processing. By the sixteenth century, an average of at least sixty whales a year were taken along the Basque coast. This seems to have been a case of overfishing, for the stock began to decline in the seventeenth century. The Basques then sailed to remote places such as Newfoundland, in what seems to be the first example of distant-water fishing. British, Russian, Dutch, Spanish, and German ships soon followed, and by the middle of the seventeenth century whaling in Spitzbergen and around Bear Island and Jan Mayen Island was big business.

From the Arctic to the Antarctic, from the Azores to Japan, from the shores of Africa to South America, off Canada, on the coast of

New England, they hunted the whale. They hunted him in open rowboats, in sailing craft, and in fast steamers; with hand harpoons, nets, blubber knives, spades, prickers, pickaxes, coshes, and grapnel. They hunted him with poisoned arrows, leaving him to suffer for days, finally to die and be washed ashore. Or, making short shrift, they slaughtered him with explosives. A bomb lance patented in the 1850s was fired from a gun: filled with gunpowder, it was designed to be exploded by a fuse in the stem only after it was inside the whale.

The bloody romance of the nineteenth century gave way to the cold steely horror of the twentieth. New whaling implements—introduced mostly by Japan, which followed Norway as the leading whaling nation, to be surpassed in turn by the Soviet Union—include electric harpoons, rocket-launching devices, helicopters, and sonar devices that indicate the whale's swimming direction and distance from the ship. The most modern whalers are 200-foot long diesel-powered ships weighing 500 to 900 tons and attaining speeds up to eighteen knots. The cannon, placed over the ship's bow, is loaded with a harpoon filled with blasting powder that explodes inside the whale. The stricken whale is winched to the surface alongside the ship and a windpipe is stuck into him, bloating him so that he floats. Then a flag is implanted in his huge body and he is cut adrift while the catcher goes after the next victim. After the day's work a floating factory picks up the raw material for processing. The factory ship is equipped with laboratory, flensing deck, blubber hacker, rotary blubber boiler, oil separators, bone saws, bone boilers, rotary meat boiler, meat-meal sack-filling plant and meat refrigeration plant, besides storerooms, recreation rooms, hospital, and so forth.

The whaling industry reached its peak in the 1930s and 1940s. In 1964, when the industry was already in decline, there still were 357 whale catchers operating from twenty-three floating factories, mostly in the Antarctic, and from thirty-nine shore stations in various parts of the world. The total whale catch was 63,001, including 318 humpback, 372 blue, 19,182 finback, 13,690 sei, and 29,255 sperm whales. The total produce consisted of 371,413 tons of whale oil, 339,045 tons of meat meal, and other products for food and fertilizer.

Clipper ships and whalers revolutionized the way fishermen felt about their occupations, altering the manner in which family and community took part in what had now become an industry. The rapid technological development of the fishing industry was the cause of its own undoing. Ruthless competitive overfishing, combined with pollution of vast stretches of water caused by the industrialization of the land and the misuse of its resources, plus the extension of similar malpractices to ocean space and resources, began to reduce the most important stocks of fish in all the oceans.

In the century from 1850 to 1950 the world fish catch increased tenfold, at an average rate of about twenty-five percent per decade. It doubled again in the decade from 1950 to 1960 and again from 1960 to 1970. Thus the protein supply from the sea would seem to have increased much faster than the world population. In reality, however, things went differently. The quantitative increase of the catch was accompanied by a qualitative change. Eventually it was not so much fish for human consumption that were taken in larger quantities as it was plankton-feeding fish swimming in large, tight schools. These were species which it had previously been impossible to fish economically; species disdained by tradition and taste, such as very fatty fish; and species that had become useful only with the development of industrial processes that could transform them into fish meal for pig and poultry feed and for fertilizer.

About thirty-five percent of the world's total catch now consists of "inferior" fish of this sort. But since it takes about ten tons of fish to produce one ton of livestock, as calculated by Sidney Holt, Director of the International Ocean Institute, clearly the real increase in the amount

WARREN D. JORGENSEN

of animal protein available to the increased world population is far less substantial than the statistics appear to suggest.

People in the rich nations fill their animal protein requirements by eating meat and drinking milk—two-thirds of the world's meat and milk production is consumed by less than one-quarter of the world's population. The other three-quarters depends on fish for the greater part of its animal protein.

Nor are the fish meal products consumed where the need is greatest, in the developing countries. They go to the rich countries, which also have the greater share of the world's fisheries. Over seventy-five percent of the world's total catch is fished by fourteen nations, including the Soviet Union, Japan, Spain, Poland, France, Norway and the United States.

Two dramatic events have contributed to, and are illustrative of, the recent decline of the world's fisheries. One is the shrinking of the whaling industry in the Antarctic and elsewhere; the other is the collapse of the Peruvian anchoveta industry.

As recently as 1937, the various species of baleen whales and sperm whales yielded as much as sixteen percent of the world's total marine catch. By 1970 the cruel depletion of these slowly reproducing creatures had reduced the catch to two percent. It is likely that even this low percentage, corresponding to one million metric tons of whale a year, will disappear over the next decades. It will disappear either because a continuation of overkill will extinguish the species, or because we stop hunting them. There is increasing recognition of the fact that whales, porpoises and other marine mammals have an almost human intellective capacity, and are of great ecological and evolutionary interest as

Oystermen of the Chesapeake Bay

These oystermen, or "watermen" as they are called, sail the beds each winter as they have done for over a hundred years; they are allowed to use power boats on only two days of the week. Scientific aquaculture could greatly increase the yield of oysters.

well. Perhaps both motives—economic failure and humane aspiration—will converge, as they so often do when "progress" is achieved in history, to make whaling a thing of the past.

The fall of the Peruvian anchoveta fishery was as precipitous and unpredictable as had been its rise. From a mere 30,000 tons in 1953, the catch of this small fish in the Humboldt Current off the coast of Peru rose to twelve million tons in 1970. Small coastal villages grew into huge, bustling fishing ports surrounded by shanty towns that tried to absorb wave upon wave of migrants coming from the country in search of work. The number of fishermen increased from 2,500 in 1957 to as many as 20,000 in 1966. Most of them did not have water, sewers, or electricity in their slums. What they had instead was the stench of fish meal from the factories that produced more than 2.3 million tons of it a year. But they could not afford the meal itself.

The small wooden boats—purse seiners—of the traditional Peruvian fishermen lost in the uneven match with the large, light, fast steel boats of the big companies. In a desperate attempt to meet this competition, many of the small shipowners wrecked their boats by overloading, and many lost their lives. To salvage something by collecting their insurance, many others wrecked their boats intentionally.

Then, in 1972, the fish began to disappear. From ten million tons the year before, the catch dropped to a 4.4 million tons, bringing the world fish catch down by ten percent. In 1973 the anchoveta catch went down again, to two million tons, and experts remained pessimistic for 1974. Environmental factors, such as changes in the Humboldt Current and in the temperature of the water, together with excessive fishing, caused the collapse. Scientists from the United Nations Food and Agriculture Organization (FAO) had warned that ten million tons was beyond the sustainable yield, but their warnings had not been heeded.

Birds which had thrived on the anchoveta, including the pelican, are also gone. Millions of them. Their droppings, called guano, which accumulated on some of the offshore islands to a thickness of 150 feet and constituted the world's best and most abundant natural fertilizer, are gone as well. And 20,000 fishermen and 8,000 fish meal workers are jobless.

These are dramatic instances, but they are not isolated. As in many areas of human endeavor, there has been tremendous technological changes in the fishing industry. And, as in many other areas, technological change unaccompanied by corresponding change in the social and institutional structure has not solved problems but has aggravated them, making the rich nations richer and the poor poorer. Without protection, the ocean's living resources, the common property of mankind, cannot survive the technological shift from artisanal to industrial fisheries. Nor will the fish heed man-made laws, laws that expand national jurisdiction over the oceans from the traditional three miles to six, or twelve, or fifty, or even two hundred. Fish migrate across the seas regardless of political boundaries.

Bigger and bigger ships, with more sophisticated and costlier gear, are depleting a supply that once seemed boundless. Irrationality is now a structural part of the system. Instead of reducing the catch or restraining the technologies, irrationality triumphs with even more gadgetry. Hydrophones make high-fidelity recordings of fish noises, which are subsequently analyzed by communications specialists who isolate the bait-eating noises, which are then played back by powerful sound projectors to lure fish over wide distances. Remote telemetering systems listen to fish calls and report suitable concentrations of fish. Completely computerized and automated fishing fleets go into action only when triggered by computers programmed to respond to the telemetric buoys. This is not science fiction. In the near future, as Brenda Horsfield and Peter Bennet Stone report in *The Great Ocean Business,* the computer of a Japanese company will "give instructions to each automatically operated fishing boat, that is, a robot ship. Complete with F.M. automatic fish detector, automatic steering devices, automatic net control systems, and various electronic equipment, the boat rushes to the instructed location. There it determines the specific location, depth, and quantity of the fish involved and automatically selects the most suitable catching method, which would be a fully automated dragging suction pump or an electric fishing net which utilizes the field effect of electric current."

Yet, with all this, the world's fisheries catch less than they could have caught with fewer ships, in fewer hours, and at less expense.

In the salmon fisheries of the United States and Canada, the same annual catch—and total revenue—could be achieved with about $50 million less capital than is currently invested each year. By 1965, the same amount of codfish could have been taken in the North Atlantic with ten to twenty percent less fishing effort than was employed, at a saving estimated at $50 to $100 million a year. The unfortunate Peruvian anchoveta fleet acquired such overcapacity that it could have harvested the equivalent of the annual United States catch of yellowfin tuna in a single day, of salmon in two and a half days. The Peruvian shore factories over-extended their facilities to such a degree that they could have processed the total fish catch of the whole world. The present government has begun to reduce this overcapacity in an orderly and rational way, and, if there is no international competition for the resource, the fishery may gradually be restored.

Another striking example of irrationality is offered by the Mexican shrimp fishery in the Gulf of California. In 1950 there were 100 boats, each catching about 100 tons of shrimp a year, the total yield thus being about 10,000 tons. With the development of this fishery, the total yield rose to 20,000 tons a year by 1960; there it stayed until 1970, when it began to decline. The number of fishing boats, however, increased to 400 in 1960—so that each boat now took only fifty tons per year. In 1970, the number of boats had increased to 800, reducing the catch per boat to twenty-five tons a year. By the middle of the decade the number of boats was still growing and the catch was well below twenty tons per boat per year.

This is obvious economic insanity, as insane as the arms race. And it is easy to predict that controls and limitations agreements for fishing will be just as ineffective as arms control attempts have been so far. Therefore, the whole basic system of harvesting the fruits of the sea must be changed.

The system is in fact changing. The technologies are at hand, and what could be achieved for the benefit of all peoples is so radical that one might call it the Blue Revolution. The scientist-fisherman of this Blue Revolution of sea farming are already at work. If the Revolution succeeds, the fishermen of tomorrow will no longer be hunters in the wild. They will be trained in the marine sciences and under their care the ocean will become a nursery for living resources.

The potential of sea farming is stupendous. If the technological changes arising from it are properly harnessed, the Blue Revolution can generate food for the rapidly growing world population.

Sea farming has historical precedents much deeper in the past than is generally assumed. Oysters were cultivated in the Orient long before the Christian era. In the Western world the first oyster farmers were probably the Romans. The Chinese developed a complex ecological fish-culture system over 1,000 years ago. They introduced a number of different species into an aquatic system, each occupying a different habitat and consuming different food. Thus, six varieties of Chinese carp could coexist in one pond: the grass carp, which consumes the large surface vegetation; two midwater dwellers, one feeding on phytoplankton, the other on zooplankton; and three bottom dwellers, feeding on mollusks, worms and the feces of the grass carp. This kind of ecological cooperation, with the bottom dwellers absorbing the "pollution" of the grass carp, is far more efficient than any monoculture system.

It is interesting that this experience—or perhaps one should call it, this philosophy—goes back so far in Chinese history. It is part of a heritage that makes the Chinese excel today in the design of industrial systems in which one factory uses the waste product of another, and where what they call the "three evils of pollution," namely waste solids, waste fluids and waste gases, are converted into "three advantages"—resources for new production. According to the Chinese, pollution can only be eliminated when waste is eliminated. Their model is organic and it is ancient.

Primitive forms of aquaculture, both in sea water and in brackish water, have been part of the traditional economy of Southeast Asia for many centuries. Farm ponds are constructed by clearing mangrove swamps and diking them with mud slabs. The ponds are then stocked with fry of various kinds, most commonly milkfish, mullet and shrimp. Initially they are fed in a "nursery pond," while a community of algae, bacteria, worms and other plankton is raised in the adjoining ponds. When the fry attain fingerling size, they are transferred to these production ponds, where they mature within a few months or, at most, a year. The average yield of such ponds is about 500 pounds per acre, which compares well with protein production on land.

John Ryther of the Woods Hole Oceanographic Institution made some encouraging extrapolations—assuming that such a simple

The modern tuna industry, active off the coast of Southern California, uses ramps to transport the daily catch from boats to land.

and inexpensive system could be applied on a global scale. "On the basis of my own calculations, I estimate that there are about one billion acres of coastal wetlands in the world. As a standard of comparison, some seven to eight billion acres of earth are now used for food production, with half of that area devoted to agriculture and half to grazing. If only one-tenth of the available wetlands, or 100 million acres, were set aside for aquacultural development, the potential yield, using improved methods of production, would be 100 million tons a year— the equivalent of the potential yield from the world's commercial fisheries. This rate of productivity is particularly impressive given the fact that it can be achieved through a relatively simple process that requires no extraneous feeding and very little labor or capital investment."

Ryther's estimate is realistic—one might even say conservative. According to the calculations of Horsfield and Stone, if it were possible to turn just the 1,000 square miles of Long Island Sound over to mussel culture, this area could produce a quantity of protein equal to three times the total world fish catch. It should be added, however, that the selection of wetlands for fish farming must be preceded by the most careful ecological investigations, for we know that wetlands, especially mangrove swamps, are the nursing grounds of most marine animals that later migrate out to sea, and must not be interfered with.

Some successes have already been recorded. When the oyster was overfished in the nineteenth century and became scarce, the French began farming it. Now the Bay of Arcachon, near Bordeaux, produces about 500 million oysters every year for the European market. Japan, which derives as much as thirteen percent of its total ocean produce from mariculture, has raised the productivity of oysters from 600 pounds per acre under natural conditions up to thirty-two tons per acre under culture—a hundredfold increase.

Green turtles are farmed by Mariculture Ltd. in Turtleland, on the Cayman Islands in the Caribbean. The ten-acre farm consists of a system of about 160 concrete pens and tanks, a huge artificial breeding pond containing a million gallons of sea water, a nesting beach, processing facilities, laboratories and offices. More than 2.6 million gallons of sea water are circulated through the pens and tanks every hour by a network of pipes. According to the company's statements, about 160,000 turtle eggs, partly "home-grown," partly collected from nesting beaches, are hatched every year at Turtleland. Whereas in nature probably fewer than two hatchlings in a thousand survive, Mariculture Ltd. is obtaining a survival rate of about ninety-five percent.

The British Oxygen Company has designed a special tank in which up to seven rainbow trout can be raised per cubic foot of water, and ten gallons a day of fresh water are required to produce one hundred tons of trout a year.

Kelp and seaweed are being cultivated for human consumption—albeit as condiments— and for a growing number of industrial and pharmaceutical uses. Carp, mullet and milkfish, plaice and whitefish, trout and salmon, shrimp, squid and abalone are already being farmed. And about 3,000 tons of farmed sea urchin roe were consumed in Japan last year.

A most beautiful and complex system, inspired by the ancient Chinese polyculture, has been devised by John Ryther at Woods Hole. Here, sewage from a nearby secondary treatment plant is used to grow plankton algae, which in turn provide food for shellfish, principally oysters. The algae remove nutrients such as ammonia, nitrate and phosphate from the sewage effluent, and the oysters remove the algae from the water. The oysters return some of the nutrients to the water in the form of excreted wastes. These are consumed by seaweeds, especially sea lettuce, which are added to the system for this purpose. The seaweeds are then fed to abalone. The oysters' solid wastes, which drop to the bottom of the tank, are eaten by sand worms, which are then circulated to a neighboring tank to serve as food for flounder.

The products of this continuous culture system are a primary crop of oysters, side crops of seaweeds, worms, flounder and abalone, and, ultimately, clean water which is returned to the sea. The sea farm in fact becomes a tertiary (or biological) sewage treatment plant. If implemented on a large scale, Ryther concludes, "such a system would be capable of producing an annual crop of one million pounds of shellfish meat from a one-acre production facility and a fifty-acre algae farm using effluents from a community of 11,000 people. The potential yield of world-wide aquaculture, based on the simplest improvements, is already an impressive 100 million tons of food. By adopting advanced culture techniques such as that developed at Woods Hole, the yield could well be multiplied tenfold within the next three decades."

Of course, there are as many problems as there are fish farms. On the biological side there are parasites, blights, epidemics. On the economic side there is the relatively high cost per unit in many cases. But the basic principle is the same throughout: to take advantage of the lavishness of nature and raise the survival rate of eggs and larvae—tenfold, fiftyfold, a hundredfold. Once the fish have been raised to the critical fingerling size, they can be released in a natural environment, provided it contains sufficient nutrients, whether natural or introduced by man.

We are only at the beginning of this development, only on the margins of the ocean. But the Blue Revolution need not stop at the water's edge. It can be carried into the deep seas. Whole ocean areas—the Mediterranean and the Caribbean—could be turned into fish ponds. The productivity of the water could be multiplied. A Soviet scientist, Boris Bykhovsky of the Academy of Sciences, has been working on schemes to improve the productivity of wide ocean areas by controlling the transfer of substances and energy. He thinks that new varieties of single-celled algae could be developed which would utilize the sun's energy more effectively.

Thus, even migratory fish could be "cultured"—raised to viable size in the desired quantity in hatcheries—and then released into wide-open fertilized spaces where there is as yet less competition with other users of ocean space than in the congested coastal areas.

The present pattern of distant-water fishing is likely to change under the impact of these developments—combined with a number of other factors, such as the extension of national sovereignty over wide coastal zones and the increase in fuel costs. Fisheries may develop in three new directions. First, fish farming of the kind just described. Second, the catch of very small organisms, such as plankton and krill, may be intensified, and these may be processed into food for human consumption. The Soviet Union is pioneering in developing such processes. The abundance of Antarctic krill alone is such that its systematic harvesting and processing could more than double present food production from all the oceans. Third, new mid-water trawling technologies, combined with new food-processing methods, make available for human consumption a wealth of mid-water creatures. One of these alone—squid—also exceed in quantity the total present world fish catch. In these technologies the Soviet Union again leads at this time.

In the light of such possibilities, does it not seem atavistic to continue to exterminate the whale, or even the anchoveta? Does it not seem futile to haggle over boundaries? The ocean is a polyculture, the manifold components of which must be understood and managed. Its uses, now conflicting, can instead be integrated. We have the tools, such as systems theories and computer sciences, to do it. We can turn wide ocean stretches into fish farms and manage them cooperatively, not competitively—some on a local, some on a regional, some on a global basis. Obviously there are great problems in projects of such scale, which require so much capital, technical skill, space and labor. It is a huge task, but it is a positive task. And we can do it if we do it together.

The Very Deep Did Rot

Samuel Taylor Coleridge

The possibilities of harvesting the sea will eventually be doomed to failure unless pollution of the oceans is controlled. In 1971 Jacques Cousteau sounded the alarm that the sea is dying. A century and three-quarters earlier, the English poet Samuel Taylor Coleridge gave us a lasting image of a dying sea in his famous poem, "The Rime of the Ancient Mariner," excerpted below.

The very deep did rot: O Christ!
That ever this should be!
Yea, slimy things did crawl with legs
Upon the slimy sea.

About, about, in reel and rout
The death-fires danced at night;
The water, like a witch's oils,
Burnt green, and blue, and white.

How to Kill an Ocean

Thor Heyerdahl

In the last ten years, a few disastrous tanker accidents and a leak under an oil platform off the coast of Santa Barbara, California have focused world attention on the problems of pollution. But such incidents account for only a small part of the growing world-wide pollution problem, as the noted Norwegian voyager Thor Heyerdahl points out in the following excerpt from an article in Saturday Review. *In 1947 Heyerdahl sailed a balsa-wood raft from Peru to Polynesia to test his theory that South American Indians had settled the islands of Polynesia (see below, p. 323). In 1970 he sailed a papyrus boat from Morocco to the West Indies, this time to test a theory that the ancient Egyptians could have sailed to the New World. His findings on the startling increase in pollution in the intervening decades is summarized below.*

My own transoceanic drifts with the *Kon-Tiki* raft and the reed vessels *Ra I* and *II* were eye-openers to me and my companions as to the rapidity with which so-called national waters displace themselves. The distance from Peru to the Tuamotu Islands in Polynesia is 4,000 miles when it is measured on a map. Yet the *Kon-Tiki* raft had only crossed about 1,000 miles of ocean surface when we arrived. The other 3,000 miles had been granted us by the rapid flow of the current during the 101 days our crossing lasted. But the same raft voyages taught us another and less pleasant lesson: it is possible to pollute the oceans, and it is already being done. In 1947, when the balsa raft *Kon-Tiki* crossed the Pacific, we towed a plankton net behind. Yet we did not collect specimens or even see any sign of human activity in the crystal-clear water until we spotted the wreck of an old sailing ship on the reef where we landed. In 1969 it was therefore a blow to us on board the papyrus raft-ship *Ra* to observe, shortly after our departure from Morocco, that we had sailed into an area filled with ugly clumps of hard asphalt-like material, brownish to pitch black in color, which were floating at close intervals on or just below the water's surface. Later on, we sailed into other areas so heavily polluted with similar clumps that we were reluctant to dip up water with our buckets when we needed a good scrub-down at the end of the day. In between these areas the ocean was clean except for occasional floating oil lumps and other widely scattered refuse such as plastic containers, empty bottles, and cans. Because the ropes holding the papyrus reeds of *Ra I* together burst, the battered wreck was abandoned in polluted waters short of the islands of Barbados, and a second crossing was effectuated all the way from Safi in Morocco to Barbados in the West Indies in 1970. This time a systematic day-by-day survey of ocean pollution was carried out, and samples of oil lumps collected were sent to the United Nations together with a detailed report on the observations. This was published by Secretary-General U Thant as an annex to his report to the Stockholm Conference on the Law of the Sea. It is enough here to repeat that sporadic oil clots drifted by within reach of our dip net during 43 out of the 57 days our transatlantic crossing lasted. The laboratory analysis of the various samples of oil clots collected showed a wide range in the level of nickel and vanadium content, revealing that they originated from different geographical localities. This again proves that they represent not the homogeneous spill from a leaking oil drill or from a wrecked super-tanker, but the steadily accumulating waste from the daily routine washing of sludge from the combined world fleet of tankers.

Netting Oil   M. Blumer

At about the same time that Heyerdahl encountered oil lumps on his Ra *voyage, oceanographers pulling skimming nets found similar evidence of pollution. The findings of one research vessel in the Sargasso Sea are reported by M. Blumer, senior scientist in the department of chemistry at Woods Hole Oceanographic Institute, in the following excerpt from* Oceanus. *The Sargasso Sea is an area of the North Atlantic with very slow-moving currents and abundant patches of floating seaweed that have both terrified and fascinated voyagers since the time of Columbus.*

Oil pollution is the almost inevitable consequence of the dependence on a largely oil-based technology. The oil reserves which have accumulated in the earth during the last 500 million years will be exhausted within a few hundred years. The use of oil without loss is impossible; losses occur in production, transportation, refining and use. The immediate effects of large scale spills in coastal areas are well known, but only through the recent introduction of skimming nets have we become aware of the degree of oil pollution of the open ocean. Thus, during a recent cruise of our R/V *Chain* to the Sargasso Sea, many surface . . . net hauls were made to collect surface marine organisms. These tows were made between 32°N–23°N latitude (corresponding to a distance of 540 miles) at longitude 67°W. During each tow, quantities of oil-tar lumps, up to 6 cm in diameter were caught in the nets. After 2–4 hours of towing the mesh became so encrusted with oil that it was necessary to clean the nets with a strong solvent. On the evening of 5 December 1968, at 25°40′N, 67°30′W, the nets were so fouled with oil and tar material that towing had to be discontinued. It was estimated that there was 3 times as much tar-like material as Sargasso weed in the nets. Similar occurrences have been reported worldwide.

A Realistic Look at Ocean Pollution

Bostwick Ketchum

The pollution of the ocean is of particular concern in the coastal zone, where the concentration of people and of industry leads to conflicting demands for navigation, waste disposal, fisheries resources, and recreation. Yet the problem of pollution is itself a complex one, involving such diverse factors as the source of pollutants, their quantity, degree of toxicity, and persistence in the environment. In the following article, Bostwick Ketchum, associate director of the Woods Hole Oceanographic Institution in Massachusetts, discusses these various aspects of marine pollution. This overview was originally presented to the United States Senate Subcommittee on Oceans and Atmosphere on July 12, 1973 and was reprinted in the Marine Technology Society Journal.

*I*t is unnecessary to elaborate before this committee the value of the coastal zone and its marine resources for the benefit of mankind. As you well know, nearly half of the population of the United States lives within the regions adjacent to our coastal waters or the shores of the Great Lakes. Historically, the reason for this concentration of population in the coastal zone has been the ease of transportation of materials and people by marine shipping.

Because of the density of people and of industry in this narrow strip of our land, the waters have long been used for the disposal of the waste products of our population and our technology. So long as population densities were low, the inshore ocean waters were able to recycle or recover from the added pollution. In many areas this is no longer true, and serious deterioration of water quality has resulted.

The oceans have also been a valuable source of food, particularly of the animal protein so essential in human nutrition, ever since our forefathers discovered and settled upon the shores. In many parts of our coastal zone indiscriminate waste disposal has depleted our fisheries resources in dramatic ways. Over 90 percent of the total harvest of seafood taken by American fishermen comes from estuaries or the waters over the continental shelf. About two-thirds of

that harvest consists of species whose existence depends upon the estuarine zone or which must pass through the estuary enroute to their spawning grounds. The salmon which used to abound in our Northeast rivers are excluded from practically all of them today, either because of the construction of dams or because of the pollution of the water itself. Many productive shellfish grounds have been closed because of pollution, and our inshore fisheries resources are less abundant, which is due, in part at least, to overfishing.

Today, there are increasing demands upon the coastal zone for many of man's activities. Navigation, disposal of pollutants, and fisheries resources remain important, but the recreational demands of our population are increasing dramatically. In the coastal waters these traditional uses are in conflict with many of the amenities and more personal uses which our population rightfully feels should be maintained for future benefits.

By hindsight we can evaluate what man has done in the past, but we do not understand well enough the ways in which the marine ecosystem works to predict the results of new or proposed engineering developments. I think that it is clear that technology exists which would prevent or ameliorate the impact of marine pollution upon the environment, but we must apply this technology and be willing to pay the cost of correcting past errors and of preserving our marine environment for the benefit of future generations.

It is worth emphasizing that large parts of our coastal zone are still relatively unmodified by man's activities and that severe deterioration has been localized in areas of large population densities. It is imperative that the natural areas be preserved in their unmodified state even as we strive to improve the quality of areas which have been degraded. . . . The Coastal Zone Management Act of 1972 recognized the need for estuarine sanctuaries which should be preserved in a natural state to permit continuing research on the ecological relationships within the area. Additional recognition of the need has also been provided in the Marine Protection, Research and Sanctuaries Act of 1972. Such sanctuaries would preserve and protect the genetic stocks of plants and animals essential for the perpetuation of the marine ecosystem.

Pollutant Characteristics

There are three characteristics of each pollutant which must be understood before one can evaluate the possible impact or hazard of its release to the environment. These are a) the quantities produced which may reach the environment, b) the toxicity of the pollutant to marine organisms and to man if it will reach him in his seafood, and c) the persistence of the pollutant in the environment. Our knowledge is incomplete; sometimes with regard to all three of these essential characteristics of pollution. . . . It is obvious that a highly toxic pollutant which reaches the environment in very small quantities may be far less important than a less toxic material which is released in massive quantities.

General categories of materials may be cited as examples of the range to be expected among these characteristics. At one extreme are the heavy metals which are produced in large quantities, are toxic at low concentrations and do not degrade biologically or chemically, even though they may be trapped in the sediments and thus

removed from the water column reducing their impact. Once added to the marine environment, however, they are there forever. Various synthetic organic compounds, particularly the chlorinated hydrocarbons such as DDT and the polychlorinated biphenyls have also been produced in large quantities and are now found even in the water and organisms of the open ocean. These compounds are not produced naturally and organisms have not evolved an ability to decompose or degrade them rapidly. We are still uncertain as to their persistence in the marine environment, but the available evidence suggests that they would be found for decades, perhaps centuries, even if all further additions to the environment could be stopped.

Oil pollution is an increasing threat to the marine environment because of our ever-increasing demands for energy and the increased sea transport of oil in tankers. Mortality of marine organisms, sometimes extensive, has been found wherever accidental oil spills have been studied, and recovery from the high concentrations produced in these accidental spills may take months, years or decades depending upon the amount of oil spilled and how rapidly the oil is dispersed and diluted to non-toxic concentrations. Domestic pollution consists of the natural products of human metabolism and can be rapidly decomposed by natural marine processes. There are two critical problems connected with the disposal of domestic pollution to the marine environment, however. First, in many of our estuaries the quantity which must be discharged exceeds the receiving capacity of the body of water to which it is added, and second, many sewage effluents contain toxic materials from industrial additions or from urban runoff. Proper treatment methods can remove most of these from the effluent, but they will remain in the sludge which also must be disposed of in some manner. The disposal of solid wastes of our civilization also poses increasing problems. Even though much of the solid waste is non-toxic, large quantities are involved. Some of this material is being disposed of in our coastal waters, such as the sewage sludge and dredging spoils disposal in the New York Bight and other areas. These substances have clearly caused deterioration of the environment where they are dumped. Each of these categories of pollutants will be discussed in greater detail below.

First, however, it may be desirable to comment in general upon the term toxicity, which is difficult to define and not always well understood. Any substance on earth is toxic if the concentration is great enough in the wrong environment. For example, a characteristic of the marine environment is the salt content of the water, but most marine organisms can survive only within a narrow range of salinity. Estuarine organisms, accustomed to brackish water, may not be able to survive in the open sea where the salinity is higher. Even pure water can be considered toxic since marine organisms cannot survive in fresh water and man cannot survive for long in pure water which is a few inches above his nose. We are all accustomed to temperature fluctuations, but organisms cannot survive in an environment that is either too hot or too cold. The substances which are of greatest concern, however, are those which are lethal in concentrations of parts per million or less. The toxic heavy metals, the chlorinated hydrocarbons and the petroleum hydrocarbons are toxic at these low concentrations, and it is because of this that they are of greatest concern.

Heavy Metals

A list of 11 heavy metals which are toxic to marine organisms and which are reaching the environment in considerable quantities is presented in Table 1. All elements reach the marine environment in varying amounts as a result of the weathering of the continents and their transport by the rivers to the estuaries and ultimately to the sea. Many elements are also present in coal and petroleum and are released in varying

Table 1: Toxic Elements of Critical Importance in Marine Pollution Based on Potential Supply and Toxicity, Listed in Order of Decreasing Toxicity

Element	Rate of Mobilization (10^9 g/yr)* A(man) Fossil Fuels	B(natural) River Flow	C Total	Toxicity † D mg/1	Relative Critical Index (10^{12} liters/year.) A/D	C/D
Mercury	1.6	2.5	4.1	1×10^{-4}	16,000	41,000
Cadmium	0.350**	?	3.0**	2×10^{-4}	1,750	15,000
Silver	0.07	11	11.1	1×10^{-3}	70	11,100
Nickel	3.7	160	164	2×10^{-3}	1,350	82,000
Selenium	0.45	7.2	7.7	5×10^{-3}	90	1,540
Lead	3.6	110	113.6	1×10^{-2}	360	11,360
Copper	2.1	250	252.1	1×10^{-2}	210	25,210
Chromium	1.5	200	201.5	1×10^{-2}	150	20,150
Arsenic	0.7	72	72.7	1×10^{-2}	70	7,270
Zinc	7	720	727	2×10^{-2}	330	36,350
Manganese	7.0	250	257	2×10^{-2}	350	12,850

* After K.K. Bertine and E.D. Goldberg, "Fossil Fuel Combustion and the Major Sedimentary Cycle," *Science,* 173: 233–235 (except for fossil fuel production of cadmium).

**Value from Third Annual Report, Council on Environmental Quality, 1972. The total includes addition to soil and thus may be an overestimate.

† Water Quality Criteria: Concentration considered to pose minimal risk of deleterious effect. After Waldichuk, NAS, 1972.

amounts to the atmosphere as we burn these fossil fuels. Estimates of the rates of supply from these two sources are given in Table 1, which also lists the toxic concentrations in sea water. A relative critical index is computed by dividing the rate of supply by the toxic concentration. Actually, this index gives the volume of sea water (in cubic kilometers) which would receive an annual increment of the element equal to the listed toxic concentration. The concentration of the element already present in sea water is not taken into consideration because, in some cases, the concentration is so low that it would not modify the calculation, and in some cases the absolute concentration in the sea water is not accurately known.

To put this index into perspective, the area of ocean surface so modified can also be computed. The mixed layer of the ocean is of the order of 100 meters in depth and the area of ocean which could be contaminated to this depth by mercury, for example, as a result of the combustion of fossil fuel would be equivalent to 160,000 square kilometers or 61,500 square miles. This area is almost equal to the land area of the state of Washington . . . It is clear that we are not talking about an insignificant problem for mercury, which presents the most serious hazard, or even for some of the other elements which would affect smaller volumes of water or areas of the sea. Another perspective is given by computing the time it would take to add this concentration of material to all of the water in the ocean. Using mercury again as an example, it would take 10,000 years to contaminate all of the oceanic waters to the indicated level of toxicity. This is not a very meaningful calculation, however, because the oceans are not uniformly mixed and the concentration at the locality where the pollutant is introduced will inevitably increase more rapidly than the average for the whole ocean.

Chlorinated Hydrocarbons

A wide variety of synthetic organic chemicals are also reaching the environment, particularly the chlorinated hydrocarbons such as DDT (and its decomposition products) and polychlorinated biphenyls (PCB's). These are not readily biodegradable, and the ocean is the ultimate sink for such compounds. . . .

[G.R.] Harvey [and others] . . . found substantial concentrations of DDT and its breakdown product, DDE, . . . and even higher levels of PCB's . . . in a variety of organisms collected from the open sea many miles from land, confirming the probability of atmospheric transport. . . .

A variety of synthetic organic chemicals, including other pesticides, detergents and pharmaceuticals are also undoubtedly reaching the marine environment with impacts which are virtually unknown. The detrimental effect of DDT on bird breeding potential is well documented and some experiments have been done on a few forms of marine life, but the information is still inadequate for a complete evaluation of the impact on the marine biota of DDT and even less adequate for the other synthetic organic compounds.

Oil Pollution

Petroleum, including crude oil, refined products and petrochemicals are now polluting the sea in large amounts. Revelle [and others★] . . . estimated the total direct oil pollution of the oceans to be 2.2 million tons annually. The sources were accidental spills, tanker operations, other ship operations, offshore production, refinery operations and industrial and automotive wastes. . . . Although accidental oil spills, such as the grounding of the *Torrey Canyon* or the Santa Barbara oil well blowout, are spectacular events and attract the most public attention, they actually contribute less than 15 percent of the total amount of oil entering the marine environment annually.

Numerous studies of toxicity and effects of oil pollution have been made, but more careful studies of selected fractions of this complex mixture of hydrocarbons are needed. It is ap-

---

★ See below, p. 200.

Domestic pollution is a major problem in coastal areas.

parent that these hydrocarbons are degraded in sea water, but little is known about the rate of turnover of this material in the marine environment. Extended studies of the spills of refined fuel oils from the *Tampico Maru* in Baja, California, and from the *Florida* in Falmouth, Massachusetts, have shown that it has taken several years for partial recovery of the biota and the data suggest that it may take a decade or more for complete re-establishment of the natural community. The initial impact of crude oil spills, such as that from the *Torrey Canyon,* seem to be less severe and the recovery more rapid. These spills generally occur in more open waters and the oil is thus more widely dispersed. There is apparently little measurable direct effect of chronic, low-level (sublethal) contamination, such as has occurred off the coast of Louisiana for several decades.

Our technology is based upon an expanding energy use which will require additional petroleum supplies including those from submarine reservoirs and increasing amounts transported in tankers from distant oil fields. If the rate of loss in our utilization and transportation of oil cannot be radically decreased by application of adequate controls wherever possible, the amount of petroleum hydrocarbons entering the sea will increase. Revelle [and others] . . . estimated that the increase would be in direct proportion to the total world production of oil without adequate controls.

Because of the increase in oil tanker traffic and of ships burning fuel oil and the resultant pollution of the high seas by oil, this has become an international problem. The Intergovernmental Maritime Consultative Organization (IMCO), a special agency of the United Nations, convened an international conference in London in 1954 which drew up the International Convention for the Prevention of Pollution of the Sea by Oil. This came into force in July 1958 and was subsequently amended by an IMCO-convened conference in 1962. Further resolutions provide for the prohibition of deliberate oil discharge from ships at sea and for the establishment of an international compensation fund for oil pollution damage. An additional conference is scheduled for the fall of 1973 to consider further regulations controlling oil pollution.*

Domestic Pollution

Human wastes are also added to the marine environment and can cause difficulties when added in excessive amounts. Sewage treatment plants have been designed primarily to reduce the organic material in the effluent which, when it decomposes in the environment, causes oxygen depletion. Complete removal of the dissolved oxygen in the water makes it impossible for most marine organisms to survive and also results in the production of hydrogen sulphide, a toxic, malodorous gas. Even when the treatment methods are successful in preventing excessive oxygen depletion, the fertilizing elements remain in the effluent and can stimulate excessive growth of objectionable plant populations in the process known as eutrophication. This plant growth can produce as much organic material as was removed at considerable cost in the treatment plant, thus partly defeating the purposes of sewage treatment.

Because of the biodegradability of domestic pollution, it is not persistent in the environment, except for the added fertilizing elements. For the oceans as a whole, therefore, domestic pollution is not a significant problem, but it is important in more confined areas where the density

of the human population is high and the recovery capacity of the system is limited. I have computed, for example, that the population of metropolitan New York discharges into the Hudson estuary about five to ten times as much domestic pollution as the system can recycle without an adverse impact. Fortunately, the mixing in the Hudson estuary is vigorous and by the time the water leaves the Harbor and enters the New York Bight the dilution is sufficient to reduce the concentrations of nutrients to acceptable levels.

A corollary of the fertilizing effects of domestic pollution is the fact that it could be used beneficially to stimulate the productivity of the sea if it were discharged within the limits of the receiving capacity of the ecosystem. This must be carefully done to avoid unfortunate side effects but it is possible, theoretically at least, to use this type of pollution for beneficial purposes rather than to dispose of it in ways that cause deleterious effects.

Solid Wastes

Solid waste disposal has become one of the most urgent and difficult problems in crowded urban centers. The types and amounts of waste materials dumped at sea in the coastal waters of the United States in 1968 is presented in Table 2. Nearly 50 million tons of waste material was dumped in United States' coastal waters, most of which was dredging spoils resulting from channel and harbor development. The Council on Environmental Quality estimated that 34 percent of these dredging spoils could be considered polluted. Pearce* has presented data to show that both the polluted dredging spoils and the sewage sludge from waste treatment plants which has been dumped in the New York Bight have caused damage to the bottom dwelling populations in the area. The Marine Protection, Research and Sanctuaries Act of 1972

Ed. note: In November 1973 this conference approved an International Convention for the Prevention of Pollution from Ships, controlling ship-generated pollution by all kinds of oil and prohibiting all discharges within fifty miles of land.

Ed. note: J.B. Pearce, *The Effects of Waste Disposal in the New York Bight,* Interim Report, Sandy Hook Marine Laboratory, U.S. Bureau of Sport Fisheries and Wild Life.

regulates the transportation and dumping of materials into the oceans, coastal zones and other waters. A permit system is established to be administered by the Army Corps of Engineers for dredging and filling and by the Environmental Protection Agency for all other purposes.

Ocean dumping is also a subject of international concern. An intergovernmental conference convened by the United Nations was held in London, 30 October to 10 November, 1972. A Convention on the Dumping of Wastes at Sea was adopted . . . [which] prohibits the dumping of some materials; requires a special permit for the dumping of other identified substances; and a general permit for all other substances.

Summary

In an effort to summarize in one figure the various aspects of marine pollution which I have been discussing, I submit a diagram [see figure] . . . showing the various processes which must be understood in order to evaluate the marine ecosystem and its capacity to accept and recycle various types of pollutants. If the diagram appears to be complicated, it is because we are discussing a complex problem.

Many, but not all, of the entries in this diagram have been discussed above. One must know the source and the amount of the specified pollutant and the routes by which this material reaches the sea, whether by runoff, discharge or by atmospheric transport. It would be desirable to have a screening mechanism established to evaluate the possible impact of new chemicals, hundreds of which are being produced annually. Where the hazard is great, these chemicals should be recycled and not permitted to enter the environment. Once the pollutant does enter the environment, its impact on the marine ecosystem and on the communities, populations and organisms which live in the ocean need to be evaluated. It is important to appreciate the fact that the impact need not be direct and immediate by causing the death of organisms but can have more subtle, sublethal effects which influence the survival or behavior of the organisms. The various chemical and geological processes which need to be considered in terms of each pollutant are listed under the "non-living" category and it is, furthermore, important to know whether two or more pollutants introduced simultaneously will augment each other's impact (synergism) or will interfere with each other's impact (antagonism). Only when the complex nature of the marine ecosystem, and the various processes taking place there, are understood can one evaluate the possible receiving capacity of a given system for a given pollutant.

Naturally, it is of concern to evaluate whether or not the pollutant can return to man in the seafood that he needs for his nutrition. Also, the

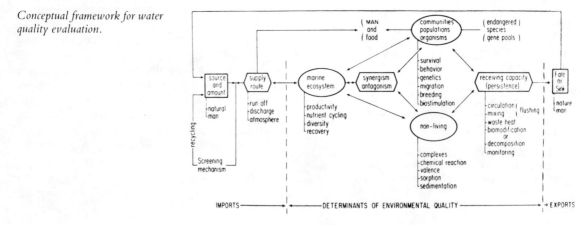

Conceptual framework for water quality evaluation.

Table 2: Ocean Dumping: Types and Amounts, 1968 (in tons)
(from the Council on Environmental Quality)

Waste Type	Atlantic	Gulf	Pacific	Total	Percent of Total
Dredge spoils	15,808,000	15,300,000	7,320,000	38,428,000	80
Industrial wastes	3,013,200	696,000	981,300	4,690,500	10
Sewage sludge	4,477,000	0	0	4,477,000	9
Construction and demolition debris	574,000	0	0	574,000	<
Solid waste	0	0	26,000	26,000	<
Explosives	15,200	0	0	15,200	<
TOTAL	28,887,400	15,966,000	8,327,300	48,210,700	100

impact on endangered species can be of special concern. An example of this type of impact is the mortality of many birds which results from oil spills. Dead, oiled birds on the beaches constitute the most immediate and obvious effect of oil spills, and this mortality can have a serious impact at certain times of year if a major part of the population of a given species is present in the area either for breeding or during their migratory passage.

In conclusion, I would like to emphasize that the only ultimate solution to the problems of pollution is to recycle the materials that we produce and use in our civilization. Discharging these materials to the environment is both wasteful of resources and causes undesirable impacts upon the marine ecosystem. Complete recycling is probably an unattainable goal, but certainly every effort should be made to recycle materials as completely as possible, not only to conserve our natural resources, but also to prevent the further deterioration of our environment, which we must preserve for the ultimate benefit of man.

Oceanic Oil Pollution

Roger Revelle, Edward Wenk,
Bostwick Ketchum, Edward Corino

Oil pollution is not confined to coastal areas; it poses an eventual threat to the ecosystems of the oceans of the world. Furthermore, as consumption of oil increases in our ever-expanding technological society, the problem of oil pollution is also likely to increase. In the following selection, Roger Revelle and three other experts analyze the extent and character of oil pollution—in which tanker accidents and offshore leaks play a relatively small part—and they suggest possible courses of action to control the problem. Revelle is director of the Harvard Center for Population Studies and former director of Scripps Institution of Oceanography; Wenk, a specialist in ocean engineering and public affairs, is a professor at the University of Washington; Ketchum is associate director of Woods Hole Oceanographic Institution; and Corino is with the Esso Research and Engineering Company.

*A*t the present time, the most conspicuously detrimental effects of oil pollution of the ocean are localized in extent and are caused by accidental spills in near-shore areas. These loci of concern, however, potentially include the coastal zones of every continent and every inhabited island so that the problem of accidental spills is of worldwide significance. Projections of future growth in ocean transport and offshore production of petroleum indicate that both the frequency and the damaging effects of local accidents are likely to increase.

Although accidental oil spills cause the most evident damage to ocean resources, they make up a small percentage of the total amount of oil entering the marine environment. At least 90 percent of this amount originates in the normal operations of oil-carrying tankers, other ships, refineries, petrochemical plants, and submarine oil wells; from disposal of spent lubricants and other industrial and automotive oils; and by fallout of airborne hydrocarbons emitted by motor vehicles and industry. The extent and character of the damage to the living resources of the sea from this "base load" of oil pollution is little known or understood. In the long run it could be more serious, because more widespread, than the localized damage from accidental spills.

The magnitude of oceanic oil pollution is likely to increase with the worldwide growth of petroleum production, transportation, and consumption. World crude oil production reached 2 billion tons per year in 1969, and production of 3 billion and 4.4 billion tons per year is predicted for 1975 and 1980, respectively.

Sources of Petroleum Hydrocarbons in the Sea

Petroleum hydrocarbons enter the sea:

1. Directly

   a. in accidental spills from ships, shore facilities, offshore oil wells, and underwater pipe lines;

   b. from tankers flushing oil tanks at sea;

   c. from dry cargo ships cleaning fuel tanks and bilges;

   d. from leakage during normal operation of offshore oil wells;

   e. from operation of refineries and petrochemical plants;

   f. in rivers and sewage outfalls carrying industrial and automotive wastes; and

2. As "fallout" from the atmosphere, probably as particles or in rain.

We shall consider all these sources except accidental spills as constituting the base load of oil pollution in the sea.

Accidental Oil Spills

At present, the average annual influx to the ocean from accidental oil spills throughout the world is probably about 200,000 tons. Most of these spills are relatively small. Out of 714 recorded accidental spills in U.S. waters in 1968, approximately half were from ships and barges, most of which were docked at the time of the accident. About 300 spills occurred from shore facilities of various types, and a few resulted from ships dragging anchor across submarine pipelines in bays.

Even under carefully controlled conditions accidental oil spills in port are negligible. Milford Haven, a relatively new British oil port, is adjacent to a national park, and great efforts have been made to control and prevent oil pollution. In 1966 the annual turnover at Milford Haven was 30 million tons with losses amounting to 2,900 tons or 0.01 percent of the total amount handled.

Accidental oil spills resulting from stranding or collision of large tankers and from accidents to offshore drilling or producing wells deservedly attract much public attention because of the extensive damage done to beaches, recreational areas, and harbors. The wreck of the *Torrey Canyon,* which discharged 118,000 tons of crude oil in the sea, is the best known example although somewhat smaller tanker wrecks have occurred elsewhere, such as off Nova Scotia and Puerto Rico. All large accidental spills to date have occurred fairly near shore, and the spreading sheet of oil has drifted or has been blown by winds onto beaches and into shallow water areas. Present efforts to contain and to dispose of the oil before it does extensive damage have been singularly ineffective. Agents such as talc, clay, and carbonized sand have been used to sink the oil. Various dispersing agents have been developed which break up the oil into minute droplets that are subsequently dispersed throughout the water. Earlier versions of these chemical dispersants were more toxic than the oil, but a number of essentially nontoxic dispersants are now available. Even with a nontoxic dispersant, dispersed oil is more toxic to marine life than an oil slick on the surface, primarily because of its increased availability to the organisms. With all our vast inventory of chemical agents, the best and safest means of disposal is apparently still absorption on chopped straw, if conditions permit.

The danger of large-scale accidents is increasing with the increasing size of tankers. Four 327,000-ton ships are already in operation; vessels of 500,000 dead weight tons will soon be constructed, and 800,000-ton vessels have been projected within the next few years. These monster ships have so much draft and inertia and are so difficult to handle that a stranding or collision is more likely to result in a destructive wreck than with smaller ships. A loss of one of the new large tankers under conditions where it would be impossible to off-load the oil would add around 20 percent to the amount of petroleum entering the oceans in a single year.

Although handling difficulties increase with size, the increase is not directly proportional to size. Moreover, larger ships means fewer ships, and, therefore, traffic can be considerably reduced. Fewer ships also means crews can be limited to highly qualified personnel, and they can be better trained. The larger tankers could also afford to install highly sophisticated navigation gear which might be prohibitively expensive for the many smaller ships.

Spectacular "blowouts" from offshore oil well drilling and production make up a surprisingly small fraction of the total influx of oil to the ocean environment. For example, the widely publicized Santa Barbara blowout has so far produced only between 3,000 and 11,000 tons of oil. Similarly, the accident to a producing well off the Louisiana coast, which began on February 10, 1970, and lasted until the end of March, released only about 4,300 tons of oil. These figures emphasize the enormous amount of damage that can be done by a relatively small amount of oil concentrated over a relatively small, previously uncontaminated area. With present drilling and production technology, accidents of this kind are nearly inexcusable. Preventing them depends on institutional changes, not technical ones.

Sources of the Base Load of Oil Pollution in the Sea

Most oil production occurs at some distance from processing and marketing areas and consequently much crude oil is transported in ocean-going tankers. In 1969, 1.3 billion tons, or about 65 percent of total oil production, was carried in tankers. Projections by the U.S. Department of Transportation indicate that the amount of oil moved by tankers will increase to 2.8 billion tons by 1980.

Normal tanker operations (ballasting, tank cleaning) were estimated to have introduced 530,000 tons of oil to the sea in 1969. Eighty percent of the world fleet used control measures ("Load on Top" or LOT). If LOT were practiced faithfully, these ships would contribute only 30,000 tons of the total losses compared to 500,000 tons from the 20 percent not using such measures. If LOT were used on all tankers, only 56,000 tons would be expected to be lost to the ocean through normal operations in 1975 and 75,000 tons in 1980. If 20 percent of the fleet continued to operate in the present fashion, total losses in 1975 and 1980 would be 800,000 and 1.06 million tons, respectively.

Nontankers, dry cargo ships of greater than 100 gross registered tons, are estimated to have discharged 500,000 tons to the ocean in 1969, primarily from pumping bilges and cleaning operations. This estimate is of low reliability because available data are very limited. The total amount, however, is comparable to that generated by the tanker fleet.

Offshore oil production is estimated to discharge during normal operations about 100,000 tons per year. At present, offshore production accounts for about 16 percent of total crude production. This percentage is expected to increase in the future, as new underwater fields are discovered and new technology permits extension of drilling and production into deeper water. Estimates of losses for 1975 are 160,000 to 320,000 tons and for 1980 are 230,000 to 460,000 tons. The smaller figures assume that offshore production will continue to represent 16 percent of world production, and the larger figures assume 32 percent. In both cases the assumption is made that no improvement in pollution abatement will occur. Many of the new wells will be drilled off the coast of nations that do not have the technological capabilities to enforce good drilling and production procedures or to deal with massive spills.

About 300,000 tons of oil are lost to the sea each year through normal operations of refineries and petrochemical plants. This estimate is based on extensive data from the American Petroleum Institute and private surveys by refineries and industry organizations. With present pollution control measures this figure could grow to 450,000 tons in 1975 and 650,000 tons in 1980. If some improvements in pollution control are made, as predicted by the U.S. Federal Water Quality Administration, oil lost to the sea from refineries and petrochemical plants could drop to 200,000 tons in 1975 and 440,000 tons in 1980.

Industrial and automotive waste oils and greases constitute a significant source of oil pollution in the marine environment. These include all petroleum products, except fuel, used and discarded in the operation of motor vehicles and industrial production, for example, spent lubricants, cutting and hydraulic oils, coolants, and solvents. Much of the disposal of these wastes occurs by dumping on land. An estimate of the quantity eventually finding its way into the ocean can be made from measurements of the hydrocarbon concentrations in river waters, multiplied by the total river discharge, plus the amounts contributed by sewage treatment plants which discharge directly to the oceans. . . . Rivers discharge . . . approximately 150,000 tons of hydrocarbons annually . . . to the oceans from the United States or about 450,000 tons for the entire earth. Perhaps as much as 150,000 tons of oil and grease are discharged to the ocean in municipal sewage effluents from U.S. cities and towns. A large fraction of oils and greases in sewage do not originate from petroleum. If we assume that one-third of sewage oils and greases are petroleum hydrocarbons and multiply by three to give the world total we arrive at 100,000 tons per year from this source. Thus all industrial and automotive petroleum wastes entering the ocean may be about 550,000 tons. This amount should increase at about the same rate as total oil production, namely, to about 825,000 tons by 1975 and 1.2 million tons by 1980.

All the preceding estimated direct losses to the marine environment made up approximately 2.2 million tons per year in 1969:

Accidental spills	0.2 million tons
Tanker operations	0.5 million tons
Other ships	0.5 million tons
Offshore production	0.1 million tons
Refinery operations	0.3 million tons
Industrial and automotive wastes	0.6 million tons
Total	2.2 million tons

The total is expected to increase to between 3.3 and 4.8 million tons by 1980. . . . Petroleum hydrocarbons entering the sea from all the above sources are about 0.1 percent of world oil production. If the possible fallout of airborne hydrocarbons on the sea surface is added, the total amount of oil and oil products contaminating the ocean may be as much as 0.5 percent of world production.

To give these figures perspective, we can make two historical comparisons.

Oil pollution of the marine environment existed long before the first oil well was drilled. This pollution came from natural seeps on the sea floor. There hae never been any measurements of the quantity of oil entering the ocean from such natural seepage areas, but two lines of evidence indicate that it must be quite small, compared to the present amounts of oil entering the ocean because of human activities. First, if much oil had continually seeped into the ocean, all of the petroleum reserves would have long since disappeared. For example, if 100,000 tons

of oil per year entered the ocean from natural seepages, within a few million years this would exceed the total estimated oil reserves of the entire earth. Second, we know from the Santa Barbara and Louisiana well accidents that any natural oil seepage producing even a few thousand tons of oil per year would have resulted in very conspicuous slicks of oil spreading over large areas of the sea surface. No such large natural slicks have ever been observed. Typically, natural seeps produce quite small quantities of oil which occasionally bubble up to the surface and produce small slicks. We estimate, therefore, that oil coming into the marine environment before the human use of petroleum began must have been considerably less than 100,000 tons per year, less than 5 percent of the present 2.2 million tons a year injected directly from land and marine sources.

Another point of comparison with today's annual influx of oil comes from the sinking of tankers and ships in World War II. . . . The total quantity of oil lost in the ocean during the six years of World War II thus may have been about twice the annual direct influx to the ocean at the present time. As far as we know, no permanent damage was done to the ocean ecosystem by these rather large releases, perhaps in part because most of them occurred far from land in relatively deep water, and in part because much of the oil may have escaped into the sea very slowly, as the sunken tanks corroded away.

A great variety of hydrocarbons is produced by marine plants. . . . [It is estimated that] about 3 million tons of hydrocarbons enter the ocean from organic activity each year.

The direct influx of petroleum hydrocarbons to the ocean is small compared to the emission of petroleum products and chemically produced hydrocarbons to the atmosphere through evaporation and incomplete combustion. The emission of petroleum hydrocarbons to the air each year is about 90 million tons, roughly forty times the amounts of these substances entering the ocean directly from ships, shore installations, rivers, and the sea floor. Most of the hydrocarbons emitted to the atmosphere may be oxidized to harmless substances within a relatively short time. It is known that others are combined with nitrogen oxides and ozone to produce substances that are highly toxic to land plants. A fraction of the petroleum hydrocarbons emitted to the atmosphere exists as, or is absorbed on, very small particles, or becomes caught in rain, just as happens to DDT and other chlorinated hydrocarbon pesticides. Much of this fraction may settle out on the surface of the ocean. If 10 percent of the petroleum hydrocarbons emitted into the atmosphere eventually find their way to the sea surface in this way, the total hydrocarbon contamination of the ocean would be about five times the direct influx from ships and land sources. This quantity should be expected to increase about as rapidly as the total petroleum production, which means more than doubling by 1980.

Physical Concentration and Distribution of Oil Pollution

Neither the base load of hydrocarbons nor the concentrated accidental sources can be expected to be distributed uniformly throughout the ocean. Obviously the intensity will be greatest near the sources and unloading points and the most heavily affected areas will be near the coasts.

It is likely that most of the oil entering the sea from ships, rivers, and the sea floor ends up in a narrow zone near shore at most only a few kilometers in width. Some of this oil will become absorbed on clay, silt, sand grains, and other particles and will settle to the bottom. The oil remaining in the water will evaporate or become oxidized. Biodegradation of the bottom-deposited oil will also gradually occur, but fractions of the bottom-deposited oil will continue to disperse into shallow overlying waters for months or years. This inshore zone is the most sensitive to severe damage to the living resources of the sea from direct pollution by oil.

Submarine reservoirs of petroleum are likely to be found on the continental shelves of almost every continent, and the incidence of local contamination from underwater drilling and production on the continental margins will ultimately be widespread.

Sources from ships as a result of tank cleaning, bilge pumping, and accidents will be expected to follow the pattern of tanker and other cargo routes, with the highest concentrations near ports and harbors and in semienclosed seas such as the Mediterranean, the Black, and North seas, the Persian Gulf, and the Gulf of Mexico. The total area of these water bodies is slightly over 2 percent of the area of the ocean, but perhaps one-fourth of the total oil pollution from ships and land sources may occur in them. The future development of oil production in the Alaskan North Slope and the Canadian Northern Archipelago may produce serious contamination in the Arctic Ocean. Regional international agreements may be the most effective way to deal with the concentration of pollution in such semienclosed seas.

On the high seas, winds and ocean currents will bring about a convergence and retention of concentrations of hydrocarbons in the subarctic and equatorial convergence zones such as the Sargasso Sea. Workers from the Woods Hole Oceanographic Institution have found that oil globules and tar balls are more abundant in the Sargasso Sea than the Sargassum weed for which the sea is named.

Probably most of the hydrocarbon fallout from the air onto the sea surface occurs in the mid-latitudes of the Northern Hemisphere. These latitudes contain the trajectories of the winds blowing from the industrialized countries. If hydrocarbons deposited from the air formed a surface film over most of the North Atlantic, its thickness might be about 1,000 angstroms.* Such a film should be detectable by suitable optical methods and might have physical as well as biological effects. It is more likely than most if the oil is in small particles, droplets, or tarry lumps, [. . .] and that much of it settles quickly below the surface. As we shall see, oil films and droplets near the surface and DDT and other oil-soluble chlorinated hy-

---

*Ed. note: One hundred-millionth (10^{-8}) centimeter.

Cleaning Up From Santa Barbara Oil Rig Leak, 1969
Accidental oil spills are conspicuous but account for less than 10 percent of the oil entering the marine environment.

drocarbons may have combined effects on the high seas which may do serious damage to open ocean ecosystems.

Modes of Hydrocarbon Removal From the Oceans

Hydrocarbons in the sea are diluted and dispersed by natural mixing and eventually disappear by microbial or physical oxidation, evaporation, and burial in the bottom sediments.

Hydrocarbons dissolved or suspended in the water column are eventually destroyed by bacteria, fungi, and other microorganisms. Some workers have found that the most toxic compounds are also the most refractory to microbial destruction, though the evidence is somewhat conflicting on this point.

No single microbial species will degrade any whole crude oil. Bacteria are highly selective and complete degradation requires numerous different bacterial species. Bacterial oxidation

205

of hydrocarbons produces many intermediates which may be more toxic than the hydrocarbons; therefore, organisms are also required that will further attack hydrocarbon decomposition products.

The oxygen requirement in marine bacterial oil degradation is served. Complete oxidation of one gallon of crude oil requires all of the dissolved oxygen in 400,000 gallons of air-saturated seawater at 60°F. (This is equivalent to a layer of water one foot deep covering 1.2 acres.) Oxidation may be inhibited in areas where the oxygen content has been lowered by previous pollution, and the bacterial degradation may cause additional damage through oxygen depletion.

The rate of oxidation is strongly affected by the temperature of the water, being at least ten times slower at 40°F than at 80°, and much slower still when the water is near freezing temperature. Estimates by ZoBell★ indicate that oil dispersed in continuously oxygenated seawater,

---

★*Ed. note:* C.E. ZoBell, "The Occurrence, Effects and Fate of Oil Polluting the Sea," *Proceedings of the International Conference on Water Pollution Research,* Section 3, No. 48 (London), Sept. 1962.

containing an abundance of oil-consuming bacteria at 80°F, may be oxidized at a rate of one gallon of oil per 400,000 gallons of water in about 2.5 days. At such a rate, 145 gallons of oil would disappear in a year. For this characteristic temperature of tropical near-surface ocean waters, the rate of oxygen uptake in the water and its nitrate and phosphate content will limit the rate of oxidation of the oil rather than temperature. Nitrate and phosphate, which are essential for the growth of oil-consuming bacteria, are present in very low concentration in the upper 100 meters of most tropical, open-ocean waters. Oxygen is continually replenished from the atmosphere in waters near the surface, but below about 10 meters this replenishment is usually extremely slow.

In the water of high latitudes, where the temperature is below 40°F, at least a month would be required to oxidize one gallon of oil in 400,000 gallons of seawater, even·if the oxygen content were continuously replenished. In bays, estuaries, and shallow coastal waters most of the oil would settle to the bottom during this period and any further decomposition would be greatly slowed by the lack of oxygen in the bottom sediments.

Consequences of Oil Pollution

Depending upon their location, character, and concentration, petroleum hydrocarbon pollutants in the ocean can produce the following unwanted consequences:

1. Poisoning of marine life filter feeders such as clams, oysters, scallops and mussels; other invertebrates; fish, and marine birds.
2. Disruption of the ecosystem so as to induce long-term devastation of marine life.
3. Degradation of the environment for human use by reducing economic and recreational values on either a short- or long-term basis and by changes of esthetics of the marine environment.

Crude oil and oil fractions poison marine organisms through different effects.

a. Direct kill through coating of surfaces.

Hundreds of thousands of oceanic birds freeze to death in winter because their feathers become fouled with oil which displaces the insulating air layer next to the skin. Many air-fouled birds are unable to fly; some lose their buoyancy and sink, others drift helplessly ashore.

b. Direct kill through contact poisoning.

c. Direct kill through exposure to the dissolved or colloidal toxic components of oil at some distance in space and time from the source.

d. Incorporation of sublethal amounts of oil and oil products into organisms, resulting in reduced resistance to infection and other stresses (one of the causes of death of birds surviving the immediate exposure to oil).

Disruption of the ecosystem may occur through

a. Destruction of the generally more sensitive juvenile forms of organisms.

b. Destruction of the food sources of higher species.

c. Possibly through interference with the communications systems of organisms.

Environmental degradation occurs on both sandy and rocky beaches and in bays and estuaries. From twenty pounds to a ton of oil and tar per mile of beach, in separate globs and in coatings on sand grains, has been observed on many beaches. Even a ton of oil per mile along U.S. beaches would represent a small fraction of the oil entering the sea near the United States each year. In bays and estuaries, boats, fisherman's nets, piers, quays, wharfs, mooring lines, buoys, and lobster pots become smeared and fouled with oil. Severe fire hazards are created near docks and piers by oil on the water surface and by oil-soaked debris.

Conclusions for Action

1. One distinguishing characteristic of the problem of oil pollution is that while most of the evident damage occurs in waters and coastal areas under national jurisdiction, international agreements and control measures are needed to

reduce or eliminate many of the causes of the pollution. Present national and international arrangements to control oil pollution are inadequate because they lack effective means of enforcement. International agreements should include provisions for easily and uniformly applied sanctions which would ensure that the costs to polluters are either greater than the benefits they receive or at least sufficient to clean up the pollution and repair the damage.

2. Perhaps a quarter of total oil pollution from ships and land sources occurs in semienclosed seas such as the Mediterranean, the Black, and North seas, the Persian Gulf, and the Gulf of Mexico, which have a total area slightly over 2 percent of the area of the ocean. . . . Regional international agreements may be the most effective way to deal with the concentration of pollution in such semienclosed seas.

3. Much more knowledge is needed about the biological effects of oil pollution. . . .

4. More information is also needed on the distributions of petroleum and other hydrocarbons at the surface of the open sea and in the water column and the bottom sediments of estuaries and near-shore areas. . . .

Particular attention should be given to the fate of oil in the ocean and to its persistence and biological degradation under different conditions of bacterial population, temperature, and oxygen.

5. Several lines of technical development designed to reduce or remove direct oil pollution in the sea should be pursued. These include:

a. Better methods of removing oil from large accidental spills such as the *Torrey Canyon* and Santa Barbara incidents. Containment of the oil where it can do least harm, followed by physical removal, is most satisfactory, but better methods to accomplish this are needed. Burning large quantities of oil at the sea surface near shore can be hazardous and may produce severe local air pollution. Use of chemical dispersants can remove oil from the surface where, in some cases, its presence may be extremely hazardous or harmful, but it does not remove the oil from the ocean since it disperses it in the water. In the dispersed state, the oil is more subject to biodegradation but also is more available to marine organisms. At some concentrations, oil dispersed even with nontoxic chemicals can poison organisms, especially filter feeders. The question whether even the limited use of nontoxic dispersants can be justified where other methods are not possible cannot be conclusively answered at this time because data on the fate and effects of oil in the sea are so incomplete.

b. More effective and economical antipollution measures for refinery and petrochemical operations.

c. More economical and easily used techniques, which do not involve the discharge of oil residues in the sea, for cleaning the bilges and fuel tanks of merchant ships.

6. Regulations and other institutional changes are needed to reduce or eliminate the discharge or leakage of waste oils and greases into rivers, lakes, estuaries, and coastal areas. Such regulations should be based on quantitative definitions of required water quality for different uses.

7. More effective international control measures for oil-carrying tankers should be developed:

a. To ensure that all tankers use "load on top" or other anti-pollution procedures.

b. To prevent strandings, and collisions: this may require monitoring and control of tanker tracks at all times, just as transoceanic aircraft are now regulated, and also provisions for tankers to load and unload only at certain specified safe locations.

8. For all ships, measures requiring that dirty ballast and washing be retained aboard and discharged to shore facilities for treatment should be investigated. Permissible limits of oil content in waters discharged to the sea should be specified.

9. Because many future offshore oil drilling and production operations can be expected off the coasts of countries that are not technically equipped to enforce effective antipollution regulations, international standards for safe procedures and an international mechanism for their enforcement should be developed.

USS *Sargo* surfacing through ice at the North Pole.

Unit Five

The law, the navy, the merchant marine—each subject is important in itself in many ways that are touched upon here. However, the subjects also overlap and become doubly important. For example, marine commerce cannot grow to meet the increasing demands of the world without laws to regulate it, sea power to enforce the laws, and ships and men to do it. What is the present status of international law of the sea? Are the navies of the world, and particularly our navy, concerned with the protection of commerce? How good are the ships and the men who man the merchant marine, and how many of them are American? Will new inventions improve the strength of our merchant marine? These are some of the questions we are seeking to answer.

Merchant shipping and naval warfare have always been governed by custom and law, but these are being revised. As the German philosopher Nietzsche wrote a century ago, "It is our future that lays down the law of our today." And at sea the future will so clearly be different from the past that we need new laws. The United Nations has been holding conferences on the law of the sea almost since the day it was established. It is still doing so as this is being written. Not much has been solved, which suggests that the issues are highly complex and continually changing.

Policy and Sea Power

The United Nations conferences are not restricted to the regulation of commerce; they are also concerned with naval warfare, freedom of passage, jurisdiction over fisheries, and the regulation and taxing of mineral resources. The self-interests of the United Nations members are highly diverse, and so are their positions on various issues. These are discussed in an article by Seyom Brown and Larry L. Fabian from *Foreign Affairs*.

Developed nations with large navies and merchant marines focus their concern on the need for laws to keep strategic straits open for free passage. Developing nations, especially those without a seacoast or ships, are more interested in debating about the division of potential benefits from mining the sea floor. The United States' position on the law of the sea is presented here in a selection from a statement by Secretary of State Kissinger.

We have already read in the article in Unit Four by Frank LaQue that the short-term benefits to a United Nations organization from mining the sea floor will be small. However, perhaps since most nations are still developing, this subject seems to dominate the conferences on law of the sea. The current conference was proposed by Arvid Pardo in 1967 when he was Maltese ambassador to the United Nations. After years of international bickering, he retains little hope that there will be agreement among the nations. Still, the conferences continue, and there is always some possibility of an agreement at any time.

World commerce is not too affected by the lack of new laws because the status quo maintains the right of free passage; but what about ocean mining, for which there is no law or custom? The capabilities of American industry to mine the sea floor have been presented in testimony before Congress. Marne A. Dubs, on behalf of the American Mining Congress, states, "There is no longer any doubt about the technical and economic feasibility of ocean mining. The technology is ready; the investment climate is not."

In order to hasten, or force, a government decision regarding ocean mining, Deepsea Ventures, Inc. has filed a mining claim for part of the floor of the Pacific Ocean. The *Notice of Discovery and Claim of Exclusive Mining Rights, and Request for Diplomatic Protection and Protection of Investment* is the first of its kind and makes interesting reading. If the validity of the claim is granted and the government accepts any responsibilities related to it, mining could begin without international agreement.

The fact that it could begin says something about naval power. "Laws are dumb in the midst of arms," wrote Cicero 2,000 years ago. If there are legal claims, whether granted by the United Nations or expressed unilaterally by any nation, an enforcing agency will be required. This will be a new role for navies, but they have assumed many new roles since the last great naval battles off the Philippine Islands in World War II.

J. William Middendorf, secretary of the navy, outlined the history of the United States Navy and its future importance to the nation in a statement included here. The most important role at present is strategic. The Polaris weapons system and its successors combine the unimaginable destructive power of nuclear rockets with the invulnerability of undetectable submarines. They are a matchless deterrent to nuclear attack.

The navy has other roles, such as support of troops ashore and protection of shipping. These roles are not always obvious. An intriguing article by George H. Quester analyzes the rules of the games navies play. For example, the degree of offense taken by a navy when one of its ships is attacked may depend on whether the ship has a name or just a number.

An increasingly important role for the navy in protecting shipping has been necessitated by the rapid growth of seagoing commerce. Perhaps the landlocked nations should turn for revenue to tolls to pass through strategic straits instead of to taxes on manganese nodules.

International commerce is particularly important to the United States because of the volume of our imports and exports and our growing dependence on foreign energy resources. A nation dependent on commerce is at the mercy of the owners of ships. When ships are owned by foreign governments, commerce can be a weapon just like the oil it transports. Nonetheless, we have allowed the American merchant marine to decline to the sixth largest by tonnage. This is somewhat misleading because American companies and their foreign affiliates own many ships under foreign registry, especially in Liberia and Panama. If these are considered as being under American control, the tonnage of our merchant marine was among the first three as recently as 1969. However, national security depends on other factors, including the availability of trained sailors and shipbuilding capacity. Among shipbuilding nations we are near the bottom.

As to qualified personnel, ships in our own registry are manned by sailors who meet the standards we set. The problems of manning ships under foreign registry are discussed in a selection from Noël Mostert's *Supership*. Some "Liberian" ships are equipped with the latest in navigating equipment, but that does not necessarily mean that anyone aboard can operate them. On one tanker that ran aground, the gyrocompass, echo sounder, radar, automatic log, speed indicator, and rudder indicator were all out of order!

Perhaps hope for the American merchant marine lies in new technology. This is discussed in a selection from a report by the National Academy of Engineering. Among the possibilities are automation of cargo handling, container ships, nuclear power, more efficient hull design, and new types of cargo ships that skim above the water or submerge completely below it. Will it take a new breed of sailors to work them?

The Common Heritage of Mankind

Resolution Adopted by the General Assembly of the United Nations, 1970

The law of the sea has become infinitely more complex than it was in 1494, when Spain and Portugal simply divided the oceans of the world equally along a north-south line through the Cape Verde Islands. In the early seventeenth century, Hugo Grotius, regarded by many as the father of international law, argued that the sea, like the air, was not subject to appropriation. But a century later, nations recognized a distinction between territorial waters and the high seas—a distinction that remained until the twentieth century.

The issue of jurisdiction was thrown into confusion by the Truman Proclamation of 1945, claiming United States jurisdiction to the limit of the continental shelf. The continental shelf, however, is not only ill-defined, but it varies throughout the world, extending 800 miles off Siberia and being virtually nonexistent off Peru, Ecuador, and Chile. The tensions created by this inequitable situation were exacerbated over the next two decades as mankind looked increasingly to the sea to supply the energy, mineral, and food resources that were being rapidly depleted on land. In 1967, in a speech before the United Nations General Assembly, Arvid Pardo, Malta's ambassador to the United Nations, emphasized that "the known resources of the seabed are far greater than the resources known to exist on land." In response to Pardo's argument that ocean technology was outstripping our rules for uses of the ocean, the United Nations created a forty-two-nation committee to study the problem. On the recommendation of the committee, the General Assembly adopted the following resolution in 1970, declaring the oceans "the common heritage of mankind." The principles set forth in this document became the basis for a series of international conferences to establish a new law of the sea.

RESOLUTION ADOPTED BY THE GENERAL ASSEMBLY

[on the report of the First Committee (A8097)]

2749 (XXV) Declaration of Principles Governing the Seabed and the Ocean Floor, and the Subsoil Thereof, beyond the Limits of National Jurisdiction

THE GENERAL ASSEMBLY

Recalling its resolutions 2340 (XXII) of 18 December 1967, 2467 (XXIII) of 21 December 1968 and 2574 (XXIV) of 15 December 1969, concerning the area to which the title of the item refers,

Affirming that there is an area of the sea-bed and the ocean floor, and the subsoil thereof, beyond the limits of national jurisdiction, the precise limits of which are yet to be determined.

Recognizing that the existing legal regime of the high seas does not provide substantive rules for regulating the exploration of the aforesaid area and the exploitation of its resources,

Convinced that the area shall be reserved exclusively for peaceful purposes and that the exploration of the area and the exploitation of its resources shall be carried out for the benefit of mankind as a whole,

Believing it essential that an international regime applying to the area and its resources and including appropriate international machinery should be established as soon as possible,

212

Bearing in mind that the development and use of the area and its resources shall be undertaken in such a manner as to foster the healthy development of the world economy and balanced growth of international trade, and to minimize any adverse economic effects caused by the fluctuation of prices of raw materials resulting from such activities,

Solemnly declares that:

1. The sea-bed and ocean floor, and the subsoil thereof, beyond the limits of national jurisdiction (hereinafter referred to as the area), as well as the resources of the area, are the common heritage of mankind.

2. The area shall not be subject to appropriation by any means by States or persons, natural or juridical, and no State shall claim or exercise sovereignty or sovereign rights over any part thereof.

3. No State or person, natural or juridical, shall claim, exercise or acquire rights with respect to the area or its resources incompatible with the international regime to be established and the principles of this Declaration.

4. All activities regarding the exploration and exploitation of the resources of the area and other related activities shall be governed by the international regime to be established.

5. The area shall be open to use exclusively for peaceful purposes by all States, whether coastal or land-locked, without discrimination, in accordance with the international regime to be established.

6. States shall act in the area in accordance with the applicable principles and rules of international law, including the Charter of the United Nations and the Declaration on Principles of International Law concerning Friendly Relations and Co-operation among States in accordance with the Charter of the United Nations, adopted by the General Assembly on 24 October 1970,[1] in the interests of maintaining international peace and security and promoting international co-operation and mutual understanding.

7. The exploration of the area and the exploitation of its resources shall be carried out for the benefit of mankind as a whole, irrespective of the geographical location of States, whether land-locked or coastal, and taking into particular consideration the interests and needs of the developing countries.

8. The area shall be reserved exclusively for peaceful purposes, without prejudice to any measures which have been or may be agreed upon in the context of international negotiations undertaken in the field of disarmament and which may be applicable to a broader area. One or more international agreements shall be concluded as soon as possible in order to implement effectively this principle and to constitute a step towards the exclusion of the sea-bed, the ocean floor, and the subsoil thereof from the arms race.

9. On the basis of the principles of this Declaration, an international regime applying to the area and its resources and including appropriate international machinery to give effect to its provisions shall be established by an international treaty of a universal character, generally agreed upon. The regime shall, *inter alia,* provide for the orderly and safe development and rational management of the area and its resources and for expanding opportunities in the use thereof and ensure the equitable sharing by States in the benefits derived therefrom, taking into particular consideration the interests and needs of the developing countries, whether land-locked or coastal.

[1]Resolution 2625 (XXV).

10. States shall promote international co-operation in scientific research exclusively for peaceful purposes:

(a) By participation in international programmes and by encouraging co-operation in scientific research by personnel of different countries;

(b) Through effective publication of research programmes and dissemination of the results of research through international channels;

(c) By co-operation in measures to strengthen research capabilities of developing countries, including the participation of their nationals in research programmes.

No such activity shall form the legal basis for any claims with respect to any part of the area or its resources.

11. With respect to activities in the area and acting in conformity with the international regime to be established, States shall take appropriate measures for and shall co-operate in the adoption and implementation of international rules, standards and procedures for, *inter alia:*

(a) The prevention of pollution and contamination, and other hazards to the marine environment, including the coastline, and of interference with the ecological balance of the marine environment;

(b) The protection and conservation of the natural resources of the area and the prevention of damage to the flora and fauna of the marine environment.

12. In their activities in the area, including those relating to its resources, States shall pay due regard to the rights and legitimate interests of coastal States in the region of such activities, as well as of all other States, which may be affected by such activities. Consultations shall be maintained with the coastal States concerned with respect to activities relating to the exploration of the area and the exploitation of its resources with a view to avoid infringement of such rights and interests.

13. Nothing herein shall affect:

(a) The legal status of the waters superjacent to the area or that of the air space above those waters;

(b) The rights of coastal States with respect to measures to prevent, mitigate or eliminate grave and imminent danger to their coastline or related interests from pollution or threat thereof or from other hazardous occurrences resulting from or caused by any activities in the area, subject to the international regime to be established.

14. Every State shall have the responsibility to ensure that activities in the area, including those relating to its resources, whether undertaken by governmental agencies, of non-governmental entities or persons under its jurisdiction, or acting on its behalf, shall be carried out in conformity with the international organizations and their members for activities undertaken by such organizations or on their behalf. Damage caused by such activities shall entail liability.

15. The parties to any dispute relating to activities in the area and its resources shall resolve such dispute by the measures mentioned in Article 33 of the Charter of the United Nations and such procedures for settling disputes as may be agreed upon in the international regime to be established.

1933rd plenary meeting
17 December 1970

Diplomats at Sea

Seyom Brown and Larry L. Fabian

Territorial disputes over land have been a major source of international conflict from time immemorial; now such disputes are being extended to the sea as well. The 1970s have witnessed growing tension between governments over fishing rights, rights of passage, mineral resources, control of pollution, and scientific exploration. In order to arrive at a peaceful solution to such problems, the United Nations convened a Law of the Sea Conference in 1973. Aimed at writing an international treaty, the conference held preliminary sessions in Geneva that year. Substantive sessions in Caracas in 1974 and Geneva in 1975 produced a single negotiating text, and the talks resumed in 1976, with more than 1,200 negotiators from 156 nations meeting in New York. As Seyom Brown and Larry L. Fabian point out in the following article excerpted from Foreign Affairs, *the positions of the delegates merely reflect other patterns of international politics. Furthermore, the various divisions among nations have their counterpart in conflicting interests within the United States. The authors, codirectors of a study of technology and international relations for the Brookings Institution, urge the need for a broad policy that recognizes the importance of a peaceful international ocean regime.*

The ocean bargaining now underway features and reinforces some of the patterns of international politics emerging in the world at large. We are referring particularly to the disintegration of the cold-war coalitions, the relative rise of non-security issues, the diversification of friendship and adversary relations, and the embittering tension between the have and have-not peoples.

Nature itself has had the heaviest hand in aligning countries in the ocean debate. Obviously, it makes a considerable difference whether a country borders the ocean or is landlocked, whether its coastline is long or short, and whether its continental margin is rich or barren, broad or narrow. Technology's impact on a country's geological and biological inheritance is an increasingly weighty determinant. Ideological inclination runs a poor third, although the cohesion of the Third World remains an important factor. And alternative visions of a world community are hardly in the picture at all on the eve of the 1974 Conference.

The broad coalition supporting pervasive national control in wide offshore zones comprises not only most of the coastal states of the Third World but also some members of NATO (Canada, Norway, Iceland), plus Spain, Australia and New Zealand, as well as China, Albania, Yugoslavia and Cuba. The United States, the Soviet Union and France (joined by Japan on some issues) are coalition leaders of the states which want to preserve as much unrestricted maritime freedom as possible. The Soviet Union and Japan, while rivals over the fishing areas in the waters between them, both are long-distance fishing powers, and therefore find themselves aligned in Law of the Sea negotiations against coastal states claiming exclusive fishing zones. The United States is a less consistent coalition partner on this subject, being responsive to its own coastal fishing interests, as well as those of other coastal states with whom the United States is negotiating. When it comes to scientific exploration, the lineup tends to correlate quite consistently with the technological prowess of the countries, with the United States and the Soviet Union partners in championing freedom for ocean research— as they also represent the dominant interests in freedom of transit through international straits.

China, only recently a full-fledged participant in ocean politics with her admission to the United Nations, is scoring points by charging

CONFLICT OF INTERESTS

Long-distance fishing powers with modern factory ships, such as the Soviet Union (right), are aligned against coastal states, such as Portugal, using age-old fishing methods (above) and wanting exclusive fishing zones.

collusion between the superpowers and identifying with the coalition of have-nots. The Chinese lose no opportunity to take the podium as spokesmen for militant anti-maritime positions (exclusive jurisdiction out to 200 miles; stringent limitations on scientific exploration), and for substantial redistributions of income and political power (direct control by an international agency over the operations and revenues of deep-sea mining; majority dominance in ocean agencies). But China has found that the putative Third-World coalition is subject to fragmentation on the many issues that lie astride the split between the landlocked states (20 of which have a per capita GNP of less than $500) and the coastal states.

Altogether the world has come a very great distance since the Law of the Sea Conference of 1958–1960, when the United States viewed itself as the coalition leader for "free world" geo-

political and commercial interests in a narrow territorial sea, against the illegitimate effort of the "Soviet bloc" to extend coastal sovereignty to 12 miles. A pluralistic world where multiple and overlapping coalitions are formed on the basis of particular interests rather than ideology is probably safer than a world in which conflicts over particular interests are inflated into grand struggles over competing ways of life. Cross-cutting coalitions of the kind now evident in ocean diplomacy are a cushion against total nation-to-nation or bloc-to-bloc hostility, and provide disincentives to the reliance on military force.

But whereas the complex intersection of interests in the ocean may be conducive to a stability of sorts by helping to contain conflicts below the level of major interstate hostility, the lack of concepts and mechanisms for attending to the continuing tasks of allocation of resources and resolution of disputes does not augur well for a stable international order over the long run. And a rampant pluralism portends a failure to conserve the resources and ecological health of the sea.

Thus the basic directions of ocean diplomacy are disturbing. Nations are greedily extending their assertions of jurisdiction seaward when even on land the system of sovereign states seems to be out of kilter with the increasing ability of politically separated peoples to affect each other's health and welfare. The preparatory period for the 1974 Law of the Sea Conference has been used by many states to grab while the grabbing's good. The prospects appear dim that the Conference will produce the coherence in policy and institutions called for by the comprehensive and interconnected character of ocean problems; nor can one be optimistic that the agreements reached will be sufficiently responsive to the continuing changes in use and knowledge of the sea.

Those countries that have seen their preferred outcomes steadily lose support during the lengthy pre-Conference negotiations now seem to be approaching the Conference convinced that international consensus can only damage their interests. And the polarization of maritime haves and have-nots shows every sign of intensifying as the Conference nears, leaving both groups skeptical about the value of the multilateral negotiating process and the concessions it requires.

To be sure, even with a complete breakdown in the negotiations we will not be at the edge of the apocalypse. Conflicts of interests still might be accommodated temporarily on a bilateral or limited multilateral basis. Yet, sometime before the end of the century the need for a more coherent global regime for the ocean is bound to become more acute, and the international community will have to try again. Things may need to get worse before they get better.

The United States—because of its prominence in maritime politics—must accept its share of responsibility for the unpromising state of ocean diplomacy. It has been several years since the American government offered an elaborate seabed resource treaty to internationalize the margins in part (making coastal states "trustees" for the world community) and to fully internationalize the deep ocean. Since then, the Byzantine maneuverings of the key players in U.S. domestic ocean politics have severely undermined the core premises of the original American draft treaty. Not without blame are the oil interests with their warnings that the Administration's announced policy endangered American energy security, and their demands for ironclad international guarantees to protect their offshore operations from coastal-state controls; the domestic coastal fishing interests with their insistence on protection against the incursions of foreign fishing fleets; the scientists who had neither the political clout nor the easily demonstrable identification with national economic objectives, but who decided that it was time to make themselves heard in the name of science; the hard-mineral mining industries with their arguments that internationalization of the deep ocean would threaten their corporate well-being and their competitive position vis-à-vis mining interests in other advanced countries; and the shipping interests with their anxieties over coastal-state interference in their navigation routes.

No less severe is the battering that the American stance has suffered on the international scene during these same years. U.S. ocean policy has taken enough twists and turns to offer almost every foreign critic a ready target. At first, the United States unequivocally resisted the very idea of a general and comprehensive law-of-the-sea conference—an attitude it shared with the Soviet Union. Then, after the furious period of domestic bargaining that produced the 1970 Nixon policy, the United States, in announcing that policy, identified itself with the internationalist notions of common heritage associated with the new perception of the ocean as a vast treasurehouse of still unappropriated wealth. International reaction to the American démarche was not, as many policy-makers seem to have expected, a rush of praise and gratitude, but a series of stinging criticisms. Former colonial nations bristled at the reincarnation of the "trusteeship" notion. The 1970 U.S. draft treaty was portrayed as at best a naïve attempt to place a tidy international organizational structure on top of the chaotic and contentious arena of international ocean politics; and at worst a cynical attempt by the Department of Defense to buy off coastal developing states whose assertiveness might hamper naval mobility. More charitably, the 1970 American policy initiative can be viewed as a gesture that conceded few real interests while rhetorically aligning the national interests with universal order. Recently, international criticism has intensified as American negotiating positions have shifted and hardened—partly to accommodate the demands of various domestic interests, and partly perhaps in a characteristic "bargaining-chip" strategy.

Not all of the policy adjustments that U.S. ocean negotiators have made on behalf of domestic interests can be faulted, and not all of the international criticism of U.S. policy positions has been fair. But it is clear that the United States needs badly to reconstitute a working domestic consensus for an ocean policy that serves an enlightened national interest and a broad conception of world community needs. Each special bureaucratic and private interest must not be free to define the national interest as its own. Nor should international policies be judged or defended merely on the basis of how much can be got for how little given.

The problem for nearly a decade now has not been primarily the pressures of domestic interests, the naïve generosity of the familiar brand of American internationalism that surfaced in the 1970 ocean treaty, or the common hypocrisy of politicians. Rather, American ocean policy-makers have simply failed to be grand enough—both in their definition of international objectives for the ocean, and in their strategies for generating domestic and international support for new policies. Being only halfway committed to only half a loaf is probably the surest way to end up with less than a quarter.

Being grand enough requires a consistent eye on the essential implications for world order inherent in the debate over the ocean's future. President Nixon, in his 1971 State of the World message—the first to prominently mention U.S. ocean policy—gave at least rhetorical content to the proposition that there is what he called a "world interest" to be served in the ocean. It is time to retrieve the spirit of this declaration and to give it direction and substance.

The cornerstone of these world order objectives should be a clear determination to promote international ocean policies and institutional arrangements that, whatever else they may do, serve to steadily increase mutual accountability among ocean users whenever they affect each other directly and whenever they affect the ocean on which they depend in common. It will not do for the United States merely to pledge a principle of mutual accountability. It must exemplify that principle in specific actions at and beyond the Conference, and must aim at having in place by this century's end a set of international structures to maintain it. A grandiose institutional blueprint is still premature. But it is not too early for the United States to commit itself to an evolving political process where it must accommodate to others, where it must share the real power of decision with others, and where it must settle for making moderate demands in return for moderate satisfactions.

The present U.S. effort to lock up even the finer details of a deep-ocean mining regime in an ocean treaty can only be regarded by others as an attempt to insulate an American-preferred mining regime from future challenge. Attempts to keep ocean environmental decisions in forums dominated by maritime powers will be interpreted by others as a device to deprive them of a voice on environmental policy in present or future institutions where their weight can be felt. U.S. enthusiasm for compulsory international dispute settlement can appear narrowly self-serving if conditions and terms are attached favoring the kinds of interests the United States expects to be under adjudication. And a persistent U.S. policy of subjecting virtually all politically and economically important ocean decisions to weighted-voting arrangements favoring the technological powers is easily portrayed as a transparent means of excluding other countries from a meaningful voice in decisions affecting them. To be sure, the United States is not alone in seeking to preserve advanced-country preeminence in ocean institutions. But an uncompromising pursuit of this objective by the United States would be impolitic, for it would stiffen Third-World reliance on the sheer weight of their numbers as a diplomatic weapon.

The second essential element of an enlightened American ocean policy must be the recognition that no resolution of conflicting international interests and perspectives is possible except via measures that bring economic justice to users of the ocean, even at some relative sacrifice to the rich states able to benefit directly from ocean-resource development. The ocean debate began as a colossal exercise in economic politics fueled by the suspicion throughout the Third World that their slice of the pie almost certainly would

219

be lost without deliberate international decisions in their favor. Even though both developing and developed nations must share some onus for the carving up of the common heritage, this fact should not lead to the conclusion that the rich are thereby divested of special obligations.

The United States, just within the past year or two, has moved demonstrably away from a precise formulation of international revenue-sharing policies, although the commitment in principle is still there. There may yet be some chance that major revenues can be realized for the world community *even if* coastal jurisdictions are extended wholesale. The United States can help by showing that it regards revenue-sharing for international-community purposes as a desirable end in itself, and not merely as a bargaining lever. It is time to press urgently and concretely for arrangements to share very substantial revenues from oil and gas exploitation on the continental margins. Offshore oil can make the real difference in near-term international returns from the ocean. Resource experts know this; politicians and diplomats have carefully tended to avoid it. Moreover, revenue-sharing schemes can be devised to place the proportionately greater obligations on the richest oil producing countries. As for the much less certain revenue prospects from any future deep-ocean mining of hard minerals, deliberate steps must now be taken to keep these prospects alive and to accept the need for a deep-ocean regulatory regime held as legitimate by the bulk of the international community. Surely the miners must have proper incentives, but it is up to political leadership to preserve a future cut for the world community.

The third ingredient in a larger American ocean strategy would be to serve the world interest by thrusting the neglected issues of ocean environmental protection to the forefront of the diplomatic agenda and to the top of the U.S. list of priorities in the years immediately ahead. To be credible, a commitment to deal comprehensively with the problem of ocean degradation must radically challenge the now-orthodox notion that the ocean is divided into national and international turf. The political logic behind this notion as it applies now to economic and security policies in the ocean probably cannot be sustained successfully for very long. Applied to environmental policies, such logic is clearly indefensible in view of the ecologically unified and four-dimensional nature of the ocean earlier described. The environmental reach of the international community in marine affairs will have to extend . . . far enough to pierce the artificial boundaries of "national jurisdiction," up to and onto the land from which so much ocean pollution comes.

A departure from prevailing nationalist concepts in the field of marine ecology will by no means eliminate the need to debate and continually refine the meaning of environmental preservation and how to implement it. Nor will it reduce the clashes of national and special interests over the nature of the regime to manage the commons, or the usefulness of less-than-global solutions to many ecological problems of the oceans. But ocean diplomatists have not yet begun to take these challenges seriously; they seem instead only to be drifting into ecological irresponsibility.

In its unfolding ocean diplomacy, the United States must face two realities: it will not get all that it wants, and it will have to cooperate intensively with many other nations in order to get what it can. The first reality suggests that the United States must carry to any ocean negotiations a clear-headed conviction of what is essential to achieve and what is merely convenient. While the United States is currently advancing and defending some very important national interests in their most unvarnished and absolute forms, none of them is rightly to be regarded as a vital interest immune from adjustment. This includes the very sticky issue of straits, for maritime military and strategic considerations do not require a universal legal right of the kind being sought—and given the political realities of the ocean debate it is quixotic to put everything on the line for this issue. On this question as well as on a host of others, reasonable adjustment to the interests of others is more likely to increase than to decrease the net overall political and economic return to this nation. And if sooner or later a truly cooperative approach to ocean management will have to be made, why not begin now?

If, as we believe, the priority task of the upcoming round of ocean diplomacy is to begin building a working political community among ocean users, then a successful Conference must be defined not merely as one that reaches agreement, but one that reaches the *right kind* of agreement. This Conference must leave future ocean negotiators with a legacy of accommodation and mutual trust in ocean affairs, not a residue of bitterness left by confrontationist diplomacy. If a general "constitution" for the ocean is to result, it must be one that is flexible and durable enough to garner the support of all major segments of the international community. A working political community requires not only a treaty laden with rights and obligations, but also a political framework of incentives that encourage joint policy-making, consultations, and collective action at appropriate levels during the decades immediately ahead.

We do not urge this approach for the United States because we think America alone can assure the success of the Law of the Sea Conference. These negotiations may fail in key respects no matter what the United States does. We urge it rather as a longer-range approach for American ocean diplomacy, and as a necessary counter, particularly within the United States, to the nationalistic and parochial pressures being catalyzed by the Conference. A country with the maritime strength and interests of the United States has little to lose by adopting an internationally responsive and less narrowly self-interested approach toward building a more peaceful and just world order. The challenge is to combine a healthy pursuit of U.S. interests in an orderly, durable ocean regime with a consistent vision of how this nation must use and restrain its power in a progressively interdependent world.

The Law of the Sea and World Order

Henry A. Kissinger

The single negotiating text of the draft treaty on the Law of the Sea, produced in Geneva in the spring of 1975, provides for an economic zone with a 200-mile limit in which fisheries and resources would be controlled by the coastal state and an extension of the territorial zone in which the coastal state exercises sovereignty from three to twelve miles. These measures are consistent with United States policy objectives as set forth in the following speech by Secretary of State Henry Kissinger, delivered before the American Bar Association in August 1975. While strongly supporting the international negotiations, Kissinger warns that the United States cannot wait indefinitely to exploit the minerals of the deep seabed and to establish control over fisheries. He sees unilateral action only as a last resort. Eight months after this speech was delivered, on April 13, 1976, President Ford signed into law an act unilaterally establishing control over fisheries within 200 miles of the United States, effective in March 1977.

*T*he United States is now engaged with some 140 nations in one of the most comprehensive and critical negotiations in history—an international effort to devise rules to govern the domain of the oceans. No current international negotiation is more vital for the long-term stability and prosperity of our globe.

One need not be a legal scholar to understand what is at stake. The oceans cover seventy percent of the earth's surface. They both unite and divide mankind. The importance of free navigation for the security of nations—including our country—is traditional; the economic significance of ocean resources is becoming enormous.

From the Seventeenth Century, until now, the law of the seas has been founded on a relatively simple precept: freedom of the seas, limited only by a narrow belt of territorial waters generally extending three miles offshore. Today, the explosion of technology requires new and more sophisticated solutions.

In a world desperate for new sources of energy and minerals, vast and largely untapped reserves exist in the oceans.

In a world that faces widespread famine and malnutrition, fish have become an increasingly vital source of protein.

In a world clouded by pollution, the environmental integrity of the oceans turns into a critical international problem.

In a world where ninety-five percent of international trade is carried on the seas, freedom of navigation is essential.

Unless competitive practices and claims are soon harmonized, the world faces the prospect of mounting conflict. Shipping tonnage is expected to increase fourfold in the next thirty years. Large, self-contained factory vessels already circle the globe and dominate fishing areas that were once the province of small coastal boats. The world-wide fish harvest is increasing dramatically, but without due regard to sound management or the legitimate concerns of coastal states. Shifting population patterns will soon place new strains on the ecology of the world's coastlines.

The current negotiation may thus be the world's last chance. Unilateral national claims to fishing zones and territorial seas extending from fifty to two hundred miles have already resulted in seizures of fishing vessels and constant disputes over rights to ocean space. The breakdown of the current negotiation, a failure to reach a legal consensus, will lead to unrestrained military and commercial rivalry and mounting political turmoil.

The United States strongly believes that law must govern the oceans. In this spirit, we welcomed the United Nations mandate in 1970 for a multilateral conference to write a comprehensive treaty governing the use of the oceans and their resources. We contributed substantially to the progress that was made at Caracas last summer and at Geneva this past spring which produced a "single negotiating text" of a draft treaty. This will focus the work of the next session, scheduled for March 1976 in New York. The United States intends to intensify its efforts.

The issues in the Law of the Sea negotiation stretch from the shoreline to the farthest deep seabed. They include:

> The extent of the territorial sea and the related issues of guarantees of free transit through straits;
>
> The degree of control that a coastal state can exercise in an offshore economic zone beyond its territorial waters; and
>
> The international system for the exploitation of the resources of the deep seabeds.

If we move outward from the coastline, the first issue is the extent of the *territorial sea*—the belt of ocean over which the coastal state exercises sovereignty. Historically, it has been recognized as three miles; that has been the long-established United States position. Increasingly, other states have claimed twelve miles or even two hundred.

After years of dispute and a contradictory international practice, the Law of the Sea Conference is approaching a consensus on a twelve-mile territorial limit. We are prepared to accept this solution, provided that the unimpeded transit rights through and over straits used for international navigation are guaranteed. For without such guarantees, a twelve-mile territorial sea would place over 100 straits—including the Straits of Gibraltar, Malacca, and Bab el Mandeb—now free for international sea and air travel under the jurisdictional control of coastal

states. This the United States cannot accept. Freedom of international transit through these and other straits is for the benefit of all nations, for trade and for security. We will not join in an agreement which leaves any uncertainty about the right to use world communication routes without interference.

Within 200 miles of the shore are some of the world's most important fishing grounds as well as substantial deposits of petroleum, natural gas, and minerals. This has led some coastal states to seek full sovereignty over this zone. These claims, too, are unacceptable to the United States. To accept them would bring thirty percent of the oceans under national territorial control—in the very areas through which most of the world's shipping travels.

The United States joins many other countries in urging international agreement on a 200-mile offshore *economic zone*. Under this proposal, coastal states would be permitted to control fisheries and mineral resources in the economic zone, but freedom of navigation and other rights of the international community would be preserved. Fishing within the zone would be managed by the coastal state, which would have an international duty to apply agreed standards of conservation. If the coastal state could not harvest all the allowed yearly fishing catch, other countries would be permitted to do so. Special arrangements for tuna and salmon, and other fish which migrate over large distances, would be required. We favor also provisions to protect the fishing interests of land-locked and other geographically disadvantaged countries.

In some areas the *continental margin* extends beyond 200 miles. To resolve disagreements over the use of this area, the United States proposes that the coastal states be given jurisdiction over continental margin resources beyond 200 miles, to a precisely defined limit, and that they share a percentage of financial benefit from mineral exploitation in that area with the international community.

Beyond the territorial sea, the offshore economic zone, and the continental margin lie *the deep seabeds.* They are our planet's last great unexplored frontier. For more than a century we have known that the deep seabeds hold vast deposits of manganese, nickel, cobalt, copper, and other minerals, but we did not know how to extract them. New modern technology is rapidly advancing the time when their exploration and commercial exploitation will become a reality.

The United Nations has declared the deep seabed to be the "common heritage of mankind." But this only states the problem. How will the world community manage the clash of national and regional interests, or the inequality of technological capability? Will we reconcile unbridled competition with the imperative of political order?

The United States has nothing to fear from competition. Our technology is the most advanced, and our Navy is adequate to protect our interests. Ultimately, unless basic rules regulate exploitation, rivalry will lead to tests of power. A race to carve out exclusive domains of exploration on the deep seabed, even without claims of sovereignty, will menace freedom of navigation, and invite a competition like that of the colonial powers in Africa and Asia in the last century.

This is not the kind of world we want to see. Law has an opportunity to civilize us in the early stages of a new competitive activity.

We believe that the Law of the Sea Treaty must preserve the right of access presently enjoyed by states and their citizens under international law. Restrictions on free access will retard the development of seabed resources. Nor is it feasible, as some developing countries have proposed, to reserve to a new international seabed organization the sole right to exploit the seabeds.

Nevertheless, the United States believes strongly that law must regulate international activity in this area. The world community has an historic opportunity to manage this new wealth cooperatively and to dedicate resources from the exploitation of the deep seabeds to the development of the poorer countries. A cooperative and equitable solution can lead to new patterns of accommodation between the developing and industrial countries. It could give a fresh and conciliatory cast to the dialogue between the industrialized and so-called Third World. The legal regime we establish for the deep seabeds can be a milestone in the legal and political development of the world community.

The United States has devoted much thought and consideration to this issue. We offer the following proposals:

An international organization should be created to set rules for deep seabed mining.

This international organization must preserve the rights of all countries, and their citizens, directly to exploit deep seabed resources.

It should also ensure fair adjudication of conflicting interests and security of investment.

Countries and their enterprises mining deep seabed resources should pay an agreed portion of their revenues to the international organization, to be used for the benefit of developing countries.

The management of the organization and its voting procedures must reflect and balance the interests of the participating states. The organization should not have the power to control prices or production rates.

If these essential United States interests are guaranteed, we can agree that this organization will also have the right to conduct mining operations on behalf of the international community primarily for the benefit of developing countries.

The new organization should serve as a vehicle for cooperation between the technologically advanced and the developing countries. The United States is prepared to explore ways of sharing deep seabed technology with other nations.

A balanced commission of consumers, seabed producers, and land-based producers could monitor the possible adverse effects of deep seabed mining on the economies of those developing countries which are substantially dependent on the export of minerals also produced from the deep seabed.

The United States believes that the world community has before it an extraordinary opportunity. The regime for the deep seabeds can turn interdependence from a slogan into reality. The sense of community which mankind has failed to achieve on land could be realized through a regime for the ocean.

The United States will continue to make determined efforts to bring about final progress when the Law of the Sea Conference reconvenes in New York next year. But we must be clear on one point: The United States cannot indefinitely sacrifice its own interest in developing an assured supply of critical resources to an indefinitely prolonged negotiation. We prefer a generally acceptable international agreement that provides a stable legal environment *before* deep seabed mining actually begins. The responsibility for achieving an agreement before actual exploitation begins is shared by all nations. We cannot defer our own deep seabed mining for too much longer. In this spirit, we and other potential seabed producers can consider appropriate steps to protect current investment, and to ensure that this investment is also protected in the treaty.

The Conference is faced with other important issues:

Ways must be found to encourage marine scientific research for the benefit of all mankind while safeguarding the legitimate interests of coastal states in their economic zones.

Steps must be taken to protect the oceans from pollution. We must establish uniform international controls on pollution from ships and insist upon universal respect for environmental standards for continental shelf and deep seabed exploitation.

Access to the sea for land-locked countries must be assured.

There must be provisions for compulsory and impartial third-party settlement of disputes. The United States cannot accept unilateral interpretation of a treaty of such scope by individual states or by an international seabed organization.

The pace of technology, the extent of economic need, and the claims of ideology and national ambition threaten to submerge the difficult process of negotiation. The United States therefore believes that a just and beneficial regime for the oceans is essential to world peace.

For the self-interest of every nation is heavily engaged. Failure would seriously impair confidence in global treaty-making and in the very process of multilateral accommodation. The conclusion of a comprehensive Law of the Sea treaty on the other hand would mark a major step towards a new world community.

The urgency of the problem is illustrated by disturbing developments which continue to crowd upon us. Most prominent is the problem of fisheries.

The United States cannot indefinitely accept unregulated and indiscriminate foreign fishing off its coasts. Many fish stocks have been brought close to extinction by foreign overfishing. We have recently concluded agreements with the Soviet Union, Japan, and Poland which will limit their catch and we have a long and successful history of conservation agreements with Canada. But much more needs to be done.

Many within Congress are urging us to solve this problem unilaterally. A bill to establish a 200-mile fishing zone passed the Senate last year; a new one is currently before the House.★

The Administration shares the concern which has led to such proposals. But unilateral action is both extremely dangerous and incompatible with the thrust of the negotiations described here. The United States has consistently resisted the unilateral claims of other nations, and others

★*Ed. note:* The bill passed the House on March 30, 1976 and was signed into law on April 13, 1976.

will almost certainly resist ours. Unilateral legislation on our part would almost surely prompt others to assert extreme claims of their own. Our ability to negotiate an acceptable international consensus on the economic zone will be jeopardized. If every state proclaims its own rules of law and seeks to impose them on others, the very basis of international law will be shaken, ultimately to our own detriment.

We warmly welcome the recent statement by Prime Minister Trudeau reaffirming the need for a solution through the Law of the Sea Conference rather than through unilateral action. He said, "Canadians at large should realize that we have very large stakes indeed in the Law of the Sea Conference and we would be fools to give up those stakes by an action that would be purely a temporary, paper success."

That attitude will guide our actions as well. To conserve the fish and protect our fishing industry while the treaty is being negotiated, the United States will negotiate interim arrangements with other nations to conserve the fish stocks, to ensure effective enforcement, and to protect the livelihood of our coastal fishermen.

These agreements will be a transition to the eventual 200-mile zone. We believe it is in the interests of states fishing off our coasts to co-operate with us in this effort. We will support the efforts of other states, including our neighbors, to deal with their problems by similar agreements. We will consult fully with Congress, our states, the public, and foreign governments on arrangements for implementing a 200-mile zone by virtue of agreement at the Law of the Sea Conference.

Unilateral legislation would be a last resort. The world simply cannot afford to let the vital questions before the Law of the Sea Conference be answered by default. We are at one of those rare moments when mankind has come together to devise means of preventing future conflict and shaping its destiny rather than to solve a crisis that has occurred, or to deal with the aftermath of war. It is a test of vision and will, and of statesmanship. It must succeed. The United States is resolved to help conclude the Conference in 1976—before the pressure of events and contention places international consensus irretrievably beyond our grasp.

Future Prospects for Law of the Sea   *Arvid Pardo*

In 1967 Arvid Pardo, then Maltese ambassador to the United Nations, issued a call for a new law of the sea, consistent with new uses of the ocean and new technological developments. In the following article, excerpted from a paper delivered at the Conference on Conflict and Order in Ocean Relations at the Johns Hopkins University School of Advanced International Studies in 1974, Pardo assesses the kind of treaty that is likely to emerge from the current Law of the Sea Conference. Whereas Kissinger is optimistic that a treaty could lead toward a peaceful ocean regime, Pardo believes that the treaty will merely recognize the self-interests of coastal states. No real provisions will be made to serve the interests of international peace, with the result that the future treaty will multiply, not diminish, conflicts between states. His misgivings seem warranted: since his paper was written, the law of the sea negotiations have been sharply challenged by forty-nine countries that are landlocked and, in their view, "geographically disadvantaged." These nations are now demanding access to the ocean through neighboring states and fishing and mining rights off their neighbors' coasts.

Pardo is now on the faculty at the University of Southern California.

*W*hatever its other provisions, . . . the future convention will revolve around the three points mentioned—international recognition of extensive coastal state jurisdiction in the seas; international recognition of the exercise by coastal states of comprehensive powers within national jurisdictional areas; and some assurances of normally unhampered commercial navigation. Agreement on these points will determine agreement on all other matters dealt with by the convention, since it is unlikely that states will fail to find appropriate formulations with regard to issues perceived as relatively secondary after having attained their more important objectives.

Perhaps I may briefly express my view on the expected outcomes with respect to each of the points which I have mentioned.

With regard to the first point—international recognition of extensive coastal state jurisdiction—I assume that each coastal state wishes to obtain for itself international recognition of the widest possible jurisdictional limits but does not wish other states to obtain more than it does. The wider the breadth of national jurisdiction, however, the fewer are the states advantaged. A 200-nautical mile exclusive jurisdiction in the seas represents an approximate political balance: if national jurisdictional limits are set at much

less than 200 nautical miles this would not satisfy the majority of coastal states which would lose the opportunity of further extending their jurisdiction. If national jurisdictional limits were set at much more than 200 nautical miles, a very substantial number of coastal states, and certainly the majority of states represented at the Conference, would not be able to avail themselves of this provision. Hence we may expect that the Conference will approve a 200-nautical mile limit to coastal state jurisdiction in the seas.

A 200-nautical mile jurisdictional limit does not, however, satisfy developed and developing oceanic states—i.e., states with broad frontage on the open oceans. There are only about two dozen oceanic states but they are very influential. These states, therefore, are likely to be accommodated by Conference approval of a number of provisions—such as vague articles on straight baselines, recognition of coastal state sovereignty up to the lower edge of a unilaterally defined continental margin, recognition that the coastal state has a special status for certain purposes in marine areas adjacent to areas under its jurisdiction, etc.—which will indirectly permit extension of the jurisdiction of some coastal states far beyond 200 nautical miles from the coast.

226

Archipelagic states are also influential and hence likely to be accommodated through the adoption by the Conference of some version of the archipelagic principles.

The question of the maritime zones of islands is troublesome. The majority view would deny to islands under colonial or foreign domination or situated in the economic zone or the continental shelf of other states the same rights as will be recognized to island or coastal states. In view of the highly controversial nature of this question, of the time constraints for the conclusion of a convention and of the large number of other issues which must be resolved, it seems likely that either no articles will be adopted on this question or if articles are adopted they will be so vague as to leave ample scope for states to interpret them as they wish. A third possibility might be explicitly to reserve the question for future negotiations.

Artificial islands will probably be dealt with on the lines of Article 5 of the 1958 Continental Shelf Convention.

In view of the sharp difference of opinion at the Conference, it is probable that a vague compromise formula will be found with respect to the question of delimitation of national jurisdictional areas between states which are adjacent or opposite each other.

The great majority of coastal states are determined to obtain international recognition not only of broad limits to coastal state jurisdiction but also to the exercise by the coastal state of comprehensive powers within national jurisdictional areas. Differences of opinion center essentially on the nature and scope of coastal state powers within the proposed exclusive economic zone. Although these differences are sharp, the terms of a compromise are foreseeable. The doctrinal dispute as to whether the waters of the exclusive economic zone are, in principle, under coastal state sovereignty or part of the high seas will not be explicitly resolved, but coastal states will receive international recognition of their exclusive resource jurisdiction and of comprehensive powers in the area. On the other hand, freedom of navigation, overflight and of the

laying of submarine cables (perhaps also pipelines) in the economic zone will be affirmed. The treaty will either not deal specifically with the possibility of conflict between the exercise by the coastal state of its rights and the freedoms recognized to the international community within the economic zone or will reserve this question to future negotiations. A third possibility could be the drafting of articles permitting decisions with regard to possible conflicts in priorities and interests between the coastal state and the international community, to be sought in the framework of a compulsory dispute settlement system that states may obligate themselves to establish at some future date.

No treaty can be adopted which does not contain provisions providing some assurance of reasonably unhampered commercial navigation through international straits and through other marine areas under coastal state sovereignty or jurisdiction. All states share this interest; at the same time, as I have already noted, states in a favorable geographical position are determined to retain ultimate control of navigation in their waters. I believe that the dilemma will be resolved, at least with regard to international straits, by the adoption of treaty articles defining as innocent all transit which meets a series of treaty defined conditions; at the same time coastal states will be recognized as having sufficiently wide regulatory powers to enable them to hamper or even prevent transit of vessels either having characteristics of which they may disapprove or belonging to states with which relations are strained.

As I have already observed, negotiations on the question of passage through international straits are being conducted on the assumption that such straits will fall within territorial waters. I have also stated that this is an erroneous assumption. Thus, whatever its provisions, the future treaty will provide no assurance at all of unhampered transit. Nor has any delegation pointed out that all the proposed definitions of international straits, largely derived from the 1958 Territorial Sea Convention, are likely to be irrelevant to the situation as it will exist after the future treaty is adopted: straits will no longer connect "two parts of the high seas" but, at best, territorial waters of riparian states or, more probably, internal waters.

I believe that it is clear from the proposals made at the Conference, many of which I have outlined, that no state has shown excessive concern for the construction of a viable legal framework for ocean space which will serve the interests of international peace and order and permit maximum beneficial use of ocean space, taking into account the implications of rapid technological advance. The immediate purpose of the future treaty will be quite different: it will be to satisfy the immediate perceived interests of coastal states in ocean space; in addition, oceanic states expect the future treaty to lay the legal foundation for a future division of ocean space which will benefit only them, but which cannot be carried out openly at the present time.

We may thus draw a few quick conclusions. First, *the future treaty will multiply, not diminish, conflicts between states.* This conclusion derives from the lack of precision of the formulations favoured by the majority of states with regard to nearly all important matters: from the simultaneous affirmation both of traditional freedoms of the seas and of wide but imprecisely defined coastal state powers within national jurisdictional areas; from the failure to establish a credible compulsory dispute settlement system; and from the inadequate provisions which will be adopted with regard to the important question of delimitation of national jurisdictional areas. A few words on this point may be useful. As national jurisdiction expands in the oceans, national jurisdictional areas of states lying opposite each other increasingly come in contact and sometimes overlap, hence jurisdictional issues will become more acute especially since the value of ocean resources is increasing. The Conference is likely to adopt (a) not necessarily compatible criteria—distance from applicable baselines for the exclusive economic zone and legal continental margin for the continental shelf—for determining the limits of national

jurisdiction; (b) imprecise criteria for drawing straight baselines; and (c) imprecise formulations with regard to criteria for delimitation of national jurisdictional areas. The combination of these circumstances, and of the fact that the exclusive economic zone of one state might easily overlap the legal continental shelf of another, guarantees confusion and conflict.

Secondly, *the future treaty will increase, not decrease inequalities between states.* In the original Maltese concept the seabed and subsequently the entire ocean space beyond reasonable national limits would have been administered by international institutions and its resources exploited for the benefit of the international community as a whole, particularly of poor countries and of land-locked countries. The thought was to utilize the resources of the vast area of ocean space originally beyond twelve, subsequently beyond 200, nautical miles from the coast to diminish inequalities between states. The future treaty will achieve exactly the opposite. Coastal states will be permitted by the provisions of the future treaty to extend their control of marine resources far beyond 200 nautical miles from the coast: international control of seabed resources in the comparatively small area which cannot be claimed by coastal states under an expansive interpretation of the provisions of the future treaty will have little or no significance.

Thirdly, *the commons of the high seas will disappear. A division of ocean space mainly between oceanic countries will be an inevitable, if delayed, consequence of the future treaty.* This is due to the combined operation of two major factors: (a) failure of the Conference to establish precise limits to national jurisdiction in the seas (imprecise baseline provisions, unilaterally definable legal continental shelf limits, recognition of special coastal state rights in areas of ocean space adjacent to national jurisdictional areas) and (b) the fact that the contemporary revolution in ocean uses and effective control of marine pollution require effective administration of ocean space both within and outside national jurisdiction. The traditional concept of the high seas subscribed to by virtually all states is obsolescent. The jurisdictional vacuum in the seas must be filled and it can only be filled either through the extension of national control or through the creation of appropriate international institutions which will preserve as much as possible the substance of freedom. A bureaucratic Seabed Authority with few real powers cannot fulfill this function.

Fourthly, *the future treaty will hamper, not improve the possibilities of effective international cooperation in the oceans.*

No recognized international forum exists for the consideration of ocean related problems in their growing inter-relationships and none will be created by the Conference. An International Seabed Authority will be added to the dozen or so United Nations agencies that already have sectoral competence in matters related to the oceans. Excessive fragmentation and overlaps in functions at the international level will be aggravated and existing gaps in competence will not be filled. All this is particularly serious since effective, not merely rhetorical, international cooperation in the oceans is ever more urgently required as the political and economic importance of ocean space increases and we acquire ever more powerful technologies.

Fifthly, *the future treaty will worsen, not improve, the prospects of continued essential transnational activities, such as scientific research and navigation, or rational exploitation of ocean resources and of the preservation of the marine environment.*

This conclusion derives from the nature of the provisions which will be included in the future treaty: the wide, imprecisely defined and essentially uncontrolled powers given to coastal states will invite abuse and selective interference with vital transnational activities in vast ocean areas; retention of the traditional concept of the high seas will also invite abuse. We may also expect, in most instances, intensified competitive exploitation of the living resources of the sea both within and outside national jurisdiction substantially unrestrained by the management provisions which will be included in the future treaty. These can be foreseen as quite ineffective since on the one hand they

assume that all coastal states have superior management capabilities within the exclusive economic zone—an assumption which is demonstrably false—and on the other hand, at best, will provide for an international organization, with neither the power to inspect fishing vessels nor to exclude new entrants in a fishery, to establish regulations with regard to highly migratory species harvested beyond national jurisdictional limits.

Above all, however, the marine environment is likely to suffer as a result of the future treaty. Present conventions deal essentially with vessel source pollution. To these the treaty will add some expressions of good intentions and some vague principles unaccompanied by the assumption of any enforceable obligations by coastal states. At the same time we may expect that the marine environment in the next couple of decades will become subject to unprecedented pressures, *inter alia,* because of expanding ocean uses and rapidly intensifying exploitation of ocean space resources. Thus we may expect the problem of ocean pollution to become increasingly serious, particularly since practical action for the control of pollution will be left in the hands of coastal states.

States have yet to learn that major activities in the oceans constitute systems, all elements of which interact, which cut across traditional legal regimes and involve ocean space in all its dimensions. Fragmented consideration of different segments of these ocean systems and the automatic application of the principles of freedom or sovereignty prevent the smooth functioning of ocean systems on which we all rely.

Perhaps I have said enough to suggest that the law of the sea treaty which will be adopted by the Conference is likely to achieve results quite different from the expectations of the Secretary General of the United Nations, who is reported, at the inaugural session of the Conference last year, to have called for "the establishment in the oceans of a viable agreed legal basis for international cooperation without conflict and in the interests of mankind."

Ocean Mining: Plans and Problems

Marne A. Dubs

Among the most important—and most controversial—issues before the Law of the Sea Conference is the creation of an international seabed authority to manage and allocate the resources of the deep sea, especially manganese nodules. While the negotiations proceed with no agreement on this provision in sight, the American mining industry is becoming increasingly impatient and is urging congressional action to permit the exploitation of these resources. In the following testimony before the United States Senate Subcommittee on Minerals, Materials, and Fuels (November 1975), Marne A. Dubs, chairman of the Committee on Undersea Mineral Resources of the American Mining Congress, summarizes the effect of the current negotiations on the large companies interested in mining the nodules. The technology is at hand, he claims, but further steps—exploring the mine sites, developing the mining and processing equipment, and investing in production facilities—cannot be undertaken unless the government is willing to guarantee protection against financial burdens arising from a law of the sea treaty and/or against interference by foreign governments.

Mr. Chairman, . . . I am director of the Ocean Resources Department of the Kennecott Copper Corp., and also am chairman of the Committee on Undersea Mineral Resources of the American Mining Congress.

As the representative of the American Mining Congress, I plan to provide your committee a broad summary of current developments in the ocean mining industry. As the representative of Kennecott Copper Corp., I will present a somewhat more parochial view of the state-of-the-art on ocean mining and future plans and problems of the industry. . . .

The American Mining Congress, recognizing the potential importance of minerals from the seabed, has had an active committee on undersea mineral resources for many years now and includes undersea minerals in its yearly major policy statement. As you know, the American Mining Congress is a trade association composed of U.S. companies who produce most of the Nation's metals, coal, and industrial and agricultural minerals. It also represents more than 220 companies who manufacture mining, milling, and processing equipment and supplies, and commercial banks and other institutions serving the mining industry and the financial community.

Mr. Chairman, I greatly appreciate the opportunity to brief the committee on current developments in ocean mining. There indeed has been substantial progress and this progress has now brought us clearly to that point, as we have previously predicted, where future progress will be determined not by the technologists and managers but by resolution of the investment uncertainties resulting from the legal and political issues of the law of the sea.

I last appeared before this committee in March 1974. Since then, there have been two sessions of the Law of the Sea Conference—at Caracas in 1974 and Geneva in 1975. A third session is scheduled for New York in the spring of 1976. I heard in testimony before you last week that these conferences achieved essentially nothing with respect to U.S. interests in the deep seabed and not much, if any, hope was demonstrated for substantive accomplishment in the approaching New York session. I certainly support that analysis and furthermore continue to believe that the prospect of agreement in the foreseeable future is very slim. Now, during this same period of 1974 and 1975 what has happened in the world of industry and technology? Where do things stand today?

The overall status can be summed up in one

230

fashion by this quote from the October 1, 1975, resolution of the American Mining Congress on "Undersea Mineral Resources":

> The technology of recovering minerals from the depths of the sea and processing them into useful raw materials has been under development by private industry for over ten years at its own cost. However, the future course of action by industry is threatened by uncertain investment conditions resulting from the failure to achieve a timely and satisfactory international regime on the law of the sea or domestic legislation which moves toward a stable and predictable investment climate.

I would translate this somewhat institutionalized wording by saying that it means that industry has progressed to the point where the next steps require very large development expenditures and they are deeply concerned about making such expenditures, since their investment may be negotiated away at a Law of the Sea Conference and there is no assurance that their own government will afford them protection against interference in the interim period before a treaty. Without a stable and predictable investment climate, progress toward commercial recovery will be halted or reduced to an insignificant rate.

It is true that industrial activity in ocean mining has continued high in 1974 and 1975 and ambitious plans are being laid for 1976. However, this activity and these plans are based on a prediction and a faith that domestic legislation will be enacted into law and will become effective in early 1976. Now, what has been happening?

Perhaps I should start with the *Glomar Explorer*.* In past committee hearings, I have extolled the probable merits of this ship and the probability that it could be a tool for achieving early success in ocean mining. I have not changed that view and believe that use of this vessel would, in fact, accelerate the pace of ocean mining. Its accomplishments, as revealed by

the press, are indeed impressive and there is little doubt that it is a further confirmation of our contention that the technology of ocean mining is at hand.

Of course, it appears that one potential ocean miner, The Summa Corp., has disappeared. However, a new one, Lockheed, has appeared and appears to have a substantive program and serious plans. Their representative is before the committee today.

In early 1974, the Kennecott Copper Corp. announced the establishment of a major consortium to work on ocean mining. It consists of Kennecott, Noranda Mines of Canada, Mitsubishi Corp. of Japan, Consolidated Gold Fields of the United Kingdom and Rio Tinto Zinc of the United Kingdom. A major program has been underway and is essentially on the projected schedule.

In this same 1974–75 period, both Deepsea Ventures and The International Nickel Co., INCO, announced consortia. . . .

In addition, the CLB group (Japanese Continuous Line Bucket System) remains active and is planning tests in the Pacific this year. The CLB group has a large number of participants including the French organization CNEXO. In Japan, an association called DOMA, Deep Ocean Mining Association, is carrying out work emphasizing exploration in the Pacific Ocean. . . .

[M]y own assessment based on informed observation of the scene is that the "collective" position of the seabed developers is as follows:

(1) They have identified nodule deposits which could provide satisfactory mine sites. Mine site definition will require a large amount of work and substantial expenditures. To carry this work out without protection of investment and assurances for the future ability to mine the site is financially very risky.

(2) They have largely solved the metallurgical problems of winning metals from nodules. They have either run pilot plants or will do so in the near future. The next steps will require much larger and more costly pilot plant or demonstration plant tests. . . .

Ed. note: The *Glomar Explorer,* supposedly built to mine manganese nodules from the ocean floor, was actually used by the Central Intelligence Agency to recover parts of a sunken Russian submarine.

231

(3) Development of the mining systems has progressed from the drawing board and computer stage and away from simple laboratory tests to large scale at sea experimentation. The costs of such work are very high and in fact are unique in industrial technology development.

In conclusion, there is no longer any doubt about the technical and economic feasibility of ocean mining. The technology is ready; the investment climate is not.

When I say the investment climate is not ready, I do not have in mind some vague generality such as stated by Secretary Maw at your October 29, 1975 hearings; namely, "We encourage private investors to develop their technology and to begin mining when they are ready." This, like other motherhood statements, is a fine sentiment. However, it provides zero assurances as to the future financial rules under which a miner would operate or as to whether he would be permitted to mine his hard chosen spot on the seabed under some distant resolution of the law of the sea treaty now being negotiated. Furthermore, the miner has no protection against the losses which would occur if his mining operations were interfered with under the present high seas regime. It is simply not feasible to invest under such circumstances and furthermore, it is not feasible to obtain funds to invest even if the entrepreneur were a high rolling risk taker.

Looking at it another way, the miner needs security of tenure over a mine site in order to spend the money necessary to explore completely the mine site, devise a mining plan, develop the expensive equipment to mine and process the minerals, and finally, to invest in the production facilities. This is the usual basis under which a mining project can be financed and undertaken. Since such security of tenure cannot be obtained under present circumstances, a substitute must be found. This substitute is protection from the risk that the industry will no longer be able to mine or would have investments impaired from financial burdens as a result of a law of the sea treaty agreed to by the United States and insurance against interference by foreign governments or firms against which there are no legal remedies available. This, Mr. Chairman, is the essence of the cure for the present unsatisfactory investment climate.

I would now like to say a few additional words about the state of the art on ocean mining as seen through the eyes of my own company. We believe we understand quite well the location, metallurgical grade, and tonnage of large manganese nodule deposits of potential commercial interest. We have tested mining equipment at 15,000 feet depth on the floor of the Pacific Ocean. . . . We believe the technical feasibility of ocean mining is clear.

Finally, we have completed our pilot plant work on the development of a hydrometallurgical process to extract metals from nodules. This unique process is low in energy consumption.

In summary, we have completed phase 1 of our program to bring nodules to commercial reality. We must now tackle phase 2 which involves the development of prototype equipment. This effort will be very expensive, requiring massive equipment and large ships. Once committed to these expenditures, there is almost no turning back from commercialization. Thus we hesitate because of the uncertainty of the investment climate resulting from the law of the sea issues. We have carried it this far on faith. We cannot go very much further. If legislation to solve the problem is not enacted into law in the next few months, we will be forced to review our plan in view of the large financial risks such inaction would impose.

Mr. Chairman, thank you for this opportunity. I hope this will assist the committee in reaching a positive decision with respect to the pending legislation. The enactment of S. 713★ into law would resolve the issue and encourage the movement of ocean mining from research and development into commercialization.

★*Ed. note:* This bill was introduced on February 18, 1975 by Senator Lee Metcalf "to provide the Secretary of the Interior with authority to promote the conservation and orderly development of the hard mineral resources of the deep seabed, pending adoption of an international regime therefor." It would permit licensing of eligible applicants to develop hard mineral resources for a $50,000 fee.

Deep-sea Claim

Deepsea Ventures, Inc.

Among the mining companies eager to force an answer to the question of ownership of the resources of the deep ocean is Deepsea Ventures, Inc. Just as the filing of mining claims with the Interior Department marked the exploitation of gold and silver deposits on the western frontier, so for the first time a mining claim has been filed to exploit the manganese-nodule deposits on the oceanic frontier. In the following letter to the secretary of state, Deepsea Ventures, Inc. lays claim to exclusive mining rights in a 60,000-square-kilometer section of the Pacific and requests United States diplomatic protection of its investment. The government's action may determine whether mining will proceed in the absence of international agreements.

November 14, 1974

The Honorable Henry A. Kissinger
Secretary of State
U.S. Department of State
2201 C Street
Washington, D.C.   20520

>*Notice of Discovery and Claim of Exclusive Mining Rights, and Request for Diplomatic Protection and Protection of Investment, by Deepsea Ventures, Inc.*

My dear Mr. Secretary:

Deepsea Ventures, Inc., a Delaware corporation having its principal place of business in the County of Gloucester, The Commonwealth of Virginia, U.S.A., respectfully makes of record, by filing with your office this *Notice of Discovery and Claim of Exclusive Mining Rights and Request for Diplomatic Protection and Protection of Investment, by Deepsea Ventures, Inc.* (hereinafter "Claim"), as authorized by its Board of Directors by resolution dated 30 October 1974. . . .

Notice of Discovery and Claim of Exclusive Mining Rights

Deepsea Ventures, Inc., (hereinafter "Deepsea"), hereby gives public notice that it has discovered and taken possession of, and is now engaged in developing and evaluating, as the first stages of mining, a deposit of seabed manganese nodules (hereinafter "Deposit"). The Deposit . . . is encompassed by, and extends to, lines drawn between the coordinates numbered in series below, as follows:

From: (1)   Latitude 15°44′ N   Longitude 124°20′W
      A line drawn West to:
      (2)   Latitude 15°44′ N   Longitude 127°46′ W
      And thence South to:
      (3)   Latitude 14°16′ N   Longitude 127°46′ W
      And thence East to:
      (4)   Latitude 14°16′ N   Longitude 124°20′ W
      and thence North to the point of origin.

These lines include approximately 60,000 square kilometers for purposes of development and evaluation of the Deposit encompassed therein, which area will be reduced by Deepsea to 30,000 square kilometers upon expiration of a term of 15 years (absent force majeure) from the date of this notice or upon commencement (absent force majeure) of commercial production from the Deposit, whichever event occurs first. The Deposit lies on the abyssal ocean floor, in water depths ranging between 2300 to 5000 meters and is more than 1000 kilometers from the nearest island, and more than 1300 kilometers seaward of the outer edge of the nearest continental margin. It is beyond the limits of seabed jurisdiction presently claimed by any State. The overlying waters are, of course, high seas.

The general area of the Deposit was identified in August of 1964 by the predecessor in interest of Deepsea, and the Deposit was discovered by Deepsea on August 31, 1969.

Further exploration, evaluation, engineering development and processing research have been carried out to enable the recovery of the specific manganese nodules of the Deposit and the production of products and byproducts therefrom. . . .

Deepsea, or its successor in interest, will commence commercial production from the Deposit within 15 years (absent force majeure) from the date of this Claim, and will conclude production therefrom within a period (absent force majeure) of 40 years from the date of commencement of commercial production whereupon the right shall cease.

Deepsea has been advised by Counsel, whose names appear at the end hereof, that it has validly established the exclusive rights asserted in this Claim under existing international law as evidenced by the practice of States, the 1958 Convention on the High Seas, and general rules of law recognized by civilized nations.

Deepsea asserts the exclusive rights to develop, evaluate and mine the Deposit and to take, use, and sell all of the manganese nodules in, and the minerals and metals derived, therefrom. It is proceeding with appropriate diligence to do so, and requests and requires States, persons, and all other commercial or political entities to respect the exclusive rights asserted herein. Deepsea does not assert, or ask the United States of America to assert, a territorial claim to the seabed or subsoil underlying the Deposit. Use of the overlying water column, as a freedom of the high seas, will be made to the extent necessary to recover and transport the manganese nodules of the Deposit.

Disturbance of the seabed and subsoil underlying the Deposit will be temporary and will be restricted to that unavoidably occasioned by recovery of the manganese nodules of the Deposit. To facilitate the United States of America's domestic policies and programs of environmental protection, Deepsea will provide, at no cost, reasonable space for U.S. Government representatives of the United States of America on vessels utilized by Deepsea in the development and evaluation of the Deposit. Deepsea does not intend to process at sea the manganese nodules from the Deposit.

It is Deepsea's intention, by filing this Claim in your office and in appropriate State recording offices, to publish this Claim and provide notice and proof of the priority of the right of Deepsea to the Deposit, and its title thereto.

A true copy of this Claim is being filed for recordation in the office of the Secretary of State of the State of Delaware, U.S.A., the State wherein Deepsea is incorporated, and on 15 November 1974 in the office of the Clerk of the Circuit Court of Gloucester County, Virginia, U.S.A., the county and Commonwealth of Deepsea's principal place of business. Copies of this Claim are also being provided to others. . . .

We ask that this Claim, and all of the annexed Exhibits, be made available by your office for public examination.

Request for Diplomatic Protection and Protection of Integrity of Investment

Deepsea respectfully requests the diplomatic protection of the United States Government with respect to the exclusive mining rights described and asserted in the foregoing Claim, and any other rights which may hereafter accrue to Deepsea as a result of its activities at the site of the Deposit, and similar protection of the integrity of its investments heretofore made and now being undertaken, and to be undertaken in the future.

This request is made prior to any known interference with the rights now being asserted, and prior to any known impairment of Deepsea's investment. It is intended to give the Department immediate notice of Deepsea's Claim for the purpose of facilitating the protection of Deepsea's rights and investments should this be required as a consequence of any future actions of the United States Government or other States, persons, or organizations.

The protection requested accords with the assurances given on behalf of the Executive Department to the Congress of the United States, including those by Ambassador John R. Stevenson, by Honorable Charles N. Brower, and by Honorable John Norton Moore, as follows:

"The Department does not anticipate any efforts to discourage U.S. nationals from continuing with their current exploration plans. In the event that U.S. nationals should desire to engage in commercial exploitation prior to the establishment of an internationally agreed regime, we would seek to assure that their activities are conducted in accordance with relevant principles of international law, including the freedom of the seas and that the integrity of their investment receives due protection in any subsequent international agreement." Letter of January 16, 1970, from John R. Stevenson, Legal Advisor, Department of State, to Lee Metcalf, Chairman, Special Subcommittee on the Outer Continental Shelf, U.S. Senate, reproduced in Hearings before the Special Senate Subcommittee on the Outer Continental Shelf, 91st Cong., 1st and 2d Sess. at 210 (1970).

"At the present time, under international law and the High Seas Convention, it is open to anyone who has the capacity to engage in mining of the deep seabed subject to the proper exercise of high seas rights of other countries involved." Statement of Charles N. Brower, Hearings before the House Subcommittee on Oceanography of the Committee on Merchant Marine and Fisheries, 93rd Cong., 1st Sess., at 50 (1974).

"It is certainly the position of the United States that the mining of the deep seabed is a high seas freedom and I think that would be a freedom today under international law. And our position has been that companies are free to engage in this kind of mining beyond the 200-meter mark subject to the international regime to be agreed upon and, of course, assured protection of the integrity of investment in that period." Statement of John Norton Moore, Hearings before the Senate Subcommittee on Minerals, Materials and Fuels, 93d Cong., 1st Sess., at 247 (1973).

The language of these extracts, and other statements similar to them made by these and other responsible officers of the Executive Branch is consistent with the Executive's continuing practice as reflected in a paragraph in President Taft's Message to the Congress of December 7, 1909, where he said:

"The Department of State, in view of proofs filed with it in 1906, showing American possession, occupation and working of certain coal-bearing lands in Spitzbergen [Spitzbergen was at that time recognized as being not subject to the territorial sovereignty of any State] accepted the invitation under the reservation above stated [i.e., the questions of altering the status of the islands as countries belonging to no particular State and as equally open to the citizens and subjects of all States, should not be raised] and under the further reservation that all interests in those islands already vested should be protected and that there should be equality of opportunity for the future." *Annual Message of the President to Congress 7 December 1909.*

Deepsea has used its best efforts to ascertain that there are no pipelines, cables, military installations, or other activities constituting an exercise of freedom of the high seas in the area encompassing the Deposit or in the superjacent waters, with which Deepsea's operations might conflict. So far as is known, no claim of rights has been made by any State or person with respect to said Deposit or any other mineral resources in the area encompassing the Deposit and no State or person has established effective occupation of said area.

Initially, approximately 1.35 million wet metric tons of nodules will be recovered by Deepsea from the Deposit per year. In accord with market conditions, this may later be expanded to as much as 4 million wet metric tons per year recovered. Deepsea's processing and refining technology, successfully demonstrated in its pilot plant, will recover copper, nickel, cobalt, manganese, and other products, depending on the market situation and competitive conditions. The recovery weight of the major four metals that the initial 1.35 million wet metric tons of nodules will yield per year will be approximately as shown in Column A below. Column B gives some indication of the dependency of the United States of America upon imports for these four metals.

	A	B
		Net U.S. Imports (1972) as a
	Production Metric Tons	Percentage of U.S. Consumption
Nodules	1,350,000	—
Copper	9,150	9%
Nickel	11,300	71%
Cobalt	2,150	92%
Manganese	253,000	93%

The importance of these minerals to the economy of the United States does not require elaboration. It has been effectively expressed to the Congress by the Executive Branch.

For your information, the capital stock of Deepsea is at present wholly owned by nationals of the United States. Ninety per cent thereof is owned by Tenneco Corporation, a Delaware corporation, and the other ten per cent is owned by individuals, all of whom are United States citizens. At this date stock options are outstanding which, if all are exercised, will result in acquisition of the following percentages of ownership of Deepsea's capital stock by others:

23.75%:   Essex Iron Company, a New Jersey corporation, a wholly owned subsidiary of United States Steel Corporation, a Delaware corporation.

23.75%:   Union Mines Inc., a Maryland corporation, a wholly owned subsidiary of Union Miniere, S.A., a Belgian corporation.

23.75%:   Japan Manganese Nodule Development Co., Ltd., a Japanese corporation.

     Respectfully,

DEEPSEA VENTURES, INC.

Counsel: By John E. Flipse, President

Northcutt Ely

L.F.E. Goldie

R.J. Greenwald

World Sea Power: U.S. vs. U.S.S.R.

J. William Middendorf II

The primary role of a navy, Mahan wrote in 1890, was to protect commerce, except in the case of a nation that had aggressive tendencies and made the navy a branch of the military establishment. But changed world political conditions since his day have demanded new roles for the navy. In the following speech delivered before the Rotary Club of San Francisco on December 3, 1974, J. William Middendorf II, secretary of the navy, reviews the traditional role of the United States Navy and outlines its current missions: sea control, projection of United States power ashore, the establishment of a political presence through the deployment of its fleet, and strategic deterrence. Arguing that our national survival depends more on seapower today than ever before, Middendorf points with alarm to the threat of a growing Soviet navy and to the declining strength of the United States Navy.

Throughout the history of the United States, seapower has been an integral part in making this nation great. A strong merchant marine has enabled the development of our economy and has transported the raw materials needed for our industrial base.

And a strong Navy has defended our coasts, protected our sea lines of communication, deterred aggression, and projected the presence of the United States throughout the world.

On 13 October 1775, the Continental Congress authorized the acquisition and construction of ships for the Continental Navy. Our country was severely challenged then. The perception of that Congress and of the peoples of those colonies, which lined the eastern seaboard, was necessarily fixed on the Atlantic Ocean and across it to Europe. For it was from Europe, across the Atlantic, that would come any foreign intervention. And it was Europe, across the Atlantic, upon which the agricultural and maritime economies of the colonies depended for our external markets and many manufactured goods.

And therefore a strong Navy was properly regarded as an essential element of defense to meet the challenges of the first decade of American history.

The War of 1812 was fought ostensibly to defend the doctrine of "freedom of the seas." At the beginning of that war, the U.S. Navy was composed of just 14 seaworthy warships. But our Navy took to the high seas against overwhelming British seapower—1,048 men-of-war—and won a number of brilliant single ship actions.

In the late nineteenth century three developments occurred which led to the evolution of the U.S. Navy's *oceanic* strategy.

First, in 1890, Rear Admiral Alfred Thayer Mahan published his internationally famous work *The Influence of Sea Power on History.* Mahan's strategic analysis rested on the military principle that the key to defeating an enemy is concentrating superior offensive force against his main force. He also delineated the *vital* links between the burgeoning American industrial production, overseas markets, the merchant marine, and the Navy.

Secondly, westward expansion and development in the United States had increased tremendously along the Pacific Coast. American interests were expanding in the Pacific and the Caribbean. And the United States found itself facing increasing European naval competition in areas thought crucial to vigorous overseas commercial activity.

Finally, steam and steel warships were replacing the Navy's wooden sailing ships. These new ships greatly increased the Navy's sea-keeping and combat capabilities.

Naval bases in Hawaii, Wake, Guam, Samoa,

and the Philippines assured the United States access to the markets and raw materials of the Pacific Basin. These bases, together with the Panama Canal and additional bases in the Caribbean, would also give the United States a protective cordon to deter attack by any major power.

Rear Admiral Mahan was the strategic thinker. President Theodore Roosevelt put seapower into action. The Spanish-American War had projected the United States into the role of a world power with overseas interests and territories requiring a strong Navy. In 1907 he sent his famous "Great White Fleet" on a two year, round-the-world cruise. This cruise demonstrated the U.S. Navy's technical proficiency, engendered goodwill, and increased American prestige.

In reflecting on the cruise of the "Great White Fleet," Roosevelt said: "In my own judgment the most important service I rendered to peace was the voyage of the battle fleet around the world."

Since their practical application by Theodore Roosevelt, American naval and, in many respects, our overall military strategy has reflected these geopolitical, oceanic concepts. And World War I and World War II represented the triumphs of that strategy. The United States Navy and Marine Corps had clearly and successfully fulfilled three missions:

(1) Sea control—primarily in anti-submarine warfare and principally in the North Atlantic, the Navy insured the unimpeded flow of men and materiel to Europe.

(2) Projection of power ashore—primarily in the Pacific theater, U.S. Marines and tactical aircraft effectively wrested strategic chains of islands from the Japanese.

(3) And the end of World War II saw a burgeoning of the Navy's mission of political presence, particularly through the continuous deployment of aircraft carriers to the Mediterranean and western Pacific.

Yet World War II ended with certainly the most significant weapon of strategic warfare.

On July 16, 1945, the United States exploded its first atomic bomb at Alamogordo, New Mexico. The following month that weapon was used twice to accelerate the end of the war with Japan.

Four years later, the U.S. monopoly of the atomic bomb ended, with President Truman's announcement on September 23, 1949, that an atomic explosion had occurred in the Soviet Union.

This event brought an entirely new equation into the broad spectrum of world affairs. Now the two superpowers had the capability to reap massive destruction on each other. And therefore the development of the military and political strategy of deterrence began to take shape.

For the Navy, this fourth mission took practical shape from 1950, when its first nuclear-capable carrier aircraft, the AJ-1/SAVAGE, was delivered. The Navy's strategic deterrence mission became more sophisticated in December 1960, when the USS GEORGE WASHINGTON went to sea with sixteen Polaris missiles.

Therefore, in the years after World War II, the Navy's three traditional missions of sea control, projection of power, and political presence and its fourth modern mission of strategic deterrence were successfully adapted to our national military strategy that encompassed potential warfare against major powers as well as a conventional naval role in smaller scale confrontations.

There is no doubt that seapower has played an integral role in the development of this country.

Today we as a nation are confronted by many crucial issues. Our ability to survive as a nation depends more on seapower than ever before.

The first of these issues is political. There has been an increase in tensions in the Middle East. As shown by previous crises in this sensitive area, U.S. seapower has been manifest in the *stabilization* of potentially violent situations and the *deterrence* of outside intervention.

U.S. seapower has inherent capabilities which can be used as instruments of national policy. Principal among the capabilities of U.S. seapower are:

(1) Ability to display a credible U.S. military commitment within range of almost any area of the world.

(2) Ability to protect or evacuate U.S. nationals in almost any area of the world.

(3) Ability to provide a military presence without an automatic commitment of forces.

(4) Ability to clearly demonstrate U.S. interest and capability, without disclosing precise intentions.

And seapower is the least susceptible to diplomatic constraints, base rights, overflight privileges or the pressures of host countries.

Today the international arena is characterized by a transition from the strict bipolarity of the immediate post World War II years. Now and in the future international issues will be increasingly characterized by political and economic issues. And U.S. seapower, within a broader national strategy, will be essential in providing international stability and in supporting our national interests.

The second issue is economic. During the past year, the monetary reserves owned by the oil producing countries have risen threefold from $13 billion to $38 billion. These reserves represent 19% of the world's monetary reserves, and are almost equal to the reserves of all of industrial Europe as recently as 1970. And Secretary of the Treasury William Simon recently noted that these accumulations of capital could exceed $500 billion by 1980, twelve times what it is now.

For example Saudi Arabia now ranks fourth among the world's countries with the largest monetary reserves. Last year at this time the Saudis held $4 billion, a healthy sum that placed them 13th on the world scale. Today their re-

U.S. NAVY VESSELS TIED UP AT
SAN DIEGO FACILITY

"Today . . . our ability to survive as a nation depends more on seapower than ever before."—Middendorf

serves total $11.5 billion. Ahead of the Saudis are only West Germany ($32.5 billion), the United States ($15.7 billion), and Japan ($13.2 billion).

This rapid accumulation of wealth, and its concentration in the extremely sensitive Middle East, threatens to create new pressures against the world's financial system.

U.S. seapower may well play a major role in the stability of this strategic area, wherein are concentrated more than 50% of the world's proven oil reserves.

Furthermore, the heaviest concentration of seaborne oil traffic in the world is through the Strait of Hormuz. Through this narrow waterway connecting the Arabian Gulf with the Indian Ocean pass an average approaching 20 million barrels of oil *per day,* representing half of world oil exports. Domination of this Strait by an unfriendly power could effectively cut off much of the free world from its prime source of oil. Similarly, domination of the Indian Ocean by unfriendly naval forces could hinder or deny freedom of passage to the oil carriers.

Soviet ships now regularly operate in the Indian Ocean with generally a missile-armed cruiser, some destroyers, an amphibious ship, and several auxiliary ships supporting Soviet interests. By contrast the U.S. Navy's Middle East Force is composed of only three ships. In addition the U.S. now periodically sends carrier task groups into the Indian Ocean.

The third issue is also economic. Today the United States is an island nation, growing increasingly dependent on foreign sources of raw materials and energy.

There are 71 strategic raw materials and metals which are integral to the economic base of the United States. We must *import* 68 of these raw materials and metals, all or in part *by sea.*

In 1973, the U.S. imported by sea more than 50 million tons of these raw materials and minerals valued in excess of $4 billion.

And in 10 years, it is estimated that half again as much of our essential imports will come by sea.

We are also dependent on energy imports. In 1973, we imported by sea about 580 million barrels of crude oil and refined products accounting for 29% of our domestic consumption.

Crude Oil and Refined Products Imported by Sea Sources

	1973
Arab Countries	10%
Indonesia	1%
Iran	2%
Nigeria	3%
Venezuela	12%
Other	1%
% of Total U.S. Consumption	29%

It is the mission of the U.S. Navy to protect the sea lanes for the transport of these critical imports. And it is the mission of the U.S. Navy to render a political and diplomatic presence in the world today in support of our national policy.

Yet today the U.S. Navy has the fewest number of ships in the active fleet since the period a year and a half before Pearl Harbor. The ships of today's Navy are indeed much more capable than their World War II–vintage counterparts.

Today's aircraft carrier ENTERPRISE (CVAN-65), with its Air Group of sophisticated F-14s, A-6s, and A-7s, packs a much more potent punch than her World War II predecessor (CV-6).

However, just six years ago we had an active fleet of nearly 1,000 ships. Because most of these were constructed during World War II and because of budget restraints we have cut our Navy in half so that today we are about to go below 500 ships.

The service life expectancy for warships is 25 to 30 years. In 1968 the average age of the active fleet had increased to 18 years. It therefore became unmistakably clear that we had to reduce the numbers of our older ships in order to build the new ships we need.

Between 1963 and 1967 the Navy planned nearly 50 new ships per year. Then came the fiscal demands of the Vietnam war with its requirements for operating funds at the expense of shipbuilding.

We also undertook the costly conversion of 31 Polaris submarines to carry the Poseidon missile, which substantially upgraded the nation's strategic deterrence capabilities.

In the last six years, although we were forced to reduce the numbers of aging ships in the active fleet, the Navy received authority to build 13 ships per year. This rate of shipbuilding— 13 ships per year—means that the U.S. Navy will never recover the strength of 900 ships but also will not even be able to sustain a fleet of 500 ships unless increases are made in the number of ships being built each year.

We have, in effect, undertaken a course of action which if allowed to go uncorrected would amount to a unilateral naval disarmament.

There are a number of solutions to this critical problem.

In our 1975 shipbuilding program, we initiated two programs which will contribute to the fleet's numerical strength because we are buying and building in *numbers.*

The first ship is the Missile Patrol Hydrofoil (PHM). Capable of operating at high speeds in heavy sea conditions, the PHM will provide the Navy a potent punch in many different operational environments. And ships of this class can readily assume the patrol and surveillance functions formerly assigned to our larger ships, thereby relieving our SPRUANCE- and KNOX-class combatants for other duties such as carrier task force operations and anti-submarine warfare. We hope to build 30 missile patrol hydrofoils.

The second ship is the Patrol Frigate (PF). We envision this ship to be a low-cost workhorse for the fleet, designed to carry out the very important escort mission. We hope to build 50 of these ships.

Additionally we are continuing to build SPRUANCE-class destroyers, LOS ANGELES-class nuclear-powered attack submarines, and three nuclear-powered aircraft carriers.

An even more advanced concept is the Surface Effect Ship, which is capable of speeds in excess of 80 knots. Once the research and development is completed, we could build and deploy these ships over the oceans of the world. Their high speed capability and sophisticated weapons systems would make them potent and nearly invulnerable ships of the future.

Congress is very much aware of the Navy's need for more ships and the problems in our shipbuilding programs. Congress approved the construction of 22 ships in the budget for this year.

However, this ghastly inflation, with which you and I are struggling, is crippling our shipbuilding effort.

For example, the price of scrap steel has nearly doubled in less than a year. And overall cost escalations will probably reach 20%. Therefore, due principally to inflation, our Navy shipbuilding programs are now $2 billion in debt.

I can assure you that people in all levels of government are aware of these critical problems and are working to resolve them. I feel everyone in the Executive and Legislative Branches recognizes the *need.* It is ultimately a question of accomplishing the task of rebuilding the fleet in the context of the greater issues facing us.

It is within the foregoing that I would address a most significant *military* issue. Having cut the size of the U.S. fleet in half and having a shipbuilding program jeopardized by inflation, we now face the challenge of the Soviet Navy.

Let me say at the outset that the Soviet Navy is *not* ten feet tall. The U.S. Navy still has tremendous staying power and real, combat-proven strength.

Yet it is the *trend* of expanding Soviet naval capabilities that I personally view as one of the most significant strategic developments since the atomic bomb.

This *trend* in burgeoning Soviet naval capabilities, when considered together with other Soviet tactical and strategic military developments, would give any Secretary of the Navy cause for concern.

With regard to the Soviet Navy:

(1) Since 1962 they have outbuilt us in every category of ship except aircraft carriers. During the period 1962–1973, they added 271 major

242

surface combatants and submarines as opposed to our 176 new ships.

Current Ship Inventories—Major Combatants and Submarines

	Soviet Navy	U.S. Navy
Attack Carriers	2 (CHG)	14
Cruisers	30	6
Destroyers	80	90
Escorts	109	66
Submarines	325	115
Total	546	291

Ship numbers and comparisons can be extremely dangerous. We cannot compare a 90,000-ton U.S. aircraft carrier to an 850-ton Soviet "minor combatant." Yet the more than 135 guided-missile firing patrol boats in the Soviet inventory do pose a significant threat.

For example, the Soviet OSA and KOMAR classes, now more than 13 years old, have been used with success in combat. An 80-ton Egyptian KOMAR sunk the 2,500-ton Israeli destroyer EILAT on 21 October 1967. The operations of Indian OSAs against merchant shipping in the war with Pakistan in December 1971 were also very successful.

The Soviet 850-ton NANUCHKA-class carries the SS-N-9 surface-to-surface missile. Its 1,100-pound warhead is twice that of the surface-to-surface missile we plan for the future.

(2) The Soviets have deployed aboard their ships highly sophisticated sensors, electronics and offensive and defensive weapons systems. They have developed an arsenal of some 20 types of anti-ship capable missiles and their variants. Having ranges from 15 to 400 miles, these missiles can be fired from aircraft, surface ships, and from submerged submarines.

I personally believe that a number of years ago we made a serious error when we chose to put all of our missile capability on our sea-based aircraft without developing an over-the-horizon shipborne surface-to-surface missile. At that time we had a force of 23 aircraft carriers, combat-proven during World War II. Now with only 14 aircraft carriers and a major Soviet anti-ship

243

missile capability we are on a crash program to introduce our first over-the-horizon surface-to-surface missile—HARPOON.

(3) The Soviet Navy has gone to sea. From a coastal defense navy in 1962, today we observe Soviet ships circumnavigating Hawaii, exercising in the Caribbean, and operating near the strategic oil routes of the Indian Ocean.

(4) They have upgraded their strategic submarine ballistic missile force. The new Delta-class submarine carries the 4,200 mile SS-N-8 missile, which is capable of hitting us here in San Francisco from its homeport a few hundred miles west of Alaska.

(5) And these *naval* developments coincide with the impending Soviet deployment next year of *four* new classes of ICBMs with warheads of one-half to 25 megatons, and a new long-range bomber, the BACKFIRE.

(6) Furthermore, preliminary analysis of U.S. and Soviet military production rates indicates that the Soviets are out-producing us in all but a few categories of military equipment by from 2:1 to 12:1.

This is where we stand today and these are the issues we face. We are more dependent on foreign sources of raw materials and energy. We have a much smaller Navy to do the job than we did six years ago, and the Soviets are building a tremendous navy at a rate faster than we are. And we are faced with a tense world economic and political situation.

For this we need the credibility of a strong U.S. Navy more than ever before.

I have never met an American who wants the U.S. Navy to become second best. Each of you in this room, either by your personal experience or knowledge, knows that the U.S. Navy was unchallenged—in Vietnam and in Korea and clearly won the battle for naval supremacy in World Wars I and II.

Yet for the United States to keep a Navy "second-to-none" we need the support of each of you. Each of you as leaders of the community can carry this message. *You* are the ones who will ultimately determine the future of the Navy and this country.

Sea Power in the 1970s

George H. Quester

Is naval warfare becoming obsolete? In the following selection, George H. Quester argues that the advent of nuclear weapons has radically altered traditional naval strategy. Whereas the navy played a major role in both world wars, naval warfare in the future is unlikely, he maintains, precisely because it would probably become nuclear and no nation wants to risk a nuclear war. As a result, naval ships function primarily as mobile military bases and as obstacles to the fleets of other nations. Their role at sea is largely symbolic: They are a trip wire of national commitment, but they are unlikely to attack another nation's fleet. Quester, professor of government at Cornell University, also analyzes the political effect of the increase in Soviet merchant shipping. This article was excerpted from a paper prepared for the Conference on Problems of Naval Armaments, convened in April 1972 by the Cornell University Program on Peace Studies to consider the emerging competition of the Soviet Union with the United States on the high seas.

Six or eight years ago, it was generally agreed that the United States together with its allies possessed an overwhelming superiority in the kinds of naval force which had been used in World War II. It was also generally thought that such force would be of limited and diminishing importance for any political and military interactions likely in the future. Such strength, to be sure, had been of immense importance in blockading or quarantining Cuba in the missile crisis, and it had facilitated interventions by United States forces in the Dominican Republic, in Southeast Asia, and in the Middle East. Yet the mere absence of a naval challenge from the Soviet bloc, and the enormous destructive force of the nuclear weapons in the air and land forces, suggested more nostalgia than real military function in the Navy's conventional firepower.

All of such reasoning has now come under challenge, if only because the Soviet Union has chosen to invest substantially in more modern versions of the very classes of ships that were being pooh-poohed, and to deploy such ships into the Mediterranean Sea, the Pacific and Indian Oceans, and at a token level, even into the Caribbean. Even if the Soviet Navy has not really grown in tonnage, it has indeed developed a capability for operating at much longer distances from its home ports.

Surface Navies

A stream of excited statements has thus emerged from various Western sources, alarmed about the substantial increase of Soviet naval power along N.A.T.O.'s "southern flank." The Soviet Navy, to be sure, has not yet invested in aircraft carriers comparable to those of the U.S., British or French navies, but it has deployed two helicopter carriers, and even these represent a noticeable expansion from what had been a zero base. More attention is devoted to destroyers and cruisers, most of them fitted with surface-to-surface missiles, an area of weaponry in which the West had lagged until recently. When aircraft based on friendly shore bases on the Mediterranean coasts are totalled in, the firepower of the Soviet fleet in the area does not exceed that of the U.S. Sixth Fleet, but the comparison is not ludicrous anymore. . . . [The] ships . . . of the United States are older than their Soviet opposite numbers, and not as well equipped in surface-to-surface missile firepower.

Yet all such comparisons of tonnage or conventional firepower, as if World War II were to be fought over again, may miss the point. The U.S. Sixth Fleet carries nuclear weapons, and so presumably does the Soviet Mediterranean fleet. It is thus very unlikely that a naval war could be fought in the area without such weapons coming into use, even leaving aside the tactical nuclear weapons that could be delivered by the aircraft of either side based within flying range of the Mediterranean. . . .

The shadow of possible or probable use of nuclears thus at least alters much of naval strategy as it would have been applied in 1944. Large concentrated convoys would be terribly inviting targets for the use of a single bomb, and thus might have to be foregone. Concentrations for amphibious assaults such as the storming of Normandy would similarly be too tempting as targets for tactical nuclear weapons. . . . Some interesting speculation has come forth on whether nuclear weapons will assist one side or the other in the murkier realm of submarine and anti-submarine warfare. Nuclear warheads on torpedoes might, as suggested above, make convoys easily destructible for any U-boat, but perhaps the same warheads will be harnessable as very effective depth charges, making it much easier for a surface vessel to destroy a submarine.

Yet any such revision of comparative firepower figures, or of naval strategy, again may miss the point of what naval confrontations are all about in the nuclear age. To use nuclear weapons at sea, or to engage in combat operations which would probably risk their use, would be to upset a very serious nuclear allergy, with risks of fallout and destruction in the countries abutting the seas involved, with great risks of escalation to World War III. Nuclear naval warfare is unlikely to happen precisely because it is so dangerous, and naval warfare is unlikely to happen because it is so likely to become nuclear.

Given all the attention that has been paid to "limited war" in the last fifteen years, this might seem excessively pessimistic on the risks of escalation in a war involving ships carrying nuclear weapons. . . .

Yet escalation threshholds are in the eyes of the beholder, and the risk of an unauthorized firing of a nuclear weapon still has some extra inhibiting effect wherever such weapons are known to be deployed. . . . If an attempt were made to sink an American ship carrying nuclear warheads, could the opposing side really be sure that such warheads would not come into use? The world has seen a surprising degree of conventional-war exemption for military forces which are nuclear equipped. Is it the defenses of the U.S. carriers in the South China Sea that protect them against surface-to-surface missile attack, or is it their "trip-wire" role, which perhaps dissuades the Russians from ever equipping Hanoi with this ability to sink such ships?

Even if nuclear weapons were not based on board the aircraft carriers of the U.S. Navy, . . . the mere size of such ships, their large crews and enormous financial costs, will make it unthinkable that an enemy could be allowed to sink one without the most serious risk of escalation to serious warfare. . . . The simple fact that large ships rather than small are required to handle airplanes ties the commitment of the United States together into a bundle that cannot be ignored.

The military strength of a naval force thus cannot so easily be fractionated when it is embarked at sea. When it is disembarked, when Marines are landed, or airplanes are launched on missions over some country, the American presence conversely can be much more fractionated. Individual planes can be shot down by some hostile protégé of the U.S.S.R.; individual soldiers or Marines can be killed, without a risk of World War III; it may even be Russians manning the surface-to-air missile (S.A.M.) sites that bring individual planes down. Back on the large ships that carry them, still escorted by destroyers and cruisers at close range, the planes and ships become part of a package that sinks or swims as a unit, so that few can dare try to sink it.

The naval forces of the nuclear powers must therefore be analyzed in terms of two distinct forms of impact. They may still be very functional for the military and political futures of the countries along whose shores they sail; they constitute a mobile military base which requires little or no acquiescence from the countries

245

exposed, or from their neighbors. Fleets can serve as a mobile artillery platform, capable of raining down shells in support of one faction or another in any civil war or international war. They serve as floating airbases which can supply the same kind of support further inland. They represent transient barracks, housing infantry which could storm ashore to help one side win a battle; they finally at least provide a floating magazine of military equipment, which might more handsomely equip any local army the fleet is supporting. Even the possibility of such intervention imposes a shadow of external influence over the countries involved.

At sea, at the same moment, the principal impact of navies will have become symbolic rather than functional, a trip wire of national commitment which perhaps may be interposed to prevent the onshore use of the other side's fleet, but is unlikely to be used militarily in attacking that force. The significance of a fleet deployed into any sea may thus lie less in the number of guns or sea-to-sea missiles it carries than in the number of ships which simply might be maneuvered into the other fleet's way, always carrying the symbolic threat of escalation to a much greater conflict. For the symbolic purpose noted here, the Russians perhaps could just as easily have deployed all their schoolchildren on board cruise liners in the Mediterranean, in a "campus afloat" program which just happened to get in the way when American carriers were meaning to launch air patrols over the Middle East, or when an American attack transport was planning to enter a harbor. To risk the accidental sinking of such a cruise ship might pose the same symbolic dangers of severe Russian reaction as in a collision with a Russian cruiser or helicopter carrier. . . .

The cruise ships carrying Russian schoolchildren are also not quite as analogous to contemporary fleets as is the U.S. garrison in West Berlin. We at least pretend that we are fixated on the fighting ability of the garrison and the carriers. The garrison maneuvers with its tanks within the confines of Berlin and tries to keep up its marksmanship. Pretending to be expecting combat makes the trip wire all the more effective, since tokens of commitment come to be less credible once they are openly discussed as such. Yet if Presidents and Generals must pretend that the West Berlin garrison is intended for the defense of the city, we know in reality that it serves mainly as a symbol to activate other retaliation in the contingency of a Russian attack. The deterrent backing the U.S. Navy similarly functions more reliably if Presidents and Admirals at least pretend that they are contemplating major naval battles with the Russians, while they may really be reminding Moscow of a more horrendous threat.

Are aircraft carriers of the U.S. Navy really as sinkable, therefore, as this analysis suggests, remaining afloat mainly due to Russian restraint in fear of escalation and retaliation? American naval officers of course are prone to deny this vulnerability, just as they are prone to support their requests for equipment, and alarm at Russian acquisitions, in terms of a war as it would have been fought in 1944. To pretend that weapons will be functional, and will be used functionally rather than symbolically, is a normal part of the international bargaining and deterrence process, and of the domestic scramble for funds. Yet Russian reconnaissance aircraft regularly find and overfly U.S. aircraft carriers to suggest the carriers' vulnerability. American aircraft are conversely dispatched to meet and escort in the reconnaissance bombers, perhaps to show that they would never have reached the carrier in a war situation to complete their attack. News reports indicate, however, that some of the reconnaissance bombers occasionally reach their "target" before being met; the use of nuclear air-to-surface missiles, moreover, would very plausibly reduce the survivability of targets as large as naval ships.

In political crises of the future, the first fleet to reach the port of the country undergoing turmoil may thus be very significant. . . . [B]eing first into port, [it] can confront the opposing navy with the choice of staying away or risking the symbolic and real collisions which might escalate to full-scale hostilities. . . . The number of ships on each side thus retains importance, if only because the side with the most ships can interpose the most trip wires in the most places, winning the most political contests by default. We have not been arguing here that no "naval arms race" is being run; rather the race is under way, but on a different racetrack from the one usually described.

As numbers still make a difference, so does location. Even when warships were being deployed in genuine expectation of combat, prior to the nuclear era, states never had total "control of the seas," in that they could muster a numerical superiority everywhere. Ever since 1945, the Russians presumably had a likely superiority in the Black Sea, and the U.S. will probably always have one within the Caribbean. . . .

The comparative size of two fleets in any particular sea, e.g. the Mediterranean, is thus significant for a series of reasons. First and foremost, the "Naval Ordinance" remains important as long as regimes along the banks of the sea regard it as important. Much of what happens in politics amounts to self-confirming propositions, and the governments on each side have to be aware of this. Second, the larger the number of ships deployed by any particular navy, the better its chance of reaching some crisis point in time to affect the political outcome, by landing marines, supplying air cover, etc. Third, the larger the number of ships on the other side, the better its chance of blocking and non-violently interfering with such operations. Finally, if a naval war were still to come, despite all the risks of further escalation, the larger navy would probably suffer fewer losses than the smaller. It is being argued here that the most important functions of navies now will rest within the first three categories, rather than the final category of actual combat. The use of navy may thus be primarily to get into another navy's way, not to shoot at it or sink it, but to block it so that fears of shootings and sinkings will divert it from its chosen course.

Some of this purpose may be linked to the tests of resolve that have occurred regularly between ships of the two fleets in the years since World War II, as Soviet destroyers have gotten in the way of American or British aircraft carriers, as American and Soviet ships have almost literally played games of "chicken" which scrape the paint, or more, off one of the ships. . . .

Lest the practice of such tactics get out of hand in some wider process of escalation or simple misunderstanding, an agreement was initiated during President Nixon's summit visit to Moscow by which each side in effect promised to desist from some of the "shouldering" and "bumping" practices of the past. Yet it is hardly clear that this agreement would be repeated on both sides if a serious crisis got under way. It might be bad "arms control" to require that the Russian and American fleets be left with only two choices, full respect for the rules of the sea or all-out combat.

The symbolic significance of navies (at sea) and the functional significance of navies (on land) are of course not totally separable. If a Russian destroyer sailed into the path of American aircraft carriers while they were attempting to launch an aircraft mission, they would in effect be applying the symbolic impact to interfere with the functional. An attempt to use ships close in to shore might similarly strain the distinction thus far drawn. Many of the ships used here might be smaller (landing craft, mine sweepers, ships the size of the *Pueblo,* etc.) and thus reflect less of any total national commitment to avenge their loss. The range of coastal artillery is not unlimited, and the sense of aggression in the sinking of such a ship would thus be less aggravated. Minefields also can be left in place in such coastal waters, which leaves the "last clear choice" to the ships which get involved, and with the responsibility for any sinkings.

Whether this coastal zone extends to three miles or twelve miles may become a function of changing notions of international law. It is interesting to note Communist Chinese behavior in 1958 during the shelling of Quemoy, when the United States decided in a very limited fashion to demonstrate its resolve to keep the island replenished with supplies. American L.S.T. ships, not the largest in any navy, but not the smallest either, carried the supplies to within three miles of Quemoy, whereupon smaller amphibious-tracked vehicles of the Chinese Nationalist armed forces went the rest of the way to the island. Confronted by this challenge, the Chinese Communists did not shell the American vessels, but only the Nationalist vehicles.

Larger American ships regularly come close to the Chinese border by sailing in and out of Hong Kong. While Peking has occasionally protested, it has never interfered with these ships. The seizure of the *Pueblo* off the coast of North Korea, probably outside the 12-mile limit, was facilitated by the small size of the ship, physically as well as symbolically. Indeed if the ship had not had a name, but only a designation such as *PC-131,* the American reaction would probably have been less severe.

The Tonkin Gulf incident which led to the first American air attacks against North Vietnam has been much debated on whether Hanoi's naval craft indeed attacked the American destroyers as much as the U.S. official account claimed. Yet the incident serves admirably as an illustration of that symbolic impact of naval presence which persists separately from the functional impact. Had Hanoi's agents merely attacked a group of American advisers within South Vietnam, it would have been difficult, albeit not impossible, to have mobilized American opinion in support of such retaliation. A destroyer is still a big enough ship, however, so that the mere prospect of its sinking can seem to call for retaliation of a qualitatively enhanced sort. The destroyer, unlike landing craft or coastal vessels, is at least an "ocean-going" warship, suggesting a discreteness and apartness which an enemy had to go out of its way to violate. It allegedly stayed out to sea far enough (certainly beyond the 12-mile limit) so that the enemy could leave it alone. Whether Hanoi thus committed a blunder in giving President Johnson an excuse for air strikes, or simply was the victim of a manufactured incident serving the same purpose, Tonkin Gulf shows the quantum jump one may be making in threatening to sink a ship of a power like the United States. . . .

Merchant Fleets

Apart from expansions of the Soviet Navy, considerable concern has been expressed in the West about the growth of the Soviet Merchant Marine. As with other commercial enterprises in the U.S.S.R., the merchant fleet is hardly a "private-sector" venture, but clearly interpretable as an arm of Soviet state policy.

In sheer tonnage, the Russian merchant fleet has grown from about 2 million tons in 1950 and 3.4 million in 1960, to almost 15 million tons in 1970, comparable to a U.S. merchant fleet nominally today at about 19 million. As is the case with the Soviet Navy, having begun later, the Russian merchant marine is equipped with speedier and more modern vessels, while a certain fraction of western merchant shipping must always verge on the obsolete.

Soviet shipping services have hardly been constricted to intercoastal Soviet domestic shipping, or even to commerce with satellites. Rather the Soviet merchant force has begun to offer its services on a number of the classic trade routes of the non-Communist world, and often at rates substantially lower than those set by the "Conferences" which govern most of the free world shippers on such routes.

The sheer tonnages figures can be misleading if seen simply in a U.S.–Soviet comparison, of course. A fair amount of American-owned shipping is registered under the "flags of convenience" of Panama and Liberia. The national tonnages of Japan, Great Britain and Norway, formal treaty allies of the United States, each exceed the figures for either the United States or the Soviet Union. American Admirals perhaps like to cite the specific U.S.–Russian comparison for its impact in shocking their public. Yet the relevant issue is obviously not parity in shipping tonnage, but the very genuinely remarkable expansion of the Soviet figure. The Soviet expansion of more than 300% since 1960 is rivaled only by the growth in Japanese merchant shipping over the same time.

How much should the free world feel threatened by a Soviet venture into what is clearly a useful and peaceful pursuit on the high seas? If goods need to be delivered, how can a believer in free trade and competition be concerned about one more competitor entering the fray?

There are military fears, of course. Each Russian merchant vessel will presumably be at the disposal of the Russian navy, and could be used as a supply vessel to keep warships on the high seas for longer periods of time. As has been historically demonstrated in the American experience, such vessels can be used also to carry troops on their way to amphibious invasions. It is even conceivable that merchantmen can be equipped with guns or surface-to-surface rockets to serve as commerce raiders or *bona fide* warships in the event of hostilities. Lacking even the most rudimentary armor, however, it is questionable how long such ships could hold up as participants in some large-scale but limited sea war.

A little away from the realm of military operations, the Soviet merchant fleet can "show the flag" in the same way as the Soviet Navy. A visible commercial presence can well amount to a more effective political presence; if "courtesy calls" have long been seen as a way to impress local rulers or citizenries favorably, the appearance of sleek modern cargo ships can do the same. On all such trade questions, the hope has been expressed on each side again and again that full commercial exchanges will lead to better understanding and closer relations between countries. Yet on each side a parallel fear has also been expressed that trade will ease hostility unevenly, perhaps reducing vigilance on our side, but not on theirs. Such fears, for example, were voiced in Poland when the first contacts with West German industrialists were opened. The same fear inhibits China from leap-

ing to seek trade benefits. And in this case, it causes Western concern at the appearance of Soviet merchant vessels.

Such fears are not totally unreasonable. In effect, the country that gets to play the role of merchant shipper gets to enter the other's home ports, while keeping such alien influences away from its own ports. Goods shipped from Australia to Russia might thus well be lured into Soviet freighters, to achieve a symbolic impact in Australian ports, and to avoid such impact in Soviet ports. Yet one can easily exaggerate how much difference this makes. We have become accustomed to the existence of Aeroflot offices in New York and London; once the novelty of such presence is behind us, the political impact becomes secondary.

Moving away from symbolism, one might express a more real fear that repeated use of cheaper Soviet shipping services within the free world will cause Western shipping capabilities to atrophy, with the result that the Soviet might suddenly deny such services during a crisis, imposing serious pressures on the Western position. The mere knowledge that a loss of regular shipping services will follow from the worsening of a crisis may inhibit Western leaderships from being resolute in such a period.

Yet the "dependence effect" in the growth of a Soviet merchant capability may not be as one-sided as suggested above. In the event of an all-out war, the Russians surely would not mind the loss of foreign-currency income in having to withdraw their freighters from routes and patronage that had been developed. Yet for more limited war, any prolonged cutoff of such revenues, after the Soviet economy has grown accustomed to them, will seem like a substantial sacrifice. If Australians come to depend on Russian freighter services, the Russians may have a dependency in reverse.

Some of the new Russian carrying capacity, moreover, tends to get tied up by Soviet political decisions to extend logistics support to distant enclaves of the Socialist bloc. The delivery of essential goods to Cuba has thus committed a considerable part of the expansion of Soviet capacity. Similar commitments have been entailed by the delivery of war material to Haiphong in support of North Vietnam and the insurrection in South Vietnam. If Chile were to become more dependent on aid from, and trade with, the Soviet Union (one of the principal fears of the free world leadership at the moment), it will at least have the incidental effect of forcing the Russians to tie up some more of their tonnage in one more new "lifeline."

This kind of reasoning can be too optimistic from the Western point of view. Losing Chile or South Vietnam or Cuba to the Socialist camp is a real loss from a free world calculus. Yet a few of such losses may be inevitable. The above argument is meant merely to remind us that few clouds are totally lacking in silver linings. The Russians cannot have their cake and eat it; if they become enmeshed in trying to support and hold ideological outposts far across the seas, their merchant marine may be taxed at close to its capacity.

One might thus be tempted to see no political effect at all in any increases in Soviet merchant tonnage, as long as the increases are consumed in overseas support of Soviet political causes. Yet there are some important effects even if the net of available and surplus Russian tonnage stays about the same. The more the Russians sail the seas, whether doing the free world's work or the Socialist bloc's work, the more the U.S.S.R. accumulates a vested interest in keeping the seas peaceful. At the very least, the investment in freighted tonnage puts a significant slice of Soviet assets into a very fragile form; ships can easily be sunk if wars break out, even if limited wars break out, unless rules are written and enforced to protect this kind of investment.

It has long been assumed that the U.S.S.R. would have a vested interest in submarine warfare in the future, even under some conditions of large-scale limited war. Perhaps the entire operation might yet be conducted clandestinely and anonymously, as various submarines and their crews "volunteer" for duty in the Med-

iterranean or the Indian Ocean, or in the South China Sea; this could come on the pattern of Russian volunteers who fly jets in Nigeria or Egypt, or the Chinese volunteers who fought in Korea.

As noted, this form of war has not occurred in the past, even when unescorted merchant ships have carried U.S. war materials to Korea and Vietnam. The explanation of this in earlier periods may have been that the U.S. was in exchange offering non-symmetrical restraints on its own armed forces, for example, no bombings of Manchuria. More recently, the exchange counter might rather have been the safe passage of Polish and Russian ships to Haiphong, just as undisturbed as U.S. ships going to Saigon. Limited wars are thus all the more likely to stay limited as the potential zones of escalation become populated with fragile assets of all the sides. The more merchant ships at sea under Communist flag, up to a point, the less likely again is any escalation of warfare from land out to the sea.

One must ask at the end how much of this argument is supported or disproven by President Nixon's decision in 1972 to bomb the docks of Haiphong, and then to place mines in the harbor to prevent merchant ships from reaching North Vietnam. Does this not show that each side is less afraid of escalation in warfare at sea than has been indicated? Has the United States since 1965 perhaps learned how to sink Russian ships in ways which do not risk a war which would wipe out the world?

To begin, one should note that Nixon's action did indeed cause a great deal of domestic criticism as to its riskiness. While the stepped-up bombing of North Vietnam was incomparably more destructive to life, it was the interference with sea-lanes that fixated American news commentators, leading them to draw comparisons with the Cuban missile crisis, and its implicit risk of general war. Yet Nixon, by using a minefield, had indeed chosen the one form of naval warfare that was much less likely to pose risks of escalation in tests of resolve. The "last clear chance" of avoiding sinkings was thus left to the Russians, who elected not to lose their merchant ships. The minefield thus resembles the blocking function as much as the shooting function of navies. Speculation was advanced on whether the Soviet Navy would send minesweepers (in some ways the most innocuous of naval ships) to try to open Haiphong, and on what the U.S. response would be. "Background" leaks hinted that the U.S. would again not sink the Soviet vessels, but merely drop more mines to replace those the Russians had removed. Again this pays off the side with the most ships, but by the rules of diplomacy more than by the Rules of Trafalgar or the Coral Sea.

Indeed, one should note how long Haiphong's docks went unbombed, and wonder whether this would have persisted if there had ever been a week in which no Soviet merchant ships were at dockside. Apart from the demands of logistic schedules, it would have been folly for Hanoi or Moscow to have ever left Haiphong empty of Russian ships in the years of aerial bombing by either the Johnson or Nixon administrations.

The Likelihood of War

What, therefore, are the prospects of war at sea? It has been contended here that we will see a complete exemption of the sea from warfare much more often than we have thought. The sea, in effect, may become everyman's sanctuary, even as one in 1967 almost ludicrously saw Communist merchant ships sailing past American aircraft carriers off the Vietnam coast, carriers that a few weeks later would be launching air-strikes directed at the very supplies the merchantmen had brought in.

Should World War III occur, of course, the seas by a certain definition would become very "active," as Polaris missile submarines of the U.S. Navy, and their opposite numbers, fired off missiles to retaliate against the cities of the two major powers.

Is there much prospect of any war in between, a World War II duplicate greater in scale than today's limited war, in which navies really are protected, but less than the all-out missile exchanges of World War III? The navies of the world obviously take such a scenario seriously enough so that it cannot be discounted. Yet the above arguments are meant to suggest that this scenario has now become a symbolic construct not to be actualized, but to be exploited for the real political and strategic uses of navies.

In both warships and merchant ships, the Russians have thus entered the race. There is indeed a race, in that great superiorities for the U.S.S.R. in either category, even if only within a particular region, can alter the political results around the shores of that body of water. Yet Russian entry into the race has done more than make it more interesting for the spectators; in an important way, it has changed the rules, with the result that the entire exercise probably will be modified.

The bolstering of the Russian fleet therefore does not "check" the U.S. Navy in any ordinary sense of the term, for the American fleet was already checked by the importance both sides attach to avoiding World War III. Except for very unusual circumstances, Soviet warships and merchant ships have been able to deliver military equipment where they pleased.

Supership

Noël Mostert

The first ocean-going oil tanker, the Glückauf, *was built 100 years ago; today, by virtue of both their size and their importance, tankers dominate modern commerce. They have been growing steadily in size since World War II, when the largest tanker had a deadweight of 18,000 tons. By 1973 tankers accounted for more than half of all the tonnage afloat, with the largest having a deadweight of close to a half million tons. Almost a quarter of a mile long, their bridges 100 feet above the surface of the ocean, requiring more than three miles to stop, demanding new navigational skills—these superships have revolutionized seafaring, much as steamships did in the days of sail. Their sailors hardly regard these depersonalized vessels as ships; they refer to them simply as Very Large Crude Carriers (VLCCs). Too expensive to remain idle, they spend less than 10 percent of their time in port. Few ports can handle these behemoths, and they often discharge their cargo offshore.*

In a world that is increasingly reliant on oil for its energy needs, these superships have come to symbolize global interdependence. But because of their unprecedented size, they also represent an unprecedented threat to the environment. Added to the unavoidable hazards of carrying a polluting, potentially explosive cargo are the dangers that result from faulty equipment and poor seamanship. These factors are discussed in the following excerpt from the best-selling Supership, *by the journalist Noël Mostert.*

*I*n a detailed study of fifty recent ship accidents published at the end of 1972, the British Chamber of Shipping said that most of the collisions involved were attributable to appalling seamanship and could have been avoided if alertness and prudence had been shown, while all the groundings were directly attributable to bad navigation. Shell Oil, in a detailed study of forty serious tanker accidents that involved pollution, found that the common link between all was that "people made silly mistakes."

A very large number of the mistakes seem to be made by ships flying one of the flags of convenience. These countries, together with others such as Greece, Formosa, and the Philippines, have dominated the marine casualty lists for some years; each year for the past five years Liberia has had the biggest total losses of any country.

Twenty years ago world shipping was largely a western European business, with Britain firmly in the lead as the biggest owner and operator of ships; outside Europe, America and Japan were the only major shipping nations. Liberia now has the world's largest merchant marine, followed by Japan and Britain, and her lead is rapidly increasing; flag of convenience fleets have regularly grown at rates more than twice those of world fleets as a whole. Liberia and Panama together now own, on paper, nearly a quarter of world shipping. Tankers dominate these expatriate fleets.

Thirty-five to 40 percent of the Liberian tonnage is American-owned, and an additional 10 percent of it is American-financed, which helps explain where the American merchant fleet, in steady decline since the end of the war, has taken itself. According to law, American-flag ships must be built in the United States and must be three-quarters manned by Americans. American shipbuilding costs used to be double those elsewhere (inflation abroad has helped make them competitive again), and American seamen's wages are still higher than elsewhere. American users of the flags of convenience, and they include Gulf, Esso, Texaco, Getty Oil, Tidewater, and Union Oil, have argued that they act not for convenience but out of necessity. Their plea has been that without the flags of convenience the American merchant fleet would have substantially vanished by now, because of costs. They have pleaded in fact that theirs is a patriotic stance in that they ensure the survival of a merchant fleet that would be vital in a war. How this squares with the fact of fewer trained American seamen, or how they would ensure the loyalty of their foreign crews and continued possession of their ships in such an emergency has never been explained.

Flag of convenience operators often say that their ships, especially many of those under the Liberian flag, are among the largest, best-equipped, and most modern in the world. This may be true. But ships are only as good as the men who run them, and the record is not impressive. Old ships traditionally have a higher casualty rate than new ones. Liberian losses between 1966 and 1970 not only averaged twice as high as those of the other major maritime nations, but, contrary to the rule, the ships they were losing were on the whole new ones, certainly newer than the ones lost by the other principal merchant marines: the average age of Liberian losses in that four-year period was 8.7 years, while that of Japanese and Europeans averaged 12 years.

To a disconcerting degree, oil cargoes have been delivered in recent years by improperly trained and uncertificated officers aboard ships navigating with defective equipment. One of the biggest of all tanker accidents involved an American-owned Liberian ship which was in the charge of an officer who had no certificate whatsoever.

After the Liberian tanker *Arrow* ran ashore in Chedabucto Bay, a three-man committee of inquiry, which was led by Dr. P.D. McTaggart-Cowan, executive director of the Science Council of Canada, found that the *Arrow*, owned by Aristotle Onassis, had been operating with almost none of its navigation equipment serviceable. The radar had ceased to function an hour before the ship struck; the echo sounder had not been in working condition for two months; and the gyrocompass, which is used to steer by and to keep the ship on course, had a permanent

error of three degrees west. The officer on watch at the time of the accident, the ship's third officer, had no license. The commission of inquiry said none of the crew had any navigational skill except the master, "and there are even doubts about his ability." In its final report the commission said: "We are well aware of the fact that no form of transportation can be 100 percent safe but from the record available to us the standard of operation of the world's tanker fleets, particularly those under flags of convenience, is so appalling and so far from the kind of safety which science, engineering and technology can bring to those who care, that the people of the world should demand immediate action."

If one judges by Liberia's recent record, it often seems to make little difference aboard a Liberian ship whether it has the newest equipment or the oldest; too often those in charge of an ultramodern bridge don't know how to use what's there, or don't know how to repair anything that breaks down, or, worse, don't even bother to report a fault when they get to port. Even in the case of well-qualified men commanding ships of the highest standard, as was the case with the *Torrey Canyon,* their judgment, responsibility, and seamanship in the long run can be affected and impaired by terms of service that would not be tolerated on any ship flying the American flag, or that of any of the other major maritime powers. When he drove the *Torrey Canyon* aground on the Scilly Isles, the ship's Italian master, who had behind him an outstanding reputation and record as a seaman, already had served 366 days on board.

As the British and French governments discovered when they sought to find someone to hold responsible for the accident, the task of trying to pin down a flag of convenience ship within any accessible frame of legal jurisdiction is well-nigh impossible. The *Torrey Canyon* was owned by the Barracuda Tanker Corporation, a financial offshoot of the Union Oil Company of California, which leased the ship and had, in turn, subleased it to British Petroleum Trading Limited, which was a subsidiary of the

British Petroleum Company. The ship, built in the United States, and rebuilt in Japan, was registered in Liberia, insured in London, and crewed by Italians. For an international lawyer any suit involving such a vessel must, one assumes, be the sort of stuff of which dreams of eternal litigation are made. The British and French, however, took a simple course. They pretended they weren't looking and, when one of *Torrey Canyon*'s sister ships, the *Lake Palourde,* ambled into the first port where the law was held to be firm, they pounced and had her arrested until the insurers, the only accessible body with responsibility, paid up $7,500,000 as a settlement for damage.

Starting with *Torrey Canyon,* most of the major oil spillage calamities of the past six years have involved Liberian ships. These have included the *Ocean Eagle,* whose wreck fouled the beaches of San Juan, Puerto Rico, in 1967; the *Arrow,* which coated sixty miles of Nova Scotia shoreline in 1970; and the *Juliana,* which in 1971 gave Japan its worst oil spill when it broke in two after hitting a breakwater off the port of Niigata. In October 1970, two fully laden supertankers, the 77,648-ton *Pacific Glory,* Chinese owned, and the 95,445-ton *Allegro,* Greek owned, both flying the Liberian flag, and between them carrying 170,000 tons of crude oil, ran into each other off the Isle of Wight. The *Pacific Glory* suffered a violent explosion and was burned out; fourteen of her crew died. Most of the oil in their tanks fortunately remained intact. The third officers of both ships were on watch at the time; the *Allegro*'s third officer, a Greek, had no certificate whatsoever. Two of her engineers, Greek as well, had no certificates either. Two of *Pacific Glory*'s engineers also had no certificates. This was, at the time, the worst maritime collision on record, but it lost this distinction in August 1972, when two Liberian-flag supertankers, the 95,000-ton American-owned *Oswego Guardian,* fully laden, collided with the 100,000-ton Greek-owned *Texanita* northeast of Cape Town in the Indian Ocean. The *Texanita,* which was empty, exploded with such violence that it rocked buildings and woke

people forty miles inland from the coast, which itself was twenty-three miles distant from the accident. The *Texanita* broke in two and vanished within four minutes. Thirty-three men died with the *Texanita,* and one aboard the *Oswego Guardian.* Both ships were traveling at high speed through fog so dense that the master of the *Texanita,* who survived, couldn't see the masts of his own ship; although they had observed each other on radar, neither ship reduced speed. *Texanita* made only two attempts to plot the course of the approaching ship, the second when it was only four miles off, and *Oswego Guardian* made no attempt whatsoever to plot the other ship.

The chief officer of a Norwegian freighter, the *Thorswave,* later provided what might be the first electronic eyewitness account of a major maritime disaster. His own ship was in the vicinity and he had watched the accident develop on his radar screen. "I saw these two ships coming closer together," he told the *Cape Argus* in Cape Town. "Then the two dots came into one. Just then we heard this terrific explosion and felt our own ship shake twice. I thought there was something wrong with our own ship because the explosion was so loud. A minute or two after this I saw the two dots coming away from each other. Then one dot suddenly disappeared from the screen."

Immediately after the collision, the master of the *Oswego Guardian* ordered his ship at full speed away from the scene. No attempt was made to pick up survivors, who owed their lives to other vessels in the area including the *Thorswave.* The *Oswego Guardian*'s SOS call gave a wrong position, which was not discovered until six hours after the accident; no correction was ever sent out. The *Texanita*'s master lost his license for eighteen months; the master of the *Oswego Guardian,* a Chinese, had his revoked.

Half the ship collisions in the world take place in the area bounded by the Elbe and the English Channel. Most of these are head-on and by far the majority of them occur in or near the Straits of Dover where, at any given moment, some forty ships usually are moving. Dodging this

SUPERSHIP: *GLOBTIK TOKYO,* ONE OF
THE WORLD'S LARGEST TANKERS

situation as well as the many wrecks and sand-
banks in the area has become the principal nightmare for all supertanker and VLCC masters; and
it is one they constantly confront because Rot-
terdam is the main tanker terminal for Europe
and the most common destination for tankers inbound from the Persian Gulf. Tankers, as one
might expect, are the ships most commonly involved in accidents there, especially flag of convenience ones, and usually because of appalling
seamanship and standards aboard them.

Between October 1970 and April 1971, for
example, ten tankers carrying among them
some 300,000 tons of crude oil were involved in
serious accidents in the area. Half of them were
Liberian and they included the *Pacific Glory* and
Allegro. . . . A Trinity House master mariner,

Captain W.L.D. Bayley, writing in *Safety at
Sea,* in its issue of December 1969, said that
supertankers with faulty VHF or radar were
so numerous that channel pilots had ceased to
report them. A further instance of almost total
inadequacy was provided when the Greek-
owned and Cyprus-registered tanker *Aegis Star*
ran aground on the Swedish coast in November
1972. A surveyor who boarded her after she had
been refloated found that her gyrocompass,
echo sounder, radar, automatic log, speed indicator, and rudder indicator were all out of
order, according to a report in the British shipping journal *Fairplay.*

A senior Trinity House channel pilot, Captain
N.R. Knowles, told me recently that, far from
improving, things were in fact getting worse,
and described an incident involving a Liberian
vessel inbound for Dunkirk which had been
advised that she would have to stay outside
because no berth was available. As pilotage is

not compulsory and many ships, flag of convenience ones especially, avoid taking aboard pilots for the English Channel run because of the extra expense, the ship in question had not asked for a pilot when she made the approaches. Fighting a gale off Dunkirk, she searched for anchorage by steaming north, and then back down the Channel to Beachy Head on the English side. Her fifty-seven-year-old master finally sent an urgent appeal for a pilot to show him to safe anchorage on the English coast. He was near exhaustion when the pilot boarded. He was the only officer on board with a mariner's certificate; his first officer had been at sea only three and a half years. Aside from the threat that such an improperly manned ship presented to tanker traffic in the area, Knowles said, she herself was typical of many tankers he'd boarded.

The menace of such vessels and their substandard operation was one of the principal factors behind the introduction of two-lane traffic in sixty-six busy maritime areas throughout the world at the beginning of this decade. Ships now are required to move in these double lanes of one-way traffic when laying course through these areas, which include the English Channel, the Cape of Good Hope, the Malacca Straits, the San Francisco and New York harbors, the Baltic, the Straits of Gibraltar. It was felt that this system would at least help minimize the risks to heavily laden supertankers. Unfortunately the lanes are ignored by many ships (referred to as "cowboys" by those who stick to their proper lane) and the results can be tragic.

On January 11, 1971, a 12,000-ton Peruvian freighter, *Paracas,* entered the English Channel and, instead of using the northbound lane off the French coast as she was supposed to do, took the shorter and more convenient downbound lane along the English coast. She struck the Panamanian tanker *Texaco Caribbean* and the resulting explosion shattered windows five miles away in Folkestone. Nine men went down with the ship.

The British coastal authorities marked the sunken *Texaco Caribbean* with three vertical green lights as a wreck warning. The following day a German freighter, the *Brandenburg,* outbound for North America, hit the wreck and sank with the loss of more than half her thirty-one-man crew. The British added a lightship and five light buoys to the green lights on the site, but on February 28 a Greek freighter, *Niki,* struck the two ships and herself went down, taking her entire crew of twenty-two. A second lightship and nine more buoys were added to the collection of wrecks, but on March 16 an unidentified supertanker ignored a barrage of rockets and flashing lamps from the guard ships, ran through one row of buoys and, to everyone's surprise, got away with it and vanished. Within a two-month period, sixteen ships were reported by British coastal authorities for having ignored the elaborate arrangement of lights and signals and entered the area of the wrecks, which have since been demolished.

It is a situation that can only get much, much worse as world trade and world fleets expand. Today's run-of-the-mill superships, the 200,000 tonners, will be tomorrow's traders of low degree. The write-off life of a VLCC is about ten years. Most of the first wave of 200,000–250,000 tonners already have seen half that. Superships aren't built to last. As they get older they begin to fall apart, to break down, and repairs and maintenance, not to speak of long tows, become too expensive to justify their retention in the service of any well-managed fleet. As the next big wave of investment starts creating the next plateau in tanker size, probably with the 500,000 tonners, the older ships will be handed down in job lots to the next generation of newcomers seeking a fortune in oil ships. So it presumably will continue, with demand and profits waxing and the oceans, alas, waning, unless some extraordinary international effort is made to control standards at sea. There seems a strange sinister touch of alchemy about it all—of black gold turned to golden gold and the lot ending up as purest dross, which will be the quality of the environment, and of life within it, we eventually will be left with.

Commerce and Transportation: Fulfilling Our Commitment

Marine Board of the
National Academy
of Engineering

Maritime commerce has played a central role in the development of the United States, as already pointed out in the selections by Samuel Eliot Morison and Alfred Thayer Mahan. Nevertheless, the position of the United States merchant fleet has been declining steadily since World War II. While the size of the fleet in deadweight tonnage has remained relatively constant, the volume of world trade has been expanding steadily. As a result, the percentage of United States foreign trade carried by United States flag ships has declined from 42.6 percent (based on weight) in 1950 to approximately 6 percent two decades later, thus increasing United States dependence on foreign shipping as well as on foreign goods. This relative decline is a cause of major concern to those who see a strong merchant marine as essential to the national security as well as to the economic and political well-being of the nation. The competitive disadvantage suffered by American ships because of higher labor costs, noted by Mostert, is only partially offset by a program of government subsidies. In the following selection from Toward Fulfillment of a National Ocean Commitment, *the Marine Board of the National Academy of Engineering assesses the importance of the merchant marine and suggests that the better application of technology might help to solve the problems facing the merchant fleet now and in the future.*

An effective national ocean commitment includes the recognition of ocean transportation as an important marine resource. Oceanborne transportation has long been a major element of world trade. The vast majority of the goods in world trade in terms of both cargo value and tonnage will continue to be carried by sea in the foreseeable future. This will be supplemented in some specialized areas by improved air transportation. Oceanborne trade is continuing to expand very rapidly and its growth makes the need for greater efficiency of ever-increasing importance.

Every nation that aspires to improving its position in world commerce regards its merchant marine as an important aid in achieving that goal. For many nations, the services rendered by their merchant ships are an important factor in their balance of payments. For instance, the Soviet Union is rapidly expanding its merchant marine as a means of ensuring that its oceanborne commerce can develop independently of political and economic pressures by other nations. Without a United States Merchant Marine able to exert a significant influence in world trade, United States shippers would find that their ocean freight rates would be based not on economics but on the changing national policies of other countries.

Development of the United States Merchant Marine

[S]hips and the shipping industry [are important] to the domestic economy and to our national security. Ships are required for the transportation of many imports that are absolutely essential to our continued existence. Many other imports could be replaced from domestic sources only at a significantly greater price and early depletion of our resources. In order to pay for these imports, it is necessary to export, and a balance over the long run is essential. These exports not only contribute to balancing the imports but create additional jobs, profits, and taxes.

Economic and Political Aspects of Development

How the needed shipping services are to be provided is a major current problem facing the country, and in particular the United States shipping companies. When the basis for our Merchant Marine was established, by the Merchant Marine Act of 1936, the country was far less dependent upon imports for essential needs. In fact, we were nearly self-sufficient. However, since 1945, there has been a tremendous increase in the importation of bulk commodities required by the domestic economy. Since the United States-owned fleet was deficient in economic carriers of the type needed, most of the new shipping services in this category have been provided by foreign flag ships manned by foreign crews. Some of these ships are entirely foreign owned and some are owned or controlled by United States companies. The exports to pay for the imports are generally manufactured products, carried in general cargo ships.

The United States emerged from World War II with a large fleet of general cargo ships, and in the years immediately following they carried a very high percentage of our general cargo. However, beginning after the Korean emergency, there was a substantial increase in foreign flag ships in the general cargo trade, as a result of postwar European and Japanese construction. This resulted in the layup of many United States ships not under subsidy, finally resulting in having most of the general cargo carried only in subsidized ships. In these latter years, the amount of subsidy has been limited, the number of subsidized voyages has been essentially constant, and the amount of cargo carried has been approximately constant. As a result, the percentage participation in the general cargo carryings by the United States fleet has significantly decreased. The present condition of our merchant shipping needs a revised merchant marine policy if it is to be improved. In establishing such a policy, both commercial needs and national security should be evaluated.

In regard to national security, alternative means of meeting military sealift requirements under various conditions of demand include the use of a Military Sea Transport Service nucleus fleet, the unsubsidized United States flag fleet,

the subsidized United States flag fleet, and the chartering of foreign flag ships. The response of the various fleets, as well as the costs incurred in the carriage of military cargo under various military emergencies, may be analyzed on the basis of historical response during previous emergencies, including those involving Korea, Lebanon, Suez, and Vietnam. It is believed that such an analysis will show that the chartering of foreign flag ships is expensive compared with the use of subsidized ships at controlled rates and that such ships are not always available when required. The minimum United States fleet essential for support of emergency military sealift requires further objective determination.

On a strictly commercial basis, there would be no significant United States fleet in foreign commerce if there were not a subsidy of some kind. Statements have been made that, through reduction in ship costs from standardization and automation, and reduction in operating costs through automation, United States ships can be competitive without subsidy. This may be possible in certain highly profitable routes under present freight rates and generally favorable conditions, but it must not be overlooked that foreign flag operators have access to all of the advanced technology available to the American flag operator and do not delay in adopting it, since they have more cooperative labor. Thus, the labor cost advantage is retained by the foreign operator and the United States flag operator will require some form of subsidy to offset the higher labor costs, even though the differential is less with improved systems and automation. Therefore, the cost of this subsidy should be considered in relation to the benefits obtained from a United States flag fleet.

Concerning the economic value of a merchant fleet, a study of the balance of payments made for the Committee of American Steamship Lines indicates the conservation of dollars that is realized from a United States flag Merchant Marine. While this study indicates the dollars that can be conserved by the Merchant Marine, the real value of the saving remains to be determined by economists. Obviously this value shifts considerably, depending upon the domestic economy of the United States and the countries with which we trade. For example, in 1948, the principal international monetary problem of the United States was to provide foreign customers with dollars to buy American goods, whereas, at present, we need to conserve dollars in foreign exchange. If a relationship can be found between the value of the balance of payments and some parameter describing the domestic economy, then the value of the Merchant Marine, as a contributor to the balance of payments, could be more precisely defined and related to the cost of maintaining a United States Merchant Marine through subsidy of the labor involved.

At various times there have been proposals both to extend the Merchant Marine subsidy as well as to reduce it. A thorough overhaul of the subsidy system is required to make it more efficient and effective, simpler to administer and to provide incentives. Different types of subsidies are required considering the types of ships essential for military needs, as well as for commercial imports and exports. This could result in a definition of the fleet that would most nearly optimize return on subsidy investment.

The Merchant Marine Act of 1970 is important in its concept of a long-range program of challenge and opportunity to revitalize the U.S. Merchant Marine. The provisions of the Act, however, will probably require early revision if its purpose is to be attained. The recommendations in this chapter could serve as a basis for this effort.

Proposals for alternative improved subsidy arrangements that provide for incentives should be obtained from experienced ship operators in the various services. In the past, too many proposals have been made by people without the necessary experience, and most of these were completely impractical. After the proposals are received, they can be evaluated by a competent, objective organization.

It has been stated that United States flag competition is necessary to maintain fair freight rates for our imports and exports. It is true that in recent years there has been little evidence to indicate that the presence of United States ships on any route has had any significant effect on import and export freight rates. However, if there were no United States Merchant Marine and if, as is very probable, ocean freight rates were strongly influenced by the policy of foreign governments such as Russia, which is rapidly developing its own Merchant Marine, then the United States might find itself at a severe disadvantage in world trade in the not distant future.

The effect of freight rates upon the quantity of goods exported should also be determined. Rough correlations of the value of goods shipped versus the freight rate have been made, but a more precise analysis is required to determine those essential commodities that are sensitive to rate fluctuations.

Another aspect of the Merchant Marine is the *flag of convenience or necessity fleet*. These are ships owned by United States interests, flying a foreign flag and manned by a foreign crew, which are considered to be always under the effective control of the United States. This assumption might not be true under all circumstances. A reevaluation of the availability of the flags of convenience or necessity ships should be carried out in light of recent experience.

As part of the determination of the essential fleet for carrying our essential imports, an analysis should be made of the potential of fully automated bulk carriers, using American crews, to determine whether these ships can compete in the open market under normal conditions and provide emergency transportation at controlled rates under limited war conditions; or whether a limited amount of subsidy should be provided to permit operation in the normal fluctuating market.

The cost of this subsidy (or United States government investment in shipping) should be considered in relation to the true value of the balance of payments and should also be compared with the return on the same investment diverted to the export production itself. While these questions cannot be answered completely, specific export trades might be studied, and the result might go a long way toward providing a better understanding of the value of alternate uses of government funds, in comparison with subsidizing the Merchant Marine.

Under Utilization of Available Technology

Many of the problems confronting the Merchant Marine do not stem from lack of technology but from the inability to utilize existing technology. This statement is still true, although over the past ten years the rising cost of ship operation has forced the industry to do some hard thinking about reducing costs. The industry has initiated programs to increase ship speed to attract more cargo and reduce the number of ships required; design specialized ships and transportation systems to reduce port time, increase utilization and reduce the number of ships and crew required for a service; reduce manning by increasing shipboard mechanization and automation; and secure more constructive labor-management agreements. . . .

Container Ships. The container ship is a significant example of recent technical innovation having a profound effect upon merchant shipping. *Its introduction raised the problem that all modern engineering faces: How to cope with the social impacts of technological innovation?* The container ship provides the marine link in a modern and efficient overall transportation system that is responsive to the rapid growth of our foreign commerce. As compared with the conventional break-bulk system, the container system provides for more rapid delivery of cargo by minimizing manual handling. Reduction in port time increases utilization so that fewer ships are required to transport the same amount of cargo as a break-bulk system. Fewer ships mean fewer crews. Cargo handling manpower in the container system can be reduced by 60 to 80 percent. In manpower decisions care must be taken to consider the system characteristics, cost differences, and the effects upon the labor forces.

Basis for Use of Available Technology. The crucial problem, therefore, is to remove the roadblocks for application of existing technology. Many of the roadblocks are in the labor-management area, and these should not be permitted to stand in the way of national progress and growth in the marine transportation industry. Labor and management in other fields have recognized the inevitability of technological progress and have accepted it; in some cases realizing that if they did not do so, the work would cease to exist because of noncompetitive costs. In other cases, provisions were made through natural attrition, advanced retirement, and retraining to provide for the displaced labor.

Advanced Technology and Its Application

Nuclear Power

Nuclear power for ship propulsion is being increasingly used by the U.S. Navy and several of the other principal navies of the world. The N.S. *Savannah* is the only nuclear powered merchant ship that has operated in regular commercial service. The Federal Republic of Germany's nuclear research ship, the *Otto Hahn,* an ore carrier, should be at sea by the time this report is issued. Other nuclear merchant ships are either under construction or being planned.

For naval purposes, nuclear power has significant advantages that easily justify its much higher cost. For merchant service, the justification must be on an economic basis.

Numerous economic studies of the application of nuclear power to merchant ships have been made in recent years. They generally indicate that present types of nuclear marine power plants are most likely to prove economic in ships of larger size and higher power than are normally required in existing world trade. A pioneer ship like the N.S. *Savannah* has been useful in developing safety standards, operating methods, and legal requirements. At the present time, it appears that the application of nuclear power to merchant ships will be very limited until there are further significant technical developments.

Unattended Power Plants

The past ten years have seen a significant development in the application of remote control and monitoring equipment to marine power plants. Pilot house control is now common. Several ships are operating today with machinery plants that do not have anyone in attendance, at least during the night hours. These developments have permitted a significant reduction in the engine department crews. There is every reason to believe that these developments will continue and that machinery plants of the future will require attendance only for maintenance and minor adjustments.

Unattended Ships

The development of unattended power plants and sophisticated navigation equipment has resulted in numerous proposals for ships which could operate between predetermined points completely unattended. As such they would be virtually crewless except for a small emergency crew and supplemental personnel who would take the ship out of port. Others would board the ship on arrival at a point outside of the harbor of destination to take it into port.

Such proposals are technically feasible but their application would require extensive development, a very high degree of reliability of the sophisticated equipment required, and a considerable change in existing laws. It must also be demonstrated that the much greater capital investment required for a ship of this type is economically justified in comparison to a minimum crew. Extensive use of unattended ships does not appear to be probable within the near future.

Resistance and Propulsion

Although ship resistance has been studied for many years, both theoretically and by experiment, our understanding of the subject is still very incomplete. Considerable progress has been made in recent years in reducing the resistance and improving the propulsive efficiency of ships by modifications to the basic hull form and by the addition of large bulbous appendages to the bow or stern or both. As the phenomena involved become better understood, additional progress may be expected. New propulsion devices such as propeller nozzles and contra-rotating propellers are being studied and applied where conditions appear favorable.

Various methods have been proposed for reducing skin friction by means of special coatings or by the discharge of air or small amounts of chemicals into the water as it passes the ship. Although progress in this area has been very limited, it is possible that a significant breakthrough could occur in the future.

Steady progress has been made in the improvement of anti-fouling coatings for ship hulls and development is continuing. Further significant progress is certainly possible in the future.

Extensive work has been done during the past decade in the study of the behavior of ships in waves and in methods of routing ships to take advantage of favorable weather and currents. Continued progress in this field can be expected although there is little likelihood that sensational advances will be made in the foreseeable future.

New Vehicles

A great variety of new vehicle types have been developed in recent years and a number of these show promise for special applications.

Hydrofoil craft are being constructed in sizes ranging from small pleasure boats to military vehicles weighing several hundred tons. Although hydrofoil craft have a promising future for certain commercial uses, particularly short-range passenger service, they appear to be unsuited for services requiring long-range or large cargo carrying capacity.

Surface effect ships of various types have been developed in recent years for military application and for short-range passenger, in which application they compete with hydrofoils. Extensive studies have been and are still being made of the feasibility of developing large size, long-range surface effect ships for carrying high-grade cargo on transocean routes. For this purpose, the captured air bubble type of surface effect ship appears to be the most suitable. However, the technical problems are formidable. The difficulties of developing vehicles of this type, which can compete with displacement type ships and the rapidly developing air-cargo carriers, are such that the prospects are not encouraging that this will become an important means of trans-ocean transportation.

A submarine operating well below the surface of the sea does not generate surface waves and can, therefore, be propelled at high speeds with less power than is required by a surface ship. Some of this saving is possible even if it is operating close to the surface with a control tower projecting above the surface. Such ships are also relatively immune to weather conditions if operated with sufficient submergence. Because of this, there have been numerous proposals for the use of large submerged or semi-submerged vessels as cargo carriers and particularly as tankers. However, the increased cost of such ships as compared with surface ships of equal capacity does not appear to be justified by the saving in power. Future possible use of such ships appears to be limited to applications in which there is a significant military advantage or a need to operate submerged, such as under arctic ice.

Conclusions

1. The volume of cargo carried in the international commerce of the United States is increasing rapidly and is expected to continue to do so. The greater part of this cargo will continue to be carried by ships in the foreseeable future.

2. An effective United States flag Merchant Marine is an essential element of our national defense.

3. A United States flag merchant fleet has a significant economic and political value to the nation.

4. The development of an efficient Merchant Marine requires close integration of ocean transport with inland transport (rail, highway, and waterway) with respect to facilities, scheduling, documentation and organization.

5. Modern, more efficient port facilities are essential for continued development of our transportation system.

6. Any real improvement in transportation systems requires the development, in co-operation with the labor unions, of a labor policy that will adequately reward seagoing and shore personnel while permitting the most effective use of manpower and technological advances.

7. Most sectors of the United States Merchant Marine cannot compete with foreign flag ships without aid in the form of subsidy or protective laws because of higher ship-building and operating costs. The most important element in the cost differential is the labor cost caused by higher shipyard and shipboard wage rates.

8. Most important maritime nations provide some form of direct or indirect assistance to their shipping industry. United States' assistance to its shipping industry is based primarily on the Merchant Marine Act of 1936 and the Merchant Marine Act of 1970. The subsidy provisions now in effect are not responsive to actual conditions.

9. The amount of support of the Merchant Marine in the form of subsidies or protective laws that is justified should be based on a broad consideration of the overall benefit to the national security and economy rather than on restricted cost-effectiveness studies. This support may be justified for only some sectors of the marine industry.

10. Adequate technology exists to permit a significant improvement in the efficiency of our marine transportation systems; however, its application requires progress in the solution of existing social, economic, and political problems.

11. Continued improvement and application

of relevant technology is necessary to improve and maintain competitive position of the United States Merchant Marine. Foreign shipbuilders and ship operators will use new technology as rapidly as in this country. Therefore, there is little likelihood that some form of government support of our Merchant Marine will be entirely eliminated in the foreseeable future.

Recommendations

1. There should be a continuing objective determination of the size and type of merchant fleet that should be maintained by the United States, considering the minimum fleet required for defense, and the minimum fleet required to meet the political and economic needs of the nation.

2. There should be a continuing objective study of how this fleet is to be provided, considering various forms of incentive subsidies and protective laws.

3. There should be a continuing objective analysis of the ships carrying United States commerce that are operating under flags of convenience or necessity to evaluate:

   a) Their availability in time of emergency.

   b) Their effect on the United States economy and balance of payments

4. Existing law should be revised:

   a) To enable implementation of the findings resulting from points 1, 2, and 3 above.

   b) To treat construction subsidies and operating subsidies separately with respect to both purpose and method of administration.

   c) To permit ship operators to order ships most suitable for their needs without unnecessary restrictions.

   d) To encourage the most efficient, competitive methods of operation, including optimum routes and schedules.

   e) To give the greatest possible encouragement to unsubsidized operation.

5. The government should support engineering research and investigation on fundamental subjects of concern to the maritime industry as a whole, and should encourage but avoid funding development which under favorable conditions will be financed by the private sector.

6. There should be studies leading to government action to provide a simplified legal and administrative framework which will encourage the integration of related functions of our entire transportation system.

7. The United States government should take a leading and active role in coordinating the planning and development of modern ports to assure their overall adequacy in keeping up with national transportation needs and cohesive development of the coastal zones.

8. A sustained effort should be made to develop an overall rate structure that will permit and encourage the most efficient utilization of all transportation facilities.

ARAN ISLANDERS

Unit Six

If there is a common theme in this unit it is that people are versatile, subtle, and complex. Without the slightest difficulty they confound all efforts to compartmentalize them and their activities. Here we have a wealth of examples concerned with the sea.

Lord Byron, during the period when he was writing the long poem *Childe Harold* that made him famous, also swam the Hellespont. At the time, it develops, Byron was far prouder of the feat than of his verse. Jack London, the novelist, describes his introduction to the sport of surfing in a lively fashion, as might be expected, but he also explains the physics of wave motion with a lucidity that may come as a surprise. We see Sir Thomas Lipton, British tea tycoon, at the end of his thirty-year quest for the grail of international yachtsmen—the America's Cup. He failed, but he had been such a good loser for so long that even his opponents were half pulling for him. We see Zane Grey, the American novelist of the arid West and author of the *Riders of the Purple Sage,* enjoying a day of triumph in his lifetime quest for world record catches of game fish.

This section is full of contrasts. The press gang is necessary to force British sailors into the crowded conditions on an eighteenth-century

Men and Women at Sea

warship. Happy volunteers endure the crowded and exposed life on Thor Heyerdahl's raft *Kon-Tiki*. Women appear in contrasting roles. Emma Roberts observes that a lady on the long passage to India in the 1840s will want to pick her cabin so she can avoid walking by the working sailors. Ann Davison is the working sailor, alone in a tiny boat, crossing the Atlantic—but she brushes her hair when a strange ship approaches.

Life at sea was, and is, harsh. The sailor endured bad food at best, scurvy and death at worst. Discipline was severe; the penalty for disobedience—again death. Still the sailors generally survived, danced their hornpipes, worked to sea chanteys, and rose above it all. Some of the passengers also maintained good spirits, or at least could give amusing accounts after reaching shore. Such a passenger was Eugenio de Salazar, a Spanish official, who took his family across the Atlantic in the sixteenth century to assume a new post in the colonies. The food was sickening and living conditions revolting, but he grew used to them and wondered how a man could "escape the boredom and misery of such a journey."

For some, boredom and misery are what sailors leave at home when they go to sea. The wives of Greek sponge divers have additional difficulties. They must manage a home for months when their men are gone but relinquish all authority instantly when they return. The ama women of Japan are relieved of this strain. They do the diving while the men stay in the boat. Elsewhere the ama approach creates new problems for women. The decision of the United States navy to accept women for sea duty is not popular with some navy wives who remain ashore.

The sea changes human lives for better or worse, but perhaps in the long run the important question is what does it do to the human spirit. How does a person who has endured a lifetime full of hardship and injustice view the sea? We may take heart from the letter to Alan Villiers from an old salt that concludes, "I still contend that the finest sight in the world is a full-rigged ship under sail."

The Medieval Sailor

Geoffrey Chaucer

Chaucer's seaman—the first to make his appearance in English literature—was a rough and dangerous fellow, skillful, disreputable, violent. This image of the sailor was to endure for centuries. It was not until the eighteenth century that sailors became more or less respectable, that the navy—in England at least—became recognized as the Senior Service, and that the public took Jolly Jack to their hearts.

The extract that follows is from the Prologue to The Canterbury Tales.

A shipman was there, hailing from the west:
For aught I know he was from Dartëmouth.
He rode upon a rouncy, as he couth,
In a gown of falding to the knee.
A dagger hanging on a lace had he
About his neck under his arm adown.
The hot summer had made his hew all brown;
And, certainly, he was a good fellow.
Full many a draught of wine had he withdrawn
From Bordeaux-ward, while that the chapman slept.
Of nice conscience took he no keep.
If that he fought, and had the upper hand,
By water he sent them home to every land.
But of his craft to reckon well his tides,
His streams and his dangers him besides,
His harbour and his moon, his pilotage,
There was none such from Hull to Carthage.
Hardy he was, and wise to undertake;
With many a tempest had his beard been shake.
He knew well all the havens, as they were,
From Gotland to the cape of Finistere,
And every creek in Brittany and in Spain;
His barge y-clepèd was the *Maudelayne*.

The Minor Horrors of the Sea

Eugenio de Salazar

We know relatively little about the details of life at sea in the fifteenth and sixteenth centuries. Sailors were practical men, little given to writing. Explorers kept journals, but rarely troubled to include information about a daily round which to them was familiar, and which they took for granted. The best accounts of ships' routine and conditions on board ship were written by landlubbers who, for one reason or another, made sea voyages as passengers. The letter which follows is the best of the surviving documents of this type. By 1573, when it was written, the trans-Atlantic crossing, pioneered by Columbus eighty-one years earlier, had become familiar to many sailors; not a commonplace, for it was always hazardous, but reasonably predictable and regular. Considerable numbers of Spanish ships—over a hundred in some years—crossed annually to the West Indies and the mainland coasts of America: out with the trade winds, with a stop at the Canaries; home with the westerlies, with a stop at the Azores. The average size of the ships had increased steadily since Columbus' day, charts and navigational techniques had considerably improved; but it is likely that the details of shipboard life were much the same.

Eugenio de Salazar was a judge who had a long and distinguished career in the Indies service. . . . He left behind a considerable volume of allegorical poetry and a number of interesting and amusing private letters, of which this, written to a friend in Spain, is one. It describes the voyage from Tenerife to Santo Domingo, which Salazar made in order to take up his appointment there.

Salazar's ship was small by the standards of the Indies trade: 120 tons, with a crew of thirty. . . . She first appears in the registers in 1552 . . . and . . . her last recorded voyage was in 1577. Salazar then traveled in a small and aging vessel, out of a minor port, away from the formal discipline of the main annual convoys, and with a new master unfamiliar with his ship. It is not surprising that he found it uncomfortable.

The extract which follows is from Cartas de Eugenio de Salazar *(Madrid, 1866), pp. 36–57: "Carta escrita al licenciado Miranda de Ron" (Translation by J. H. Parry).★*

To the Licentiate Miranda de Ron:

Qui navigant mare, enarrant pericula ejus. Those who go to sea may speak of the perils of the deep; and since I have just had to make a sea voyage, for my sins, I write to tell you about my maritime sufferings; though I must admit that they included (thank God) no pirates or shipwrecks.

I was in the island of Tenerife when my new appointment came through, and I had to make my own arrangements for getting to Hispaniola. I inquired about sailings, and eventually booked passage, at great expense, in a ship called the *Nuestra Señora de los Remedios*—better by name than by nature, as it turned out. Her master assured me that she was a roomy ship, a good sailer, seaworthy, sound in frames and members, well rigged and well manned. Accord-

ingly, on the day we were to sail and at the hour of embarkation, Doña Catalina[1] and I, with all our household, presented ourselves on the bank of the Styx. Charon, with his skiff, met us there, ferried us out to the ship, and left us on board. We were given, as a great privilege, a tiny cabin, about two feet by three by three; and packed in there, the movement of the sea upset our heads and stomachs so horribly that we all turned white as ghosts and began to bring up our very souls. In plain words, we were seasick; we vomited, we gagged, we shot out of our mouths everything which had gone in during the last

---

★*Ed. note:* This headnote is from J. H. Parry, *The European Reconnaissance: Selected Documents* (1968).

[1] His wife.

two days; we endured by turns cold depressing phlegm, bitter burning choler, thick and heavy melancholy. There we lay, without seeing the sun or the moon; we never opened our eyes, or changed our clothes, or moved, until the third day. Then . . . I dressed as well as I could, and crawled out of the whale's belly or closet in which we lay. I discovered that we were riding on what some people call a wooden horse, or a timber nag, or a flying pig; though to me it looked more like a town, a city even. It was certainly not the city of God that the sainted Augustine talked about; I saw no churches, nor courts of justice: nobody says mass there, nor do the inhabitants live by the laws of reason. . . . In sum, from bow-spirit to bonaventure, from stem to stern, from hawse-holes to tiller-port, from the port chains to the starboard top-gallant yardarms, from one side to the other, there is nothing for which a good word can be said; except indeed that, like women, it is a necessary evil.

There is in the city a whole community of people, all with their duties and dignities in strict (if not angelic) hierarchy. The wind is the real owner and master; the navigator governs as his deputy. The captain is responsible for defense; . . . the master has charge of the general work of the ship; the bo'su'n, of stowing and breaking out the cargo. The able seamen work the ship; the ordinary seamen help the able seamen; and the boys wait on the able seamen and the ordinary seamen, sweep and scrub, chant the prayers, and keep watch. The bo'su'n's mate . . . has charge of the ship's boat, sees to the water supply, and looks out for ways of cheating the passengers. The steward is responsible for the food. The caulker is the engineer who fortifies the city and secures the posterns through which the enemy might enter. The city has a surgeon-barber, to scrape the sailors' heads and bleed them when they need it. In general, the citizens of this city have as much faith, charity and friendship as sharks encountering in the sea.

I watched the navigator, the wind's lieutenant, seated in all his dignity upon his wooden throne; there he sits, an imitation Neptune, claiming to rule the sea and its waves. From time to time the sea unseats him with an unexpected lurch, so that he has to hold on to the pommel of his saddle to avoid a ducking in salt water. From there he rules and governs; "since Lanzarote out of Brittany came," no knight has been more faithfully served. Certainly I have never seen a gang of rogues obey more promptly, or earn their wages better, than these sailors; for when he shouts, "up forrard there," they come tumbling aft in a moment, like conjured demons, all their eyes on him and mouths open, awaiting his command. . . .

I was fascinated to watch the city and the activities of its people, and intrigued to hear the marine (or malign) language, which I could follow no better than heathen gibberish; and I doubt whether Your Honor, for all his cleverness, has understood all the words and phrases I have written. If any have escaped you, look them up in Antonio's word-book; and if you don't find them, ask the sailors in the town of Illescas to translate, for this jargon is much used there; but don't ask me. . . .

I would pass the time listening to the master giving his orders and watching the sailors carrying them out, until the sun was high in the sky; and then I would see the ship's boys emerge from the half-deck with a bundle of what they called table cloths; but alas, not white or handsomely embroidered. They spread out these damp and dirty lengths of canvas in the waist of the ship, and on them piled little mounds of broken biscuit, as white and clean as the cloths, so that the general effect was that of a cultivated field covered with little heaps of manure. They would then place on the "table" three or four big wooden platters full of beefbones without their marrow, with bits of parboiled sinew clinging to them. They call the platters *saleres,* and so have no need of salt-cellars. When the meal is laid out, one of the boys sings out, "Table, table, Sir Captain and master and all the company, the table is set, the food is ready; the water is drawn for his honor the captain, the master and all our good company. Long live the King of Castile by land and by sea! Down with his enemies, cut off their heads! The man who won't say 'amen' shall have nothing to drink. All hands to dinner! If you don't come you won't eat." In a twinkling, out come pouring all the ship's company saying "amen," and sit on the deck round the "table," the bo'su'n at the head and the gunner on his right, some crosslegged, some with legs stretched out, others squatting or reclining, or in any posture they choose; and without pausing for grace these knights of the round table whip out their knives or daggers—all sorts of weapons, made for killing pigs or skinning sheep or cutting purses—and fall upon those poor bones, stripping off nerves and muscles as if they had been practicing anatomy at Guadalupe or Valencia all their lives; and before you can say a *credo,* they leave them as clean and smooth as ivory. On Fridays and vigils they have beans cooked in salt water, on fast days salt cod. One of the boys takes round the mess-kettle and ladles out the drink ration—a little wine, poor thin stuff, not improved by the baptism it receives. And so, dining as best they can, without ceremony or order, they get up from the table still hungry.

The captain, the master, the navigator and the ship's notary dine at the same time, but at their own mess; and the passengers also eat at the same time, including myself and my family, for in this city you have to cook and eat when your neighbors do, otherwise you find no fire in the galley, and no sympathy. I have a squeamish stomach, and I found these arrangements very trying; but I had no choice but to eat when the others were hungry, or else to dine by myself on cold scraps, and sup in darkness. The galley—"pot island" as they call it—is a great scene of bustle and activity at meal times, and it is amazing how many hooks and kettles are crowded on to it; there are so many messes to be supplied, so many diners and so many different dinners. They all talk about food. One will say, "Oh for a bunch of Guadalajara grapes!"; another, "What would I give for a dish of Illescas berries?"; somebody else, "I should prefer some turnips from Somo de Sierra"; or again, "For me, a lettuce and an artichoke head from Medina del Campo"; and so they all belch out their longings for things they can't get. The worst longing is for something to drink; you are in the middle of the sea, surrounded by water, but they dole out the water for drinking by ounces, like apothecaries, and all the time you are dying of thirst from eating dried beef and food pickled in brine; for My Lady Sea won't keep or tolerate meat or fish unless they have tasted her salt. Even so, most of what you eat is half-rotten and stinking, like the disgusting fu-fu that the *bozal* negroes eat. Even the water, when you can get it, is so foul that you have to close your eyes and hold your nose before you can swallow it. So we eat and drink in this delectable city. And if the food and drink are so exquisite, what of the social life? It is like an ant-heap; or, perhaps, a melting-pot. Men and women, young and old, clean and dirty, are all mixed up together, packed tight, cheek by jowl. The people around you will belch, or vomit, or break wind, or empty their bowels, while you are having your breakfast. You can't complain or accuse your neighbors of bad manners, because it is all allowed by the laws of the city. Whenever you stand on the open deck, a sea is sure to come aboard to visit and kiss your feet; it fills your boots with water, and when they dry they are caked with salt, so that the leather cracks and burns in the sun. If you want to walk the deck for exercise, you have to get two sailors to take your arms, like a village bride; if you don't, you will end up with your feet in the air and your head in the scuppers. If you want to relieve yourself . . . you have to hang out over the sea like a cat-burglar clinging to a wall. You have to placate the sun and its twelve signs, the moon and the other planets, commend yourself to all of them, and take a firm grip of the wooden horse's mane; for if you let go, he will throw you and you will never ride him again. The perilous perch and the splashing of the sea are both discouraging to your purpose, and your only hope is to dose yourself with purgatives. . . .

We sailed on alone for the first six days; for the eight other ships which left Santa Cruz harbor in Tenerife in our company all disobeyed the instructions which the judge of the *Contratación de Indias* sent us, and each went off on his own during the first night. What pleasure can a man have on board a solitary ship at sea? No land in sight, nothing but lowering sky and heaving water; he travels in a blue-green world, the ground dark and deep and far below, without seeming to move, without seeing even the wake of another ship, always surrounded by the same horizon, the same at night as in the morning, the same today as yesterday, no change, no incident. What interest can such a journey hold? How can he escape the boredom and misery of such a journey and such a lodging?

It is pleasant to travel on land, well mounted and with money in your purse. You ride for a while on the flat, then climb a hill and go down into the valley on the other side; you ford a running river and cross a pasture full of cattle; you raise your eyes and watch the birds flying above you; you meet all kinds of people by the way and ask the news of the places they have come from. You overtake two Franciscan friars, staves in their hands, skirts tucked into their girdles, riding the donkeys of the seraphic tradition, and they give you "Good-day and thanks be to God." Then, here comes a Jeronymite father on a good trotting mule, his feet in wooden stirrups, a bottle of wine and a piece of good ham in his saddle-bag. There will be a pleasant encounter with some fresh village wench going to town scented with pennyroyal and marjoram, and you call out to her, "Would you like company, my dear?" Or you may meet a traveling whore wrapped up in a cloak, her little red shoes peeping below the hem, riding a hired mule, her pimp walking beside her. A peasant will sell you a fine hare to make a fricassee; or you may buy a brace of partridge from a hunter. You see in the distance the town where you intend to sleep or stop for a meal, and already feel rested and refreshed by the sight. If today you stay in some village where the food is scanty and bad,

tomorrow you may be in a hospitable and well-provided city. One day you will dine at an inn kept by some knife-scarred ruffian, brought up to banditry and become a trooper of the *Santa Hermandad;* he will sell you cat for hare, billy-goat for mutton, old horse for beef and watered vinegar for wine; yet the same day you may sup with a host who gives you bread for bread and wine for wine. If, where you lodge, tonight, your hostess is old, dirty, quarrelsome, querulous and mean, tomorrow you will do better and find a younger one, clean, cheerful, gracious, liberal, pious and attractive; and you will forget the bad lodging of the previous day. But at sea there is no hope that the road, or the host, or the lodging will improve; everything grows steadily worse; the ship labors more and more and the food gets scantier and nastier every day. . . .

We ran with a stiff northeast wind for the next four days; and the navigator and the sailors began to sniff the land, like asses scenting fresh grass. It is like watching a play, at this time, to see the navigator taking his Pole Star sights; to see him level his cross-staff, adjust the transom, align it on the star, and produce an answer to the nearest three or four thousand leagues. . . . They always went to great pains to prevent the passengers knowing the observed position and the distance the ship had made good. I found this secretiveness very irritating, until I discovered the reason for it; that they never really knew the answer themselves, or understood the process. They were very sensible, as I had to admit, in keeping the details of this crazy guesswork to themselves. . . .

In the middle of all these vain conflicting arguments among masters, navigators, and sailors who claimed to be graduates in the art, on the twenty-sixth day out, God be praised, we sighted land; and how much lovelier the land appears from the sea than the sea from the land! We saw Deseada—appropriately named—and Antigua, and set our course between them, leaving Deseada to the east. We ploughed on; Santa Cruz hove in sight to windward, and we passed it at a distance; we reached San Juan del Puerto Rico and coasted along the shore some way, keeping a careful watch on Cape Bermejo, which is a notorious haunt of pirates. We recognized Mona and the Monitos—easy to identify, even at a distance—looked for Santa Catalina but failed to see it; and eventually came in sight of Saona, the land of the blessed saint, and blessed sight to us. All this time we were repeatedly soaked by downpours of rain; but we made light of them, and thought ourselves lucky to have been spared hurricanes. . . .

But I will weary you no more with the perils and miseries of the sea; except to ask you to imagine, if life can be so uncomfortable with fair winds and a relatively calm sea, what it must be like to experience contrary winds, encounters with pirates, mountainous seas and howling gales. Let men stay on firm ground and leave the sea to the fishes, say I.

Next day at dawn our city came to life, with much opening of trunks and shaking out of clean shirts and fine clothes. All the people dressed in their best, especially the ladies, who came out on deck so pink and white, so neat, so crimped, curled and adorned, that they looked like the granddaughters of the women we had seen each day at sea.

The master went ashore, and I sent my servant with a message of greeting to the president of the court. Boats began to put out to the ship; and since there was a head wind and the ship had to be warped up the river, my family and I went ashore directly in a launch which they sent for us. So we reached the longed-for land, and the city of Santo Domingo. We were kindly welcomed; and after a few days' rest I took my seat on the Bench, and here I stay for as long as God wills, without any desire to cross the sea again. I hope soon to hear that you also have the appointment which you deserve. Doña Catalina and the children send their respects and best wishes.

Rotting Ships, Rotting Men

J. H. Parry

The appalling conditions suffered at sea—bad food, crowded conditions, lack of sanitation, and disease—did not improve much from the sixteenth to the eighteenth century. Disease aboard naval ships claimed far more victims than battles. The poor health conditions among eighteenth-century seamen and the discovery of the cure for scurvy, the dreaded disease that particularly afflicted sailors, are the subjects of the following excerpt from J. H. Parry's Trade and Dominion.

The activities of navies . . . were prominent among the factors which enabled merchant ships to reduce their crews; but any surplus which resulted was promptly snapped up, in wartime, by the navies themselves. As navies increased in size, their demand for men increased in direct proportion. Hence, in England, the press; the French managed better, by maintaining, from 1689 onwards, a register of seamen; but compulsion, in one form or another, was an essential feature of recruitment in all major navies. Hence, also, as more and more resentful and potentially mutinous recruits were dragged on board, the increasing severity of naval discipline towards the end of the century and throughout the French wars. This discipline involved not merely more frequent and more savage floggings, but also—and probably much more resented—long, indeterminate commissions, and the close incarceration of men on board ship when in harbour, to prevent their running away.

Navies not only absorbed men, they destroyed them; and that not only—not even mainly—in battle, but by disease. Like the ships, the men in them rotted. In the best of circumstances, ship-board life was unhealthy. It was difficult to keep ships dry; wet clothes and bedding encouraged rheumatoid complaints which, if they did not kill men, slowed their movements. In bad weather it was difficult to keep fires alight to cook hot meals. Extra spirit issues, though doubtless welcome, were an unsatisfactory substitute. Significant improvements were made in both these respects in the course of the eighteenth century: flush-decked ships were drier than waisted ones: permanent brick-built hearths between decks were vastly more reliable than the old, primitive fire-box in the break of the fo'c'sle; and much ingenuity was exercised in devising more efficient galley equipment. There were worse dangers, however, than rheumatism and general discomfort. Throughout most of the eighteenth century ships' companies, especially the crowded crews of warships, were repeatedly devastated by killing disease. During the American War the British navy raised 171,000 men. Of these 1,240 were killed in action, 42,000 deserted and 18,500 died of disease. The main killers were the mosquito-borne diseases, malaria and yellow fever, which attacked men in tropical harbours, especially in the West Indies; typhus in home waters; and scurvy everywhere, in ships which spent long periods at sea. Quinine had long been used as specific for malaria; for yellow fever there was no known cure; and nothing could be done to prevent either disease in harbour until, more than a century later, the role of the mosquito as carrier came to be understood. Meanwhile, the best remedy was to put to sea.

Putting to sea gave no immunity from typhus. This, the dreaded "gaol fever," often devastated fleets after a year or two of war, when the reservoir of merchant seamen was exhausted and the press-gangs began to bring in verminous recruits from the slums of big cities. There was an appalling epidemic of typhus in the British Channel fleet in 1780, when the flood of new entry overwhelmed the rudimentary hospital organization. The fleet could scarcely be got to sea. Everyone in England concerned with manning problems was thoroughly scared; and perhaps for that reason, a striking and continuous improvement in the handling of fleet epidemics began after that calamitous year. Much of the improvement was due to the influence of the great naval physician James Lind, who, though he had no notion of the significance of lice and fleas, did understand the importance of cleanliness. Lind insisted in 1781 on the establishment of Receiving Ships, where new entries could be examined, issued with new clothes, and quarantined.

Scurvy, the worst killer of all, presented problems of a different kind. No one in the eighteenth century had heard of vitamin C; but everyone knew that men fed on fresh food did not suffer from scurvy. It was impossible, with eighteenth century methods of storage, to keep most kinds of food fresh for more than a short period; and in many parts of the world where European ships went, it was difficult to obtain fresh supplies. Some officers, therefore, shrugged off scurvy as an inescapable hazard of the sea; others pursued the search for anti-scorbutics which could be preserved. . . . There were endless experiments with half-measures and useless nostrums. Sauerkraut and "portable soup" probably did some good, extract of malt little, elixir of vitriol none. Cook achieved a resounding triumph in 1775, when he returned from his second world voyage without losing a man from scurvy, and proved that careful victualling could eliminate the disease. Cook, in his determination to keep his men healthy, had tried virtually everything; it was not clear, even to him, which of the many reputed antiscorbutics carried in his ships had really been responsible for his success. Probably several of them contributed. During the twenty years after his return, however, informed opinion gradually fixed upon the best of the various remedies, a remedy so simple that it seemed too good to be true; namely lemon juice. (Orange juice would do equally well. West Indian limes were an inferior substitute.) The therapeutic value of citrus fruits had been known at least since the beginning of the seventeenth century. Fruit was expensive, and whole fruit was difficult to keep in good condition; whole fruit, moreover, even whole juice in cask, was bulky; it was difficult to carry enough of it to meet the needs of a heavily-manned ship on a long voyage. The problem was, therefore, to find a method of concentrating lemon or orange juice without impairing its efficacy; to induce naval authorities to insist on its issue and use, regardless of its cost; and to persuade the owners of merchant ships that dead men cost more than lemons. So far as the British navy was concerned, the Admiralty, persuaded by Sir Gilbert Blane, who had been Rodney's physician, ordered, in 1795, a regular issue of lemon juice concentrate, daily to all hands. Within five years, scurvy was almost unknown in the fleet.

The conquest of scurvy was a striking example of the gap between scientific recognition of a technique and its administrative adoption. However tardy, it was probably the most important single advance in nautical management in the whole period with which we are concerned. It made possible the long blockades which broke the naval power of Napoleon. . . . Progress in merchant shipping is more difficult to trace, depending there as it did on the initiative and enlightened self-interest of owners. On the whole, the East India companies were ahead of the navies in this matter; but they were not consistent, and sometimes tried to economize on the wretched soldiers they transported. . . . By 1800 no one doubted that scurvy could and should be prevented. For sailors, soldiers and emigrants alike, long sea passages lost much of their horror (though none of their tedium). One more limitation on European trade and dominion overseas had been removed.

A Naval Officer's View of Naval Discipline

Frederick Chamier

The conditions described in the selections by Salazar and Parry were hardly such as to attract sufficient numbers of sailors to the navy, so all the major navies relied on compulsion to recruit seamen and harsh discipline to keep them. Impressment—conscription that frequently amounted to kidnapping—was widely practiced, and the British habit of stopping American merchant ships on the high seas and impressing their crews was one of the major causes of the War of 1812. The following account by Frederick Chamier, a British novelist who served in the navy from 1809 to 1856, presents an officer's view of naval discipline and impressment.

And now, as I have come to anchor after a long cruise to the Western Indies, I will give my opinion about corporal punishment and impressment of seamen; because every now and then I run foul of a kind of sea-lawyer, one of the devil's attorneys, and he is always prating to those who will listen to him, and trying to make them believe that they are, like dogs, taken, stolen, whipped, and kicked by every man with a pair of epaulettes, who happens to walk a quarter-deck.

As for corporal punishment then—which means a little back-scratching—I think I may say that it could not be abolished without injury to the service. When the wind is whistling, the rain pouring down, the sea getting up, and the after-guard, main, mizzen top-men, and marines are lugging away at the weather fore-topsail-brace, with their eyes all running over with rainwater, and their tails blowing over to lee-ward—I say, when it is dark, a dirty, murky, rainy, windy, snowy night, there is many a man who prefers a hammock to a wet jacket. Of course, if these men remain skulking below, the duty will fall the heavier upon the men aloft. Now, it is nothing but the fear of the cat and her tails that keeps such fellows from sleeping, and if you put them in irons, why you only encourage their idleness. You may make them pick oakum, and that's all you can do. You may stop their grog, and they will get more than their allowance from their shipmates. You may clap them in the black list, but that is a bad remedy: nothing breaks a good man's heart more than being mixed up with fellows on that list—and perhaps he may have dozed a bit on the lookout, or the officer of the watch may have seen a stranger before him.

Let them be educated, say some: I say, no, you'll make them worse. Instead of talking of the good old times, spinning a yarn about the Nile, running up one's memory about Nelson, and such like, they would all be squatting about the decks like a set of Turks, with newspapers before them, settling the affairs of the nation, and talking about that which none of them understand! Let them alone; they are used to it, they think less of the disgrace than the pain; and whilst we have officers who are as humane as they are brave we have little to fear from tyranny, and that tyranny can always be stoppered.

Well, then, as to impressment, without being the least personal, I take the liberty of saying, that none of the great people understand the subject at all. Who is to know so much about it as we, who have been on shore, and lugged the man out of his warm bed to make a sailor of him? And, curse their ungrateful souls! some of them try to run away afterwards! as if it was nothing to have board and lodging; to peck and perch at the King's expense; to be allowed to fight their enemies; and to sing a jolly song in the forecastle, when the ship's under a close-reefed main-topsail, rising over the waters like a duck, with as jolly a gale of wind for a chorus as ever seaman can wish: and then, on Saturday night, to have "sweethearts and wives," and to know that

> "There's a sweet little cherub which sits
>   up aloft,
> To look out for the life of poor Jack!"

Well, there are ungrateful people in the world, and we sometimes find them in these pressed men. . . .

A Sailor on Impressment

The sailors, as was to be expected, did not necessarily share their officers' benign view of impressment. At least some of them, however, regarded "the press" as a necessary evil. The following selection first appeared in The Naval Sketch Book, *published in London in 1826 by "An Officer of Rank."*

You may talk o' the hardships of pressing—your man-hunting—and the likes of such lubberly prate; but if there's never no ent'ring, how the h—ll can you help it? Men-o'-war must be mann'd, as well as your marchanmen—marchanmen must have their regular convoys, for if they havn't, you know, then there's a stopper-over-all upon trade—so take the concern how you will—"by or large"—there's not a King's Bencher among you can mend it. Bear up for Blackwall—ship aboard of an Ingee-man, and see how you will be badgered about, by a set o' your boheaing-hysun-mundungo-built beggars! Get hurt in their sarvice—lose a finger or fin by the chime of a cask in the hold—or fall from aloft and fracture your pate—then see where's your pension or "smart." I'm none o' your arguficators—none o' your long-winded lawyers, like Paddy Quin the sweeper, or Collins the "captain o' the head"; but d——n it, you know, there's never no working to wind'ard of truth.

Daily Routine in a Sailing Warship   *Bill Truck*

Life aboard a crowded nineteenth-century warship, carrying some 850 officers and men, had to be carefully regulated, and the discipline of work was governed by the striking of the bell at half-hour intervals. Work was relieved by rations of grog (watered rum), which helped to make the men forget the scum on their water and the beetles in their biscuits. The typical daily routine of the sailor is set forth in the following excerpt from The Man-o'-War's Man *(1843) by Bill Truck, Senior Boatswain of the Royal Naval College of Greenwich.*

Discipline was somewhat more relaxed in port, where the naval seamen were kept aboard lest they desert. Wives, however, were allowed, sharing hammocks slung between the guns with their men—and giving birth to the phrase "son of a gun." Some women quietly slipped out to sea when the ships weighed anchor again.

The nautical day commences, either by observation or account, at the sun's meridian, generally supposed to be our twelve o'clock—noon—on shore. At that moment (meridian) the officer of the watch, or more commonly the master of the ship, orders the marine sentinel to turn an half-hour sand-glass (which he has always in charge, and which has been previously run out), and strike eight bells forward: which is accordingly done, and the dinner is piped. No sooner is this glass run out than the sentry calls, "Strike the bell *one,* forward!" and again turns it, when the grog is immediately piped. When it runs out a second time, he again calls, "Strike the bell *two,* forward!" which is no sooner done than the boatswain's mate calls the afternoon watch. Thus he proceeds until he comes to the eighth bell; which is no sooner struck than the watch expires, and the grog is again piped. Previous to this, however, in order to relieve the quartermaster, the helmsman, the look-out at the masthead, and the sentinel at the glass or elsewhere, an individual of each of these classes of the watch below goes to the purser's steward when the seventh bell has struck, gets his quartern of rum unmixed, takes his supper, and is ready, as soon as the bell strikes, to relieve his man with the rest of his watch. All

hands now take supper, and, when the bell again strikes, the first *dog-watch* is called. This is only a watch of two hours, and, of course, when the fourth bell has struck, the second *dog-watch* is called, which lasts other two hours, and brings the supposed time pretty accurately to our eight o'clock at night. By this time, however, the hammocks having been piped down, the watch relieved generally retire to rest. The watch on deck, therefore, execute all the necessary duties of the ship until their eighth bell has struck, when the *middle watch* is called; and this, again, is relieved, after the same time, by the morning watch, who do the ship's duty during other eight bells, which brings the account of time to eight o'clock in the morning, when breakfast is always piped. At one bell, after breakfast, the forenoon watch is called, who do the duties on deck, while the watch below are scrubbing or fumigating the lower deck, or probably mending their clothes; and thus they continue until a fresh observation of the sun is again taken, and the necessary correction made on the time lost or gained. All this being accomplished, the eighth bell is struck, the day at sea is completed, the glass is turned to commence a new one, the dinner is piped, and the watch called as before.

Sea Language

J. H. Parry

One of the characteristics that for centuries set the sailor apart from other men was his specialized use of language. The extract that follows is from J. H. Parry, "Sailors' English," The Cambridge Journal (Cambridge, England, November 1949).

To describe the English as a seafaring race is an exaggeration. Though ships and sailors have traditionally played a leading part in shaping England's fortune, no more than a very small proportion of Englishmen have ever earned their living at sea. Yet both the slang and the technical terms of the sea are known, either through literature or through direct contact with sailors, to a wide English public, and the English language is full of nautical words and phrases, many of which have become so familiar that their origin is forgotten. As far as literary evidence is to be trusted, this enrichment of our ordinary daily speech is comparatively modern. Sailors have usually been treated sympathetically in English literature (Chaucer is an exception); but the dramatists and novelists who used the sea and ships to elaborate their plots have not always taken the trouble to portray the sailor himself as a character shaped by his profession, with customs, superstitions and dialect all his own. In the whole great range of Elizabethan drama there is hardly a character who is clearly distinguishable by his speech and manner as a sailor. There are a few nautical touches in Restoration drama. Wycherley, who had seen service in the Fleet, convincingly sketched a group of seamen in *The Plain Dealer,* but they are minor characters in the play. Apart from these, Congreve's Ben Legend, who first trod the boards in 1695, was almost the first genuine and recognizable sailor on the English stage. Ben was sympathetically handled by his creator, but it is clear from the dialogue that to be "half sea-bred" was at that time far from a social recommendation. His nautical slang provoked derision and not kindly interest—certainly not a desire to imitate.

The novelists lagged behind the dramatists in exploiting the characteristics of sea-faring men. Defoe's characters constantly went to sea and suffered piracy or shipwreck; but that is almost the extent of their nautical colour. It was not until the middle of the eighteenth century, in *Peregrine Pickle* and *Roderick Random,* that the sailor with his virtues, eccentricities and peculiar dialect became at last a literary feature and attraction.

The French wars naturally created a wave of interest in the doings of sailors, which was reflected in novels then and later. In 1805 appeared *The Post Captain: or Wooden Walls well manned,* by John Davis—a light and frivolous performance, with little attempt at plot or regard for probability. It was written by a sailor, however, and is full of lively humour and authentic nautical slang. *The Post Captain* enjoyed great popularity at the time. It has been called the parent of all our nautical novels, though its effects were not at once apparent. Other early nineteenth-century novelists continued to introduce seafaring characters into their books without much attempt to reproduce seafaring manners and speech. There is nothing particularly nautical about Captain Wentworth in *Persuasion,* except the business-like way in which he sets about making his fortune out of prize-money. Miss Austen has, it is true, drawn two convincing seafarers in *Mansfield Park*—the open-hearted midshipman William Price and his father, the ex-lieutenant of marines with his smell of spirits and his interest in dockyard details; but even the elder Price is not made to reveal his profession by his choice of words. A good many years were still to pass before Marryat, seconded by lesser writers like Howard and Chamier, gave the sea-novel a well-deserved vogue and made the sailor, with all the details of his dress, his customs and his speech, a familiar figure in English fiction.

The Englishman's acquaintance with the language of sailors, however, is not derived solely from fiction. Seafaring is not a closed profession. . . . In no country in Europe is the seafaring population more widely spread, both geographically and socially, than in England, and that fact also helps to account for the prevalence of nautical terms in the general English vocabulary.

The words and phrases here described as "sailors' English" fall roughly into four groups. The first group comprises technical terms in current use at sea, words used in giving orders, making reports, and in general carrying out the work of a ship. The second group comprises slang terms in use at sea—words used among sailors in ordinary conversation to describe the familiar surroundings of daily life, their food, their clothes, their shipmates, and so on. Thirdly, there is a surprisingly well-defined group of sea-metaphors habitually used by sailors to describe objects and actions ashore. Obviously there is a good deal of overlapping between these groups, and in course of time words pass from one group to another. In particular, words from the first and second groups are used with different meanings in the metaphors of the third, and these uses lead on to the fourth group, comprising words and phrases which originated in the nautical technicalities or nautical slang of a former period, but which have passed into ordinary use ashore and have in some cases entirely lost their nautical connections.

The technical language of the sea is remarkable for its terseness and its accuracy, as might be expected; for it is used for giving orders to considerable bodies of men, often under conditions of bad weather or of sudden emergency, concerning the handling of a valuable, extremely intricate and often perverse mechanism—a ship. There is a word or phrase for every action and every object, and usually the word or the phrase is short and unmistakable. Often, too, the wording of conventional orders and reports is crisply descriptive. Let anyone who doubts these statements try to express, in less than three times the number of words, the meanings of "anchor's a-cock-bill," "at short stay," or "foul hawse"; or consider the sequence of orders for hoisting a boat: "Hook on—haul taut singly—marry the falls—hoist away—high enough—ease to the life-lines—light to." Sailors in general are extremely sensitive to any misuse of the technical terms of seamanship, and are reluctant to accept new terms. The capstan, for instance, when turned in the reverse direction, is "walked back," though it is many years since capstans were worked by tramping sailors shoving at the bars. . . .

But technicalities are only the dry bones of nautical English. It is the slang of the ward-room and the mess-deck (in some respects the two languages are quite distinct) which reveals the inventiveness and peculiar humour of the sailor. Slang terms are necessarily less definite and lasting than technicalities, and although many current slang sea-phrases have a good antiquity, their history is more difficult to trace. The first lexicographer upon whom the historian can rely was a nineteenth-century one. Admiral W. H. Smyth's *Sailors' Word-book* was published in 1867, in the decade, that is, when the victory of iron over wood was won. It is a mighty monument of nautical erudition. Like Doctor Johnson, the admiral was a staunch conservative, and allowed no finical scruple of impartiality to prevent his expressing his own opinions in his definitions; so we have:

"Boatswain-captain: an epithet given by certain popinjays in the Service to such of their betters as fully understand the various duties of their station."

"Followers: . . . the young gentlemen introduced into the Service by the captain, and reared with a father's care, moving with him from ship to ship; a practice which produced most of our best officers formerly, but innovation has broken through it, to the great detriment of the Service and the country."

"Shipmate: a term once dearer than brother, but the habit of short cruises is weakening it."

"Lime-juice: a valuable anti-scorbutic included by act of parliament in the scale of provisions for seamen. It has latterly been so much adulterated that scurvy has increased three-fold in a few years." It is curious that Smyth makes no mention of "limey," the name given by American sailors to Englishmen, and still current in New England. . . .

Sailors have a considerable variety of names for those who shirk their work; but here one must avoid hasty conclusions. "Idlers" are not, or were not, lazy men, but men whose duties did not include watch-keeping. "Send all the idlers up," vociferates Midshipman Echo, upon "all hands" being piped. "Daymen" is a more usual synonym today. The "mess-deck dodger," also, is not a man who avoids the work of the mess-deck, but one whose work lies there, as a sweeper; all that he dodges is bad weather on deck. "Waister" is a deceptive word. In wooden ships it described a man employed in cleaning the waist, the mid-ship section of the upper deck; the waist-party was composed of men too slow or too incompetent for work aloft, and "waister" thus came to mean good-for-nothing. A lazy fellow is a "skulker," a "proper ullage." He used to be called—until another service appropriated the word and changed its meaning—an "urk."

Two of the best examples of long-lived slang phrases are "warming the bell" and "tapping the admiral." To "warm the bell" is to do something prematurely. When watches were measured by the half-hour glass, an unscrupulous watch-keeper could shorten his watch by warming the glass between his hands, and so making the sand trickle through a little faster; and the metaphor, it is said, has become transferred from the obsolete glass to the bell which still marks the passage of the half-hours at sea. "Tap-

ping the admiral" is a more uncommon phrase, but it is still used occasionally to describe the habits of those who will drink anything. Smyth mentions it—it was old in his day—and says it alludes to "the drunkard who stole spirits from the cask in which a dead admiral was being conveyed to England." These are good hoary old phrases; and a true account of sailors' English must set beside them the long list of new slang terms, invented by sailors of the present day but already established throughout the service: "flat-iron" for an aircraft carrier, "tin fish" for a torpedo, "ping" for ASDIC [British sonar], and "skimming dish" for that delightful toy, the motor dinghy. It should in fairness be recorded, as a sign of the times, that a larger type of boat, decorously known in the Royal navies as a motor whaler, is dubbed by the other English-speaking navy a "gasoline gig."

Sailors' slang, then, no less than sailors' technical jargon, is characteristic and vigorous, and although much of it dates from the days of sail, there is little reason to suppose that the "habit of short cruises is weakening it." What happens to sailors' English when it goes ashore?

The habit of using nautical metaphor when ashore has long been a recognized characteristic of the sailor, both in literature and in real life. With the growing popularity of the Navy during the French wars, this habit began to attract more frequent literary notice and to provoke affectionate caricature. Most of the humour in *The Post Captain* is of this kind. The First Lieutenant proposes marriage in nautical language:

"Divine Flora, the havoc committed by shells thrown into the seaport of an enemy, is a mere trifle in war-time, compared, queen of queens! to the destruction of my heart from the fire of your eyes. Yes! goddess of goddesses! a shot from either one or both of those heavenly bow-chasers has raked my heart fore and aft and knocked it into splinters; splinters that no carpenter can repair but the magic of your smiles. Alack! alack! every time I lie down in my hammock, I fairly make the clews strand, conceiting I hold you, beautiful Flora, in my arms; and if this be not a proof of my most ardent love, I know not in which point of the compass it lies. Lowering my top-gallant sails to you, I am your dying lieutenant, Henry Hurricane." . . .

The Post Captain, in the words of a laudatory review of the period, reproduced "that correct sea language which since the time of Smollett has been seldom found in works of imagination." The same might be said of the works of Marryat, Chamier and the rest. But it did more; it showed sailors—or at any rate naval officers—exaggerating their jargon in ordinary talk ashore, because it was expected of them. The public loved them for it, and they have gone on doing it ever since. For Jack ashore, a lawyer is a "land shark" and an empty bottle a "dead marine." "Backing and filling," a sailing-ship phrase, may describe the habit of changing one's mind, or even the operation of turning a motor car round in a narrow space. To excuse or justify one's self is to "square one's yard arm"; and the attitude of mind of a selfish person is concisely expressed in the sentence "pull the ladder up, Jack—I'm inboard."

Language is an important part of professional *esprit de corps.* When the administrative depots, stores and training establishments formerly housed in wooden hulks were moved into barracks and offices ashore, those barracks and offices were commissioned under the names of ships and provided with gravel "quarter-decks," brick-built "mess-decks" and "ward-rooms." The practice even deceived the German high command, who during the recent conflict claimed to have sunk a naval airfield situated fifty miles inland. While the Admiralty creates "ships" ashore, the sailor carries the language of the sea, as he has always done, into his home and the public house where he meets his friends. . . .

Hundreds of nautical expressions have found a permanent place in colloquial English. "Taken aback" is an obvious example. So is "son of a gun," a reminder of the days when women lived on board the king's ships in harbour, and sometimes at sea too. To "cut and run" once meant to "let run" the furled sails of a ship by cutting the yarns which secured them. How many novelists who use that curious phrase, "to the bitter end," suspect a possible nautical origin? A *Seaman's Grammar* of 1653 explained that "a bitter end is but the turn of a cable about the bitts, and veare it out little and little, and the bitter's end is that part of the cable doth stay within board." The modern equivalent, also used in conversation among sailors, is "out to a clinch," referring to the steel forging in the cable locker to which the inboard end of the cable is shackled. A "nipper" was a stopper upon a cable, and by a simple transference became the ship's boy by whom in former times, the stopper was worked. "Loggerheads" were iron bars, used when hot for caulking seams with pitch—handy and dangerous weapons, apt to be snatched up in the heat of a mess-deck dispute. The process of incorporation still goes on. The submarine service is already adding slang terms to the English language, and no doubt aircraft-carriers and landing-craft will have their contributions to make. . . .

Sailors' English is not a mere specialists' jargon. It is a vital and living branch of English; a speech of clearness, precision and—for all its oddities—of beauty. Conrad, who of all sailor-authors understood its beauty the best, compared it to an anchor—"a forged piece of iron, admirably adapted to its end; and nautical language is an instrument wrought into perfection by ages of experience, a flawless thing for its purpose." It was of nautical *English* that this adopted Englishman was thinking; for he added, a little later: "If I had not written in English, I would not have written at all."

Handing Sail in a Squall

Richard Henry Dana

A fine example of sailors' English is provided by Richard Henry Dana in the following excerpt from Two Years
Before the Mast. *Blending the techniques of literature and journalism, Dana gives a vivid account of his
voyage in the mid-1830s as a common sailor aboard the* Pilgrim, *a two-masted ship that sailed from Boston around
Cape Horn to California. The book, published in 1840, achieved great popularity and has remained a
classic. Like John Masefield in "Dauber," Dana describes the situation that supremely tested the sailor's skill and
courage: handing (furling) sail in a storm.*

We met with nothing remarkable until we
were in the latitude of the river La Plata. Here
there are violent gales from the south-west,
called Pamperos, which are very destructive to
the shipping in the river, and are felt for many
leagues at sea. They are usually preceded by
lightning. The captain told the mates to keep a
bright lookout, and if they saw lightning at the
south-west, to take in sail at once. We got the
first touch of one during my watch on deck. I
was walking in the lee gangway, and thought
that I saw lightning on the lee bow. I told the
second mate, who came over and looked out
for some time. It was very black in the southwest, and in about ten minutes we saw a distinct
flash. The wind, which had been south-east, had
now left us, and it was dead calm. We sprang
aloft immediately and furled the royals and top-
gallant-sails, and took in the flying jib, hauled
up the mainsail and trysail, squared the after
yards, and awaited the attack. A huge mist
capped with black clouds came driving towards
us, extending over that quarter of the horizon,
and covering the stars, which shone brightly
in the other part of the heavens. It came upon us
at once with a blast, and a shower of hail and
rain, which almost took our breath from us. The
hardiest was obliged to turn his back. We let
the halyards run, and fortunately were not taken
aback. The little vessel "paid off" from the
wind, and ran for some time directly before it,
tearing through the water with everything flying. Having called all hands, we close-reefed
the topsails and trysail, furled the courses and
jib, set the fore-topmast staysail, and brought
her up nearly to her course, with the weather
braces hauled in a little, to ease her.

This was the first blow, that I had seen, which
could really be called a gale. We had reefed our
topsails in the Gulf Stream, and I thought it
something serious, but an older sailor would
have thought nothing of it. As I had now become used to the vessel and to my duty, I was
of some service on a yard, and could knot my
reef-point as well as anybody. I obeyed the order
to lay aloft with the rest, and found the reefing
a very exciting scene; for one watch reefed the
fore-topsail, and the other the main, and every
one did his utmost to get his topsail hoisted
first. We had a great advantage over the larboard
watch, because the chief mate never goes aloft,
while our new second mate used to jump into
the rigging as soon as we began to haul out the
reef-tackle, and have the weather earing passed
before there was a man upon the yard. In this
way we were almost always able to raise the cry
of "Haul out to leeward" before them, and
having knotted our points, would slide down
the shrouds and back-stays, and sing out at the
topsail halyards to let it be known that we were
ahead of them. Reefing is the most exciting part
of a sailor's duty. All hands are engaged upon it,
and after the halyards are let go, there is no time
to be lost—no "sogering," or hanging back,

then. If one is not quick enough, another runs over him. The first on the yard goes to the weather earing, the second to the lee, and the next two to the "dog's ears"; while the others lay along into the bunt, just giving each other elbow-room. In reefing, the yard-arms (the extremes of the yards) are the posts of honor; but in furling, the strongest and most experienced stand in the slings, (or, middle of the yard,) to make up the bunt. If the second mate is a smart fellow, he will never let any one take either of these posts from him; but if he is wanting either in seamanship, strength, or activity, some better man will get the bunt and earings from him; which immediately brings him into disrepute.

The Morn of Life.

Scrimshaw often occupied what idle hours the sailor enjoyed at sea.

A Clipper Captain's View on Steam   *John Masefield*

Dana wrote Two Years Before the Mast *in 1840, during the Golden Age of Sail. The next decade witnessed the development of the clipper ship. Built for speed, these slender sailing vessels carried up to thirty-five sails as they raced from the East Coast around South America to China, to bring back tea, or to California, carrying fortune seekers and supplies. The clipper ship became the queen of the seas during the mid-1800s, but even then the days of sail were numbered. For almost 5,000 years, since the Egyptians invented the sail, shipbuilders had concentrated on increasing the size of ships and the efficiency of the rig. Then in 1807, Robert Fulton built the first successful steamboat, and in 1838 the British side-wheeler* Sirius *became the first ship to cross the Atlantic under steam power alone. By the 1860s, screw-propelled steamers were regularly crossing the Atlantic, and the opening of the Suez Canal in 1869 gave the steamers a further advantage. The romance and beauty of the great days of sailing were just about gone by 1900, and with them went a way of life that pitted man against the elements. The contempt of the sailor of this earlier era for the seaman of the new steamship is bitingly portrayed in the following excerpt from John Masefield's* The Bird of Dawning *(1933).*

Shall we walk a little, Mr. Trewsbury?" the Captain asked. "Athwartships, though, if you please, even if it be 'three steps and overboard.' An old seaman once warned me never to walk up and down a poop, fore and aft, always athwartships. 'It will be less easy for them to knock you on the head, so, from behind.' He had been mate in a Western Ocean packet, where an officer had to look out for these things. However, after a week of calm, such as we've been having, a sudden death might seem a pleasure."

Cruiser wondered at the Captain's sudden gentleness; however, it was a pleasant change; he had had no conversation since leaving the Min except an occasional chat with one of the boys at night. He turned to walk with him in the mist, in the few feet of space between the break of the poop and the mizen mast.

"So you were not in the China tea race before, Mr. Trewsbury?" the Captain said. "Were you in the Navy, by any chance?"

"No, sir, never. I was in Gloucester Brothers'; the Blue and White Line they call them, big, full-rigged ships of over 2,000 tons, the *Talavera,* the *Vittoria,* the *Bidassoa:* they were all named after the battles in the Peninsula."

"What were they? Colliers?"

"We took general cargoes to Australia or San Francisco, and came back with grain, or hides, or hides and tallow. I served my time in them, and went one voyage in the *Bidassoa* as second mate."

"A South Shields firm, you say?"

"No, sir, a Liverpool firm, Gloucester Brothers, with a house-flag of blue and white stripes. The *Fuentes d'Onoro* holds the record to San Francisco."

"Did you say that you took convicts to Australia?"

"No, sir, general cargoes."

"Ah, yes, And you were never in the Navy?"

"Never, sir."

"Strange," the Captain said. There was a pause after this, till the Captain suddenly asked:

"May I ask how you came to the East, if you were never in this tea race before, and not in the Navy?"

"Certainly, sir. When I had passed for mate, I passed in steam, and tried to get a berth in a steamer."

"I should have thought that little effort would be needed," the Captain interposed. "There is surely little competition for a place in the Black Guard."

"Young men are taking to steam, sir," Cruiser said. "Steam is the new thing, and the new men turn to it."

"I will admit that they may be new, Mr. Trewsbury, but let us not agree to call them men. Men master the elements; in that there is beauty and fitness. The new scheme is that men should become the slaves of machines: and in that there is neither. But I interrupt, I fear. May I ask if you were unsuccessful in your attempt?"

"No, sir, I was successful. I was given a second mate's berth in a steamer bound to Sydney: a ship called the *Thunderbird*."

"I am sorry that you should call a tank moved by a machine a ship. May I ask if you felt a little ashamed, when you stood upon the coal or oil platform to empty the cinders?"

"I was very proud, sir, to be helping to master the elements."

"Proud, did you say? But Life is continually astounding one, with the conditions which cause pride in others. However, in this case, I ought not to feel surprise. And your pride, as you call it, maintained itself, perhaps; you grew to love your oil-rag and what do they call it, your coal-slice?"

"Sir, she was a ship twice the size of this: and the last word in man's advance in mastery. But I had a bad time in her, for other reasons."

"Very often, a berth is not what one thought it would be," the Captain said. "So this blackened strumpet of your love was not all that she should have been as a wife?"

"The ship was all that one could ask, sir. She was as steady as a rock: and logged her seven knots day in, day out. No one could have complained of the ship."

"Yet you left her?"

"Yes, sir, I left her in Sydney."

"So the collier's joy was widowed. Perhaps you repented of the intercourse. I think you did not come all black from a bunker into this beautiful ship."

J. and F. Tudgay, *Flying Cloud*

At Sea in Sail: The 1880s

A Foremast Hand

The pride in seamanship noted by the captain of the clipper in John Masefield's story was often shared by the crews, who vied with each other in establishing speed records. The Champion of the Seas *made a run of 465 miles in twenty-four hours in 1854, a record that was not to be beaten by a steamship for more than twenty-five years. Yet despite some improvements in technology, such as the addition of the steam winch, conditions on board sailing ships at the end of the nineteenth century were in many respects quite similar to what they had been 150 years earlier. Poor food and scurvy remained common complaints, as evidenced in the following extract. Describing conditions in the 1880s, this letter was written by a foremast hand to Alan Villiers, who quotes it in full in his book* The Way of a Ship *(1953).*

The writer spent more than ten years at sea in sail. I have doubled both capes. Four times around Cape Horn, and six times around the Cape of Good Hope. Three voyages around Cape Horn I made in Lime-juicers, and one voyage in an American hell ship. Although an American by birth, most of my deep-water sailing was done in English ships for the very good reason that the "Limeys" were always better manned, and although the food in American ships as a rule was better than in any others, American deep-water ships were usually sent to sea short-handed with a mixed crew of "soldiers and dishwashers," and at the time of which I speak, in the early eighty's, we still had brutal, bucko mates on many American deep-water ships.

I was an ordinary seaman on Dicky Green's famous Aberdeen clipper, *Thermopylae,* when she made her famous run from London to Melbourne in 60 days. In the *Thermopylae* we carried 30 before the mast. Six apprentices, 8 ordinary seamen, and 16 A.B.'s. Although she was only of about 1000 tons, we surely needed all we had because the old man would carry on until her lee rail was almost under water before he would take a stitch off her. As soon as it began to blow great guns, we would take in the kites and stow the head sails, except the foretopmast stays'l, and the rest of them had either to stand

the gaff or blow out of the bolt ropes. She also had a steam winch, a great blessing to sailors, and the only ship I ever sailed in that had such a luxury.

I can see her now in memory with her slender tapering tall masts and yards which many a time I helped to paint white, swinging in a bosun's chair. Her hull was painted a bright green as were all of Richard Green's ships, and she looked more like a yacht than a cargo carrier. One of my shipmates that was in her when she first came out, told me that she logged more than 300 knots many a day.

Two voyages I made in another famous sailing ship, the *Sir Lancelot,* a famous tea clipper in her day, but she wasn't in the China trade when I was in her. I made one trip in her for wool from Melbourne, and one to Calcutta for jute. I left her in 1895, the year before she was lost, somewhere in the Indian Ocean, I believe. The two ships above mentioned were probably the fastest ships that ever sailed, with the possible exception of the *Flying Cloud, Sovereign of the Seas, Red Jacket,* and one or two others built by Donald McKay.

Every red-blooded boy at some time feels the urge to go to sea, and it isn't a matter of geography either, as it don't make any difference whether he was raised inland or on the coast. I have been shipmates with men that were raised

286

in the corn belt of the west that were excellent sailors. My first experience was in fishermen out of Boston to the Grand Banks, than which there is no better training for a man that intends to follow the sea. Next in 2, 3, and 4-masted schooners carrying ice from Rockland, Maine, to Cuba, lumber from Brunswick, Georgia, and coal from Norfolk to Boston and Providence.

Any boy that makes the voyage around the Horn will have something to talk about for a long time afterwards, but there will be mighty few of his listeners that will know what he *is* talking about. Most of the men that have sailed on any of the old time clippers are, like myself, pretty well up in the sixties, and a considerable number are in Sailors' homes, and before many more years have passed most of these old "Shell-backs" will have weighed anchor for their final voyage. [This was written in the 30's.]

There has been a lot of controversy about what country produces the best sailors. I have sailed under four flags: American, British, Italian, and Spanish. I have had nearly all nationalities as shipmates, and I don't believe *any* country produces the best, as I have seen many excellent sailors of *all* nationalities.

I have had experiences at sea that I would not care to live over again. I had some pretty good times also, but taking it as a whole it was mostly grief. Sailors, as a general rule, are a pretty good class of men. In the American ship referred to earlier, we had a hell afloat in our voyage from New York to San Francisco, via Rio Janeiro. As you know, sailors have a "chantey" for every job aboard ship, but there were no "chantey's" on her after we passed Sandy Hook. The third day out we went aft in a body and complained about the food which was the very worst any of us had ever had. The Captain told us the food was good enough for a bunch of wharf rats, and he said some other things that I cannot put on this page, ordered us forward, and as we did not move fast enough to suit the mates they waded into us with belaying pins and beat us up plenty. If we hit back, that would be mutiny, of course.

However, in the second dog watch that eve-

ning we passed the word that we would teach that Skipper a valuable lesson. A sailor usually loves his ship because it's his home. But we said: "To hell with her. The first time we get into a gale and have to shorten sail, we will make a darned slow, sloppy job of it and let the rags blow out of the gaskets." We got it about the 8th day out, off Hatteras. We passed the gaskets so loosely nearly half the sails were blown to ribbons. We bent new ones, and lost them before we got to Rio. Two days before we got to Rio we had another gale. The second mate struck the man at the wheel in the face because he was half a point off the course. The helmsman knew better than to do it, but he let go of the wheel and she came up "all standing" and shook the fore and main to'gallant masts out of her. Hell broke loose, but we went about the work of cutting away the gear just as calmly as if she was tied to the dock. We tried to jump her in Rio, but could not get away as we lay out in the harbor. The Skipper disrated the second mate for striking the man at the wheel and the consequent loss of the upper sticks. The old man sent him forward, and when a couple of the boys got through with him he wasn't worth picking up. He never turned to with the watch he was in but was paid off in Rio.

We sent new sticks aloft in Rio, bent new sails that we had made there, and before leaving port the Skipper called us to the break of the poop, threatened to put us in irons, said he would have us put in prison for insubordination and mutiny, threatened to shoot us, hang some of us from the yardarm as an example to the others; but as the grub did not get any better, neither did the crew. We had a hell of a time getting to the Horn, got off our course through compass error and came near piling up on Staten Island, about 60 miles east of Cape Horn on July 4th, 1886. Twelve days we "laid to" off the Horn in a westerly gale that would blow the hair out of your head, lost a few more sails, and to make a long story short, we limped into Valparaiso 104 days from New York, with nine of the crew down with scurvy and the Skipper with a broken leg. We jumped her in Valparaiso and glad of the chance to do so.

Many a spar has been lost through the sullen response of a badly treated crew. Many a sail has been blown out of improperly tied gaskets. With a willing crew that jumps to it and knows their business, a ship may carry her royals and to'gallants'ls to the last moment. With a sullen, ill-treated half-starved crew it means take everything in and reef down as soon as the glass falls, or you will lose sails and spars. As in our case, this means longer, more expensive voyages lengthened by many weeks, without mentioning the added cost of the new sails and cordage, and the risk of losing the ship.

In my experience I have found that deep-water sailors treated like *men* are the most loyal, active, dare-devil bunch in the wide world. They will put up an awful fight to save a ship they love, in which they get decent food and fair treatment. By this I do not mean that they must be coddled, because sea discipline must be maintained at all times. Treated as dogs, like we were, they can be sullen and unresponsive, caring less than nothing either for their own lives or for that of the ship. We had a *good* crew on the ship above mentioned, amongst which we had five Liverpool Irishmen that, to use a sailor's expression, "were hard men to shave." I really believe that after the Liverpool gang declared themselves that the remainder of the crew were more afraid of them than they were of the Skipper and the two mates. She was a 2100 ton ship. Her complement was 24 men before the mast. We sailed from New York with 17. Bad as she was, she sure was a beauty, and I still contend that the finest sight in the world is a full-rigged ship under sail.

Haul Away, Joe

As pointed out in the previous selection, the harsh conditions and hard work of the sailors were often lightened by the singing of chanteys. These work songs helped pass the time and also established a rhythm that enabled the crew to synchronize their tasks. The chantey reprinted below was an old favorite during the days of sail.

(Solo:)
When I was a little lad,
And so my mother told me,
(Chorus:)
Way, haul away,
We'll haul away, Joe.
(Solo:)
That if I did not kiss the girls
My lips would grow all mouldy.
(Chorus:)
Way, haul away,
We'll haul away, Joe.

King Louis was the King of France
Before the revolution.
But then he got his head cut off
Which spoiled his constitution.

Oh, once I had a German girl
And she was fat and lazy,
Then I got a Brooklyn gal,
She damn near drove me crazy.

So I got a Chinese girl
And she was kind and tender,
She left me for a Portugee,
So young and rich and slender.

Way, haul away,
I'll sing to you of Nancy.
Way, haul away,
She's just my cut and fancy.

Oh, once I was in Ireland,
A-digging turf and praties,
But now I'm in a Yankee ship
A-hauling on sheets and braces.

The cook is in the galley,
Making duff so handy,
And the captain's in his cabin
Drinkin' wine and brandy.

Way, haul away,
The good ship is a-bowling.
Way, haul away,
The sheet is now a-blowing.

Way, haul away,
We'll haul away together.
Way, haul away,
We'll haul for better weather.

Life on a Trawler

Jeremy Tunstall

Except for a few pleasure ships and training ships, the large sailing vessels have passed into history. Yet certain aspects of the seaman's life today remain as they have always been: The ship is still a tight community, cut off from the rest of the world for varying periods of time; it has its own social structure and its own demands for work, which tend to mold the seaman's whole life. The following selection from The Fishermen *by Jeremy Tunstall, a sociologist at City University, London, portrays the crew of a modern fishing trawler off the coast of England. Vessels such as these remain on a far more human scale than the superships that now dominate international commerce.*

The Hierarchy

There are two separate eating places on a trawler—the "cabin" for the "officers" and the mess-deck for the men. The officers who eat apart are the skipper, mate, bosun, chief and second engineer, the radio-operator and the cook. There are also usually separate sleeping arrangements for them though these vary according to the design of the ship. The ordinary deckhands on an old trawler sleep for'ard in very cramped conditions in a single fo'c'sle, and on a new ship in several cabins aft. On an old ship only the skipper has individual accommodation. But on a new ship, not only the mate, bosun, and chief but also the "sparks" and cook have separate cabins.

Differences in rank are based on jobs done, and are maintained by physical distance when the men are not on duty. But there are certain anomalies in this ranking system—and the officers are not in all respects superior to the others. For instance the higher social status of the cook and the sparks is not reflected in higher pay than that of the men. Another anomaly is that the chief engineer, though clearly belonging to the officer category, is not in the line of succession for command of the ship. He is paid only slightly less than the bosun, who is of course third in command, but it is the third hand (only really first among equals with the other deckhands) who, despite being paid much less money than the chief, is next in command. The chief is one of the skipper's most vital assistants—indeed in a way he is nearest in status to the skipper, because he is in charge of a department where the skipper has no competence and by tradition will never interfere. Ultimately, however, the chief is inferior to all of the deckmen, since the deckmen always take command.

These anomalies in rank, pay and status, are only part of the subtle confusion of interests which shapes the social life of the senior men on a trawler. Another, of course, is the formal hierarchy itself. The skipper does not eat alone, but with his fellow officers. However, the very power and authority of the skipper casts a shadow over the table. The others talk to him, swear at him, disagree with him sometimes. But always at the back of the talk, nobody can forget that he is the skipper—that he is in command, his orders will be supported in a law-court, and that ultimately his decisions may affect the life and death of the crew. Especially if he is an established skipper, he can on his own initiative sack any of the officers with whom he shares the eating-table.

Quite apart from the divergence of interests between men who do different kinds of jobs—the cook, the sparks, the chief and the second engineers, and the deckmen—there is much room for conflict between the three senior deckmen—the skipper, mate and bosun. The minor anomalies in pay just referred to are so small that the frustrations which centre specifically around them are comparatively trivial; but the gradations in pay between the top three men in the hierarchy are relatively big. When the skipper is making £126 a week, the mate £79, and the bosun £38, there is room for *major* frustrations and resentment. It is often the case that the bosun has seen the very same mate or the skipper, under whom he is now working, come up from an ordinary deckhand, junior to him, during the years. Similarly with mate and skipper. Many mates hold skippers' tickets and have been skippers off and on themselves. This means that the man who finds himself mate may previously have commanded a trawler on which the present skipper was only the mate. One man described to me how he was mate and was fired by a particular skipper because of a certain disagreement. Later he found himself on a ship as skipper with the same man as mate under him.

But even if grounds for such resentments are not present between their respective personalities, the skipper's relations with his mate and bosun are still likely to be uneasy. The personal whims of the skipper can decide to a great extent the relative comfort of their lives. For instance, on one trawler I went on, the skipper was impatient during the fitting of a new trawl. The job took an hour and the sparks, on the bridge, commented that it was quite quick. But the skipper, who evidently was in tacit agreement with this opinion, was impatient to be catching fish again, and said: "I'm going to give that bosun a bollocking for being slow, just the same." And he did shout and curse at his bosun. Later these men must meet at table and on the bridge. The bosun has been out on the deck, out in the weather, while the skipper has been standing, or even sitting, on the warmed bridge. The bosun has been forcing himself and cursing his deckmen to get the job done as fast as possible—straining, in fact, to please the skipper and to keep his own job. Meanwhile he has, he feels, been helping that "man in a glass cage" to make £126 a week. Naturally the mental strain on the skipper will be under-estimated by the

bosun, who may himself be a man in his forties and under great physical strain in not only keeping himself going for eighteen hours on deck but also in keeping the other deckmen at it. What can such a bosun do to express his resentments? Nothing apart from cursing the other deckhands—who in turn, he will be aware, regard him as a surly miserable bastard like most bosuns in this world. With the skipper, the bosun and the mate must be wary, and indeed the safest thing is to say as little as possible.

This explains why other members of the crew advise the visitor not to eat with the officers in the cabin—"they never talk to each other in there, but just eat in silence." . . .

The Need for Sleep

"And bed, he thought. Bed is my friend. Just bed, he thought. Bed will be a great thing"—ERNEST HEMINGWAY: *The Old Man and the Sea*

"But sleep, on the *Uganda,* passed only briefly over the crew, like a few drops of rain over a parched land."—B.J. TAYLOR: *Steam-Trawler Uganda*

"Think not, is my eleventh commandment, and sleep when you can, is my twelfth." (Stubb.)—HERMAN MELVILLE: *Moby Dick*

The cook and the sparks are different from all men in a trawler's crew (except the galley-boy) in that they usually have a regular night's rest. The cook has to be up at about five to start making the breakfast but since the main part of his duty is over after the six p.m. meal, he can usually get eight hours sleep in a row. The sparks similarly has a full night's sleep, and it is more than a coincidence that the two men who are most likely to be unpopular are the only two who always get adequate sleep.

Lack of sleep is normal on a trawler and it is a powerful force in the social life of the crew. The engineers work six on and six off, meaning that they must usually sleep twice during the twenty-four hours. They have a split-shift at work and a split-shift at sleeping.

The deckmen, as we know, are desperately

short of sleep, during fishing. But not only then. On the way back to port most men take a day or two to catch up on sleep. By this time they are half-way home. Some find it difficult to sleep the last night out—they call this having the "channels"—and lie awake thinking of what they will be doing when they get ashore. Thus some men arrive ashore already short of sleep and those who go to parties get little sleep at night, frequently staying up all next day as well. Such a man at the end of two and a half or three days ashore goes back on to his trawler with a lot of sleep to be caught up on. Time on shore has been too precious to be wasted on sleeping. This means that all the time off watch for the first day or two is likely to be taken up with sleeping. A man often only has about two days when he is relatively well off for sleep before he plunges again into the ten days' stretch of fishing and inadequate sleep.

If a man works an average number of trips on a fairly successful trawler he is probably short of sleep for well over half of his days. This is a fact in the trawlerman's life which again marks him off from most other men. It is difficult to say what exact consequences follow from the fishermen's consistent lack of sleep, but some can certainly be suggested.

When on a trawler for the first time, one is at once struck by the extraordinary repetitiveness of the talk. The same man will repeatedly make the same point, and different men will each come up to the visitor and tell him all the usual trawler clichés as if they were fresh and interesting opinions. It is easy to suggest one reason why this happens. A man can continue to work when he is very tired, he can perform great physical feats and (as in war) can be extremely obedient and disciplined. Lack of sleep does, however, dull the mind. The body and its muscles develop a tired "automatic" routine, and the mind does the same thing, repeating to itself dull and dreary patterns. It is notoriously difficult to think clearly or freshly with a tired mind. When one is short of sleep the mind turns to the most immediate need for rest. Frustrations and aggressions build up around this immediate issue and those persons immediately present.

An obsession with lack of rest is a short-term interest. It can eventually be relieved by one long session of sleep. When the familiar state of tiredness is shaken off there tend to disappear with it also all those attitudes and feelings of aggression which centre on the now-sated need. From this point of view the fisherman's normal state of tiredness can be seen as a reinforcing agency in his normal tendency to a short-term view of things.

Crew as a Social Environment

A fisherman spends a high proportion of his time at his place of work—on board trawlers—which is one reason why his job influences him so deeply. While the ordinary shore worker can be regarded as making brief excursions to his place of work for 40 or so of the 168 hours in the week, the fisherman's life is the other way about—he makes brief excursions to the shore from his trawler, his place of work, where the majority of his time is spent. This means that his job has a much greater proportion of his total hours in which to shape him, quite apart from the extremely unusual kind of influence that life on a trawler represents.

Like the sailor, the fisherman tends to describe everything in terms of his job. One fisherman was talking to me when I was at sea, and trying to discover whether my trip to sea was a holiday from my usual form of activity. "So this is a trip off for you, is it?" he said. On another occasion a man said: "She went astern, got two bags," when explaining that his wife had had twins. (When a trawler has enough fish in the trawl to make more than one "bag," the skipper usually goes astern to move the net further for'ard along the ship's side.) Of course, there are "occupational languages" in other jobs. But the fisherman's special language is so all-pervasive that outsiders, who know nothing about fishing, find it difficult to follow a fisherman's conversation. The fisherman's language thus tends to cut him off from non-fishing people and to reinforce the social isolation from which it springs.

Most fishermen start off in this life in their 'teens and one of the forces which motivate them to stick at it is a particularly eager desire to be manly, to copy the men with whom they live and work at sea. He is cut off from shore-bound boys and their shore-bound ideas and ambitions, and he submits to this influence over an extremely high percentage of his total hours, days, weeks. As we have seen in the last section, most fishermen spend much of their time in a state of extreme physical tiredness, and with a deckie-learner (who requires more sleep and is unused to the hours anyhow) this state borders on almost total exhaustion some of the time. The boy keeps going, dizzy from want of sleep, and in this state he drinks in the influence of the trawler's social groupings and attitudes. He half sleeps and dreams, half experiences the trawler's life as the fishing continues day and night.

When he is a deckie-learner of sixteen or seventeen, even the very brief hours that he has ashore often seem to pass slowly for him. Shore is like a piece out of his childhood, a reversion from the heightened manly world of fishing. Even when he is ashore he lives uneasily in the shadow of the world of trawling. The real world for him is at sea.

What then are the special social features of this crew-life on a trawler to which the young fisherman submits himself? We have already seen the important differences between new successful ships, and old badly-paying ships. These differences are reflected in the make-up of the social life aboard. On a top ship the deckhands, especially, will be mainly efficient quick workers, experienced without being too old. Most deckhands on top trawlers are married men between the ages of twenty-five and forty. On a very old ship the less good deckhands are normally found and they will be mostly inexperienced youths under twenty-two or so and old men, aged up to perhaps fifty-five. On an old ship

there are usually markedly less married men. Moreover the deckie-learner himself is usually a different type, the stronger more efficient lads, of course, being on the top ships. On a bottom ship a young inefficient lad typically finds himself adjusting to a group of mainly very old and very young deckhands. The latter group are the ones with whom he identifies himself particularly. On a top ship an older, stronger, more efficient deckie-learner adjusts to a set of deckhands with a smaller age-spread, and probably a quicker, more determined way of working.

Nevertheless, despite these differences there are much more important similarities between the social set-ups on all trawlers. The crew number is the same, twenty, the ship is away the same length of time, three weeks, the hours are the same and, most of all, the specialized jobs are the same. A boy may move from ship to ship but always the same life meets him. One notices, watching an almost entirely new crew together on a trawler for the first time, that they fall into place very quickly. Even though they do not know one another, each man knows what social reflexes to expect from the others.

On a trawler if you do your work you always know where you are with the other men. The actual faces change, but the things they say, the language they use, and the roles their particular jobs force them into remain the same. This accounts for the fact that some men find, despite the extremely long hours, a curious sense of relaxation on a trawler. There is never an unexpected psychological challenge.

The absolute and unchanging quality which pervades group-life on a trawler is made all the more marked by the fact that, with minor variations, it is the same as old fishermen remember. This gives the fisherman a certain feeling that whatsoever exists in the life on a trawler is somehow permanent and ordained.

Ashore the boy is never at the work-place for more than a few hours at a time. In Hull a boy will usually even go home for his midday meal, and domestic values are never far away. If he moves from one job to another he may find a different social atmosphere. The boy may be confronted at work with women fellow-workers or he may have regular contact with people of a different social class. In different jobs ashore the same person may be called on to perform quite different social roles.

But on a trawler, by comparison, there is a social atmosphere of inevitability, of a changeless hierarchy and specialization of jobs. Talk on a trawler often turns to bitter complaints, savage denunciations of both trawler-owners and the union, but it never passes on to restrained discussion or careful deliberation about how action might be taken to alter such things. What happens on trawlers can only be altered ashore. This is where all the vital long-term decisions and changes are made. But these moves are made as a result of concerted discussion, moderate statements, careful manoeuvres, the manipulation of conflicting attitudes and personalities. Social life on a trawler, however, militates against a fisherman being able to move easily in the more subtle, shifting, tangential social atmosphere of the shore.

All this helps to explain the fisherman's conservatism and fatalism which in terms of shore values sometimes seem difficult to understand.

Sea Nymphs of Japan

Luis Marden

The relationship of women to the sea in Western cultures has traditionally been indirect —through fathers, husbands, sons, or brothers. But in some non-Western societies, women have a direct working relationship to the sea that dates back thousands of years. Their roles, however, are strictly defined and generally limited to gathering shellfish and seaweed rather than hunting large fish. In the following excerpt Luis Marden, chief of the foreign editorial staff of National Geographic, *describes the ama divers of Japan, whose tradition is at least 2,000 years old.*

More than any other nation in the world, Japan, crowded on mountainous islands with little arable land, looks to the sea for sustenance. From earliest times artists and poets have celebrated the ama, most curious of Japan's fisherfolk.

The ama dive for food—shellfish and edible seaweeds—never for pearls. Some of the things they have been doing for 2,000 years (a venerated Japanese work tells of ama diving before the time of Christ) appear to go against the most basic rules of modern diving. Ama plunge without breathing apparatus of any kind to depths as great as 75 feet many times a day. Now they have begun to attract the attention of scientists as well as of poets and painters.

There are ama all round Japan's coasts, except in the far north, but most live along the east and west shores of Honshu. . . .

Elite Among Ama Work From Boats

The ama of Hekura, like those of the rest of Japan, belong to two classes: The *kachido,* "walking people," dive in shallow water, usually from the shore, and toss their catch into a floating

JAPANESE AMA DIVER

wooden tub; the *funado,* "ship people," older and more experienced, dive in deeper water from an anchored boat. . . .

I set out for the diving grounds in one of the 50 ama boats. Most carried one ama and a crewman—a husband, father, brother, or other close relative who steered the boat and tended the diver while she was on the bottom.

A friend in Kyoto once said to me, "In Japan we envy the ama's husband above all men." When I pointed out that with all the pulling and hauling, the husband often finished the day more tired than the wife, he said, "Ah, but we refer to the kachido's husband. He just sits at home." . . .

I rode with Nakamichi-san and his two diver-daughters, Hideko, age 16, and Toshiko, 18. Except for a few older women, the ama of

Hekura no longer dive semi-naked, and the girls wore black leotards. Most others wore all-enveloping suits of black neoprene, the diver's wetsuit. . . .

Finally the girls tied a string of egg-shaped weights round their waists and slipped over the side. Hideko took a deep breath, expelled some of it, and dived. Nakamichi-san passed her lifeline over his right forefinger. "I feel her moving, like a big fish," he said.

A strong tug: Swiftly he pulled the line in, hand over hand, until Hideko's head broke water, blowing like a surfacing seal in a corolla of cascading water. As Hideko handed up four *sazae* (*Turbo cornutus,* a species of marine snail), Toshiko exhaled a valedictory sigh and slipped beneath the surface.

The ama of Hekura dive in two periods: from ten in the morning until noon, when they go ashore to eat lunch and warm themselves round a fire on the beach; and an afternoon period from two to four o'clock. In the afternoon I searched for abalone *(Haliotis)* in deeper water with Masako. This 40-year-old woman was, like most ama, short and stocky, with well-developed legs and a deep chest. . . . The ama, naked except for tight-fitting shorts, . . . dropped over the side and, holding to the descending line, began to hyperventilate—that is, to breathe slowly and deeply. She expelled her breath through pursed lips in a plaintive whistle, like the distant crying of curlews on a wind-swept shore. Japanese poets call this *iso nageki,* the elegy of the sea. . . .

Our ama made nearly fifty dives that afternoon; she remained on the bottom almost exactly sixty seconds each time. Halfway through she climbed into the boat and put on a short cotton jacket to get warm.

When I surfaced, I asked her the classic question: Why do women, rather than men, dive in Japan?

She smiled shyly and replied with some diffidence; Japanese women hesitate to question male superiority: "Because I can stay in the water for three or four hours. Men cannot stand more than an hour."

True; women have a thicker layer of subcutaneous fat than men. I am sorry, ladies, but that is the way the physiologists put it.

Our ama had taken nearly thirty pounds of awabi. That night—and for nearly a week afterward—we dined on awabi and sazae, raw, steamed, and sun-dried. . . .

Factory Jobs Tempt Women of the Sea

[The next] afternoon I sat in the headquarters of the fisheries cooperative drinking tea with a robust ama of 48. Two gold teeth flashed when she smiled.

"I started to dive when I was 16," Hatsue said, "and I've done it ever since. Our season is short, from the middle of May until about the 10th of September; the rest of the time I pack sardines in the cannery. Most of us don't live beyond 60; I suppose it's the cold water and the hard breathing.

"Many of our daughters don't stay in the village. They go to Tokyo to work for the big companies like Sony and Matsushita, so they can have weekends off and buy the things they see on television. Some people say ama won't last another generation. I don't know; we still have 20 young girls in our village, so we can keep going for a while."

Although the ama lose some of their daughters to industry, they are still far from a relict population. By recent count, there are as many as 7,000 women divers in Japan.

Singlehanders

Francis Chichester

A few bold women have always dared to defy both tradition and the elements to meet the challenge of the sea. In the following selection from Along the Clipper Way, *Sir Francis Chichester, who himself sailed around the world alone at the age of sixty-five in the ketch* Gipsy Moth IV, *pays tribute to another solo voyager Ann Davison. Having lost her husband to the sea, she completed alone the voyage that they were to have made together, crossing the Atlantic from Plymouth, England to Antigua in the West Indies in 1952. Her reactions to the ceaseless work, to the solitude, and to the sea are presented in excerpts from her own account of her voyage.*

*B*efore leaving the north-east trades I want to introduce you to the only woman who has sailed across an ocean alone, Ann Davison. In 1952 she made a passage from Plymouth to Antigua in the West Indies. This is not properly along the clipper way, because for the first half of her voyage—although she was following the general direction of the route down to the Canary Isles—she made short hops; and her next stage, the passage from the Canaries to the West Indies, cut across the clipper way instead of continuing along it.

What a comparison—the thoughts and feelings of a young woman sailing across an ocean alone and those of the big crews of the big clippers! Ann Davison and her husband had previously set off to cross the Atlantic in a converted old fishing ketch of 70 feet overall. This ship was untried and unready, and the difficulties they immediately got into, combined with fa-tigue, were more than Davison could stand. His mind could not bear the strain. Eventually they were wrecked on Portland Bill, escaped in a liferaft but were swept into the Portland race. The raft capsized time after time and finally Ann found that her husband was dead. Fourteen hours after leaving the wreck, the raft was swept ashore.

She climbed the cliffs to start life again, alone. She bought a 23-foot sloop built by Mashford Brothers in Plymouth. It was 19 feet on the waterline with a beam of 7½ feet and a draught of 4½ feet. Total sail area was 237 square feet. It was called *Felicity Ann*. In this she set off alone to cross the Atlantic.

Here she is writing about the passage from Casablanca to the Canary Isles. *Felicity Ann* took 29 days for the passage of about 530 miles—an average of 18¼ miles per day.

It was an extraordinarily pleasant voyage, certainly the nicest so far, and I enjoyed the sort of lazy-hazy lotus-eating sea life one dreams of walled up in a city. . . .

Most of the time, however, there was a huge swell in which *FA* rolled abominably and flung her boom from side to side with a viciousness that threatened to wrench it clean out of its fastenings. She rattled her blocks and everything not immovably fast below with an aggravating irregularity, so that I was driven to a frenzy of restowing and rigging preventers in an effort to restore peace. An intermittent blop—rattle—crash on a small boat at sea is the nautical version of the Chinese water torture. . . .

For the first nine days out of Casablanca there was not a ship to be seen, and I missed them, grizzling quietly to myself at the loneliness; then we joined the north- and south-bound shipping lane and two steamers appeared on the horizon at the same time, whereon, embarassed by riches perhaps, I perversely resented their presence. "What are you doing on my ocean?"

Being in the shipping lane again meant the resumption of restless, sleepless nights. I figured out it took twenty minutes for a ship invisible over the horizon to reach us, and as a big ship was extremely unlikely to see me I had to see her, so any rest below was broken every twenty minutes throughout the hours of darkness. Enough practice since leaving England had endowed me with a personal alarm system which rang me out of a comatose condition at the appropriate intervals. Occasionally it let me oversleep, and once I awoke to find a south-bound steamer twenty-five yards astern of us. She was deep in Aldis conversation with another vessel, northward bound to west-ward of her, and utterly oblivious of our existence. A miss is as good as a mile maybe, but twenty-five yards is a narrow enough margin in the ocean, and it gave the required jolt to the personal alarm clock. On these ship-watching nights I used to get two hours of genuine sleep at dawn, when it could be assumed that *FA* was reasonably visible, and I couldn't care less by then anyway, but the overall lack of sleep did not improve the general physical condition, already much lowered by dysentery. The thought processes, never on Einstein levels, were reduced to a positively moronic grasp, and I had some rare hassels with navigational problems. However, the balance of nature was somewhat restored in that I was eating better on this trip than on any of the previous ones—the voyage from Douarnenez to Vigo was made almost exclusively on oranges—and there are several references to cooked meals in the log book. I may say that anything mentioned in the log book at this stage outside of navigation notes was a sure indication of an *Occasion,* though they were simple enough meals in all conscience, consisting mainly of scrambled eggs, or an omelette and coffee, or weird mixtures of cheese and onions. One reads of explorers and other isolated people dreaming up extravagant concoctions they are going to eat on returning to civilisation, but quails in aspic were not for me. I had an uncomplicated yearning for plain boiled potatoes and cabbage. As these do not represent a normal taste on my part, I concluded it was a deficiency desire, and stepped up the daily dose of vitamin tablets: a strict necessity for ocean voyagers, as I discovered on the nineteen-day Vigo to Gibraltar run, when I tried to do without them and broke out into reluctant-to-heal sores. The only canned goods whose vitamin content survives the canning process are tomatoes, which probably explains why canned foods lost all appeal for me as soon as I went to sea. Very practically I was learning what stores would be required for the long passage.

One supper was especially memorable, though not for the menu. At 1750 hours, Sunday, 5 October to be exact, I was fixing some cheese nonsense on the stove, for it was a flat calm and I was in an experimental mood, and whilst stirring the goo in the pan I happened to glance through the porthole over the galley and spied a steamer way over on the horizon, the merest speck to eastward of us, going south. A few minutes later I looked out again and to my surprise saw she had altered course and was making towards us. Coming out of her way specially to look at a little ship. Thrilled to the quick, I abandoned supper, brushed my hair, and made up my face, noting with detached amazement that my hands were trembling and my heart was beating, and I was as excited as if I was preparing for a longed-for assignation.

She was a tall, white-grey Italian liner, the *Genale* of Rome, and she swept round astern of us, the officers on her bridge inspecting *FA* keenly through their binoculars. As she had so kindly come many miles out of her way, I had no wish to delay her needlessly, for minutes are valuable to a ship on schedule, so I made no signals, but waved, and the whole ship seemed to come alive with upraised arms waving in reply. She went on her way satisfied that all was well with her midget counterpart, and the night was a little less lonely from the knowledge of her consideration. . . .

Here are her views on the passage across the ocean from Las Palmas to Antigua which now faced her.

Preparing for this voyage had been in a way unlike preparing for any of the others, although naturally it was a projection and development of them all, and the preparations were more in the nature of a mental strengthening of the skipper than a material provisioning of the ship, a feature with which I was now pretty well acquainted. It was to be a much longer passage than any of the others, and it would be much lonelier; there would not be the comforting knowledge of vast continents only a hundred miles or so to the eastward, and I could not expect to see any ships en route. Once the busiest sea lanes, the trade wind belts are now the most deserted, for steamships have no need of following winds, but sail great circle courses direct to where they want to be. And there could be no turning back on this voyage. No change of plan. This sort of certainty was sobering. When there is no way back, no way out, you must be very, very sure of what you are doing. I did not know how I would react to absolute solitude. It is an experience few of us are ever called upon to undergo and one which few of us would voluntarily choose. It is almost unimaginable, because solitude is something that normally can be broken at will. Even being on one's own in undeveloped country, popularly supposed to epitomise loneliness, is not true solitude, for one is surrounded by trees and bush and grass and animals, all part of the substance of one's own living. But the sea is an alien element; one cannot live in it or on it for long, and one survives that little time by one's own wit and judgment and the Grace of God. When a man says he loves the sea, he loves the illusion of mastery, the pride of skill, the life attendant on seafaring, but not the sea itself. One may be moved by its beauty or its grandeur, or terrified by its immensity and power of destruction, but one cannot love it any more than one can love the atmosphere or the stars in outer space. . . .

On Sunday, January 18th, she makes a landfall, Barbados. She writes in the log:

We round Harrison Point, well off shore, towards the end of the afternoon and I am fascinated by the occasional glimpse of a red roof and a stone wall. The island looks peculiarly English from the little I can see . . . None of this is true . . . We make south towards Carlisle Bay, Bridgetown, and HAVEN.

We cannot make it before dark, though, and I have hove-to on the starboard tack, about 8 miles off shore, so I can fill the lamps, tidy the ship and myself, get supper as have neither fed nor drunk today and feel pretty whacko. . . . Then we can go in in the morning all fresh . . .

And, anyway, *we have crossed the ruddy ocean.*

Women as Passengers: A Voyage to India

Emma Roberts

Perhaps far more typical than Ann Davison's attempt to master the sea was the rather passive role of Emma Roberts, a passenger on a sailing vessel in the mid-nineteenth century. In the following account from The East India Voyage, or The Outward Bound *(1845), she relates how a "delicate person" can make the best of a sea voyage. East Indiamen were the aristocrats of the sailing merchant fleet, the only ships that made a business of regularly carrying passengers and that gave serious attention to their comfort. Even so, a passage to India was not without hardship for the modest or the squeamish.*

*T*o ladies, whether married or single, the upper or poop-cabins are certainly the most desirable, the disadvantages of the noise overhead being more than counterbalanced by the enjoyment of many favourable circumstances unattainable below. In the first place, these cabins are much more light and airy: it is seldom, even in the very roughest weather, that the ports are compelled to be shut; and it is almost inconceivable to those who have never been at sea, how great a difference it makes in the comforts or discomforts of a voyage, whether a delicate person can have the enjoyment of light and air in bad weather, or be deprived of both, condemned in illness to a dark, close cabin, without the possibility of diverting the mind by reading, or any other employment. There is also another great advantage above stairs, which is the comparative degree of seclusion attainable in these cabins. A few steps lead from them all to the cuddy, or general apartment: there is no necessity to go out upon deck, or to go up or down stairs to meals; thus avoiding much of the annoyance of a rolling vessel, and all the disagreeables attendant upon encountering persons engaged in the duties of the ship. It may seem fastidious to object to meeting sailors employed in getting up different stores from the hold, or to pass and repass other cabins, or the neighbourhood of the steward's pantry; nevertheless, if ladies have the opportunity of avoiding these things, they will do well to embrace it; for, however trivial they may be in a well-regulated ship,

very offensive circumstances may arise from them. The two after-cabins on the lower [i.e. middle] deck are generally considered to be the best in the ship. . . . They are certainly more free from noise than any others in the vessel; but there is a greater difficulty in keeping them clean, and a much greater danger of their being infested with rats or other vermin. The upper cabins, on the other hand, may with a little care be always neat and comfortable; nor are they liable to have the sea wash into them, which may be the case in fine weather below, if by any awkwardness in the management, the ship should make a sudden dip: but they are certainly noisy. Neither during the night nor the day can the inmates of the poop-cabins expect peace: persons on duty are always stationed above their heads, and it is a favourite walk with the passengers; added to this, the hencoops are usually placed upon the poop. . . . In bad weather, or during the working of the vessel, the noises made by trampling overhead, ropes dragging, blocks falling etc. etc. are very sensibly augmented by the cackling, chuckling, and screaming of the poultry, while throughout the day, whether fair or foul, they are scarcely ever silent. In those ships in which the comfort and convenience of the passengers are paramount considerations, the hen-coops do not occupy a place upon the poop, and it is probable that a general doom of banishment will shortly be pronounced against them.

Women on Warships: Navy Wives React

Linda Charlton

Today there are indications that the traditional taboos against women participating in an active life at sea are beginning to break down. In Russia, Scandinavia, and a few other areas of the world, women are beginning to serve on merchant ships, and increasing numbers of women oceanographers are serving aboard research vessels in the United States and elsewhere. In 1972 Admiral Elmo R. Zumwalt Jr., United States Chief of Naval Operations, shattered tradition by lifting the ban against women serving with men aboard warships at sea. While some feminists welcomed this end to job discrimination, the order aroused a storm of protest among many navy wives who feared sexual involvement and the disruption of the disciplined community of the ship. Their reactions are presented by reporter Linda Charlton in the following special report from The New York Times, *August 27, 1972.*

*E*ver since ships were powered by the winds and guided by a flotilla of deities, women have been considered bad luck at sea—by men. Now that the man at the top of the United States Navy has reconsidered this notion, the bad-luck warnings, and worse, have been sounded again—by women.

Specifically, the alarm is being sounded by a small group of women in this steamy port city, which is essentially a wholly owned subsidiary of the Navy. They are, in their own proud term, "enlisted wives," living in Government quarters with their husbands away at sea for months at a time.

What has angered them out of their traditional Navy-wife silence in public is the recent directive—"Z-Gram Number 116," which seems destined to live in infamy in their minds—from Adm. Elmo R. Zumwalt Jr., the Chief of Naval Operations, lifting the ban against women serving aboard warships at sea.

"I'll be a Canadian citizen before I'll let my daughters go to fight a war," said Mrs. Barbara Stone.

"I just don't think it's right they take our husbands away from us so many months at a time—and then put other women with them," said Mrs. Sally Bedgood, 33 years old, the wife of a commissaryman. Mrs. Bedgood was cradling the youngest of her six children, six-month-old Denise, and drinking coffee in the kitchen of Mrs. Sue Jackson, another leader of the irate women.

The four other women in the group, all, like Mrs. Bedgood, "enlisted wives" with at least four children each, interrupted each other to enunciate their agreement.

Their outrage, which has prompted them to start a petition-signing campaign—they have between 300 and 400 signatures—focused on several aspects of the admiral's message besides a number of variations on Mrs. Bedgood's theme.

Some, such as Mrs. Jackson, were afraid that passage of the equal rights amendment, followed by implementation of the admiral's women-at-sea directive, would lead inexorably in the direction of women being drafted into the Navy. Mrs. Jackson, who has four sons, said: "I wouldn't want my daughter going out in hand-to-hand combat in the combat zone."

Relaxation Stressed

Mrs. Stone, who declared her intention of leaving the country rather than allowing any daughter of hers to "fight a war," said that, besides, she thought having women aboard ship would be unfair to the men: "They're not going to be able to relax—they can't run around in their skivvies. . . ."

"And curse," interrupted Mrs. Marge Thebarge, whose husband is a boatswain's mate. "When men go out to sea, they're animals. They live like animals."

Mrs. Stone was frank about her motives in becoming involved in the petition campaign: "The first thing that prompted me was jealousy," she said. "But please don't stress that point, don't think that's the only thing. I don't believe that it's a woman's place to be aboard ship. I don't think they can handle the job for one thing."

"I'd like to see a woman pick up 200 pounds," said Mrs. Lucille Kennell, who lives just across the street from Mrs. Jackson, in identical "married quarters"—frame row houses built, the women say, sometime during World War II, shabby now and with low ceilings that trap the sultry heat of a Norfolk summer.

The women say that their husbands are in complete agreement with them.

Threat to Efficiency

Women aboard ship would be a hazard, if not to navigation, at least to efficiency, the wives said. "The men would be more or less looking out for the women, when they should be looking after their own jobs," Mrs. Kennell added.

"I think they ought to be treated as equals, but I don't think they will be," said Mrs. Stone.

"Why not?" a visitor asked.

"Because the men are going to be men," said Mrs. Kennell.

Which returned the conversation to that other subject. With their husbands away routinely for months at a time, none of the women seemed alarmed about the matter of fidelity—in por that is. "There, they can turn away from it," sa Mrs. Bedgood.

And the others, too, seemed to make a curiously fine distinction between the accepted shore revels of sailors on liberty and the possibility, which they saw as more than likely, of shipboard sexual adventures, leading to

THE *KRISTALL*'S SKIPPER

Nadezhda Artysh is captain of the Russian passenger ship Kristall. The idea of active roles for women at sea has met strong opposition from the wives of many U.S. sailors and navy officers.

hostility and perhaps "hand-to-hand combat" among the men.

The tradition of no-women-afloat that the Norfolk wives are intent on saving dates back, according to some authorities, to ancient beliefs that the immortals who protected ships at sea were women—and that having another woman on board would make the goddesses jealous.

The women's views are not universal among "enlisted wives." One who welcomed the "Z-gram," as Admiral Zumwalt has dubbed his often surprising messages, is Mrs. Dorcas Carriglitto. Mrs. Carriglitto, 28, served two years and seven months in the Navy and is now the wife of a hospital corpsman and the mother of two children.

"I Totally Disagree"

"I totally disagree," said Mrs. Carriglitto. "I wish they'd had it [the possibility of sea duty] when I was in, and now I'm talking about going into the Reserves." Mrs. Carriglitto, who said she believed that women should hold any shipboard rating that they are "physically and men-tally qualified for," has had some angry words with other wives. She added:

"My neighbor told me they [Navy ships] would be floating whorehouses, and that was a slap in my face. They'll both have a job to do and they're going to be doing it."

She also disagrees with the contentions of the angry wives that making the necessary changes in shipboard quarters would be costly.

The Norfolk Navy wives have collected between 300 and 400 signatures and they speak vaguely of many other Navy wives in other places who agree with them. They have no hard evidence of a growing hue and cry anyplace else. Norfolk, of course, is the world's largest naval base, and there are more United States Navy wives here than anywhere else.

The group gathered at Mrs. Jackson's house got angrier the more they talked and Mrs. Thebarge spoke of what they might do if their protest elicits no response from the brass: "If all these Navy wives said "Take the Navy or me . . ."

The Fisherman and His Wife

There is a certain universal quality in the problems faced by the wives of seafarers that transcends national boundaries and cultures. All must learn to live for long periods of time without their men and must, therefore, achieve a certain measure of independence, coping alone with minor annoyances and major problems. Yet in some traditional cultures, these problems are greatly increased by societal sanctions against independence for women. The resulting conflicts and the carefully circumscribed role of women in one such society, that of the sponge fishers of Kalymnos, Greece, are the subject of the following selection by H. Russell Bernard, an anthropologist at West Virginia University. This article was originally presented at the Third Mediterranean Social Anthropological Conference in Athens in 1966.

This paper discusses some aspects of male-female relationships on the island of Kalymnos in Greece. Kalymnos is a rather barren-looking rock in the Dodecanese chain. It is only 49 square miles, of which about 82% is nonarable. The economic prosperity of Kalymnos is directly linked to its native sponge-fishing industry. The sponge fleet of approximately 30 vessels leaves Kalymnos around Easter. It spreads out through the Aegean and, until recently, many boats travelled all the way to the North African coast. The boats remain at sea for about half the year, until the weather and cold water drives them home in early November. During the winter months the divers, crew members, and captains remain on the island. They work on repairing the boats and preparing for the next expedition. There is much to do. The captains negotiate loans from banks and from the buyers to underwrite the extremely high costs of an expedition—about $35,000, on the average, per captain. (Each captain runs two or three small boats and a single mother ship to carry supplies.) The divers and crew negotiate with the various captains to obtain the best *platika,* or advance payment on their earnings. Most of these complex negotiations take place in the coffee houses around Kalymnos, especially in the port city of Pothea where over 70% of the island's population resides.

Just before Easter, the island honors its folk-heroes, the divers, with a formal banquet; the streets are lined with bunting and signs wishing the men safe return; there are visits from mainland politicos and displays of boats in the harbor; and then they are gone, leaving their families for six months.

Sponge fishermen are variously held in high esteem and disdain by the majority of the island's population. They constitute a scant 10% of the labor force, yet they account for more than 40% of the goods and services of Kalymnos. . . . They risk their lives (indeed, there are men killed and crippled nearly every year from diving accidents) to make this significant contribution to the Kalymnian economy. Their efforts yield a total of more than a million dollars per annum in foreign currency flow to Greece, and their internal spending supports dozens of shops and hundreds of families beyond their own.

Divers earn two and a half times in six months what a similarly educated laborer earns in a year. Their high earnings and death-defying job behavior account for their folk-hero status among lower class Kalymnians. Their economic importance accounts for the banquet held in their honor by middle- and upper-class Kalymnians. The banquet, however, belies a harsher reality. The merchants and landed classes of the island despise the fisherman for the latter's antisocial behavior. Indeed, the raucous antics of young sponge divers are famous throughout Greece. Not knowing if they will return from a trip, they demand and get their high wages in advance. Many of the younger divers spend their money freely, often leaving Kalymnos penniless or in debt to the captain against bonuses for large catches.

When the divers return in the winter, they borrow money from the captains against the next year's wages, and so on. A diver may borrow from several captains, promising each that he will ship out at the proper time. The captain, who eventually signs a man, must pay off his fellow captains if they are holding notes against the diver.

The exploits of the younger divers and their seemingly reckless fiscal philosophy are tolerated by the outraged gentry. The divers have a kind of license for antisocial behavior. This antisocial behavior is largely directed against women. It is the young wives who suffer when their husbands openly throw money away in the taverns; and it is the young wives who suffer through the summer when their husbands leave them without funds and force them into debt to local merchants. When the men return from the sea they must borrow from the captains to pay off their wives' debts. The older, more stable divers are conscientious family men; and not all the young divers are irresponsible. But the exploits of a few are enough to reinforce the image of divers held by the gentry: "They are a cancer of this island."

The irresponsible image of divers is transferred to their wives. Everywhere on Kalymnos the divers' wives are described as "loose women, waiting for their husbands to leave so they can run around behind their backs." A very small number of cases of adultery serve to reinforce this prejudice. An abortionist told me that the peak time of year for these operations was soon after Easter. "The men like to leave their wives pregnant so they can be sure they will not make them cuckolds. These women will not make them cuckolds. After all, how would they get away with it on a small island? But neither do they want so many children. So they come to me for help."

The essential relationship of distrust between men and women in the Mediterranean culture is drawn in high relief in this example. On Kalymnos, because the men are gone for long each year, the picture is drawn more sharply than usual. The basic quality of the male/female relationship is one of battle—a game of resource manipulation whose prize is personal power over the opponent. There are certain crucial rules in this game: (1) the status of man is defined as superior to that of woman and everyone must acknowledge this, at least superficially, in order to be eligible to play the game; (2) final and formal authority for legitimate use of decision-making power within the family is vested in males; a corollary of this rule is that a man succeeds to this authority formally when he transfers his primary allegiance from a nuclear family of orientation to one of procreation; and (3) the resources for control of this authority and power within the conjugal domestic unit are not the same. Essentially a man controls decision-making by force of tradition. His resources for maintenance of that control are, among others: (1) the sanction of custom and tradition; (2) the threat or actual use of physical coercion; and (3) the control of finances as breadwinner. The resources of the wife are, among others: (1) the legitimate use of cajoling techniques; (2) argument; (3) the threat of infidelity; and (4) the withholding of sexual favors. The husband's objective is maintenance of control. For the wife, the goal is the usurpation of power (both physical and that of decision-making) from her husband in order to attain her own goals.

Men and women each have a private realm within which they may exercise considerable control. "The house is the woman's domain (so long as she runs it to suit her husband)" and "the outside world belongs to men" are two cultural clichés that sum up this polarity. At the same time, they demonstrate that, ideally, the man gives his wife power in the house at his tolerance.

A Kalymnian woman must understand that a man's *philotimo* (honor) is at stake every time he deals with women. The side of family life which faces on public display must demonstrate the husband's control of the situation. A woman's own *philotimo* depends to a large extent on her not doing anything to harm her husband's *philotimo*. Of primary importance is the notion that shame rather than honor is the independent variable in *philotimo*. A man's honor depends on the shame *possessed* by his wife, sisters, and mother. If a woman does not possess shame, she brings dishonor on her family, specifically on the men of the family. The dishonor throws open to public questioning the men's authority to control social, political, and economic life.

Given this, the wife of a Kalymnian sponge fisherman leads an extraordinary existence. She is not exempt from the primary rules outlined above, neither is she free to play her role in the orthodox manner prescribed by Kalymnian (and Greek) society. For the sake of her own *philotimo,* and that of her children and of her family, she must preserve her husband's by not appearing to usurp his authority during his absence. Yet, the simple facts of life dictate that she must act in his stead. For six months of every year she must ask herself if she is indeed free to act in her husband's name, or whether she should defer action until his return.

Consider the use of financial credit. Ordinarily only men can incur debts with local merchants. On Kalymnos a woman whose husband is away at sea may buy what she requires for herself and her family on credit. These debts become debts of honor for her husband. To fulfill his honor, the sponge fisherman must go back into debt to a captain for cash against his next year's earnings.

The use of this power cannot be taken for granted, however. In one case, a diver returned to find himself in debt to a beauty parlor. He refused to honor the debt and sued for divorce.

Registering the children in school is considered to be a masculine activity. Since the sponge fishermen are still away during Sep-tember, this task falls to their wives. One more example may suffice. Until recently, women did not shop at the central market place for fresh fish and meats. This was reserved for men. The wives of sponge fishermen, however, purchased these items between May and October while their husbands were working at sea. When the men returned, they took up the shopping duties.

Because the sponge fisherman's wife acts in traditional male roles for part of each year, she is the subject of gossip and suspicion by other Kalymnians. A sponge fisherman's wife must take every opportunity to protect her husband's *philotimo*.

In one case, for example, a woman was approached by a distant relative and was asked to sell off a piece of property so that the relative might put together a decent *prika* (dowry) for his daughter. The price was fair, the property was legally hers (she brought it with her to the marriage in her own dowry), and there was some urgency in the sale. The prospective groom, it seemed, was demanding to know exactly what land would be included in the dowry lest he break the engagement. Still, the woman did not sell but told her relative that he would have to wait until her husband could decide. A letter was dispatched to North Africa where her husband was fishing at the time.

A month later during a shore leave, the letter caught up with him. The letter contained a full exposition by the wife of why she thought the offer a good bargain; but her husband was left to make the final decision. The groom was told what the problem was and agreed to wait. Two months after she had first been approached, the woman sold the land with her husband's consent.

Her public display of obedience to her husband's authority in this matter went a long way toward maintaining her own reputation as a "good woman" and the reputation of her husband as a male who had his family firmly in control.

The role of the Kalymnian "business widow" is ambivalent, difficult and precarious. She is held in contempt, in admiration, in distrust,

Women watch their men go off to sea.

and in envy by other Kalymnians, male and female. We may briefly examine each of these characteristics.

The morals of a sponge fisherman's wife are universally suspect by non-sponge fisherman males. During the months prior to the departure of the sponge fleet, gossip becomes rife on the island concerning the impending orgiastic behavior of the women after their husbands leave. During investigation of various problems dealing with infidelity, I was told many times: "Wait until the men leave; then you'll really see something." Another favorite explanation was that the young men of Kalymnos had to seek sexual exploits somewhere and, given the lack of prostitution, the sponge fishermen's wives were the most likely prospects. "What do you think? They all just put themselves in deep freeze?" was another favorite saying among some males.

At sea, anxiety runs high concerning infidelity. For one thing, being a cuckold is a very unenviable status in Greek society. For another, a sponge fisherman rarely finds out directly about his wife's indiscretions. Comfreres find out first in letters from the island, and they, in turn, tell him. On a boat 33 x 11 feet where fifteen men live out half a year the thought of being known as a cuckold is almost intolerable to many individuals.

During the several days each season when the sponge boats find port shelter for one reason or another, many men go to prostitutes. One informant summed up the double sexual standard as follows: "When I went to a prostitute last time, I didn't even ask her name. And other times, if I knew it at the moment we were in bed, I forgot it the next morning when I left. For a man it is a purely superficial act, one which one does with sexual organs, not with his heart and soul. But for a woman—a woman must love a man to go to bed; she will remember him and compare him to you the next time you go to her."

Mediterranean men are culturally conditioned to be jealous of female sexuality capacity. They seek to restrict female sexuality because they view it as threatening to their self concepts. Most men will not discuss these anxieties; but at sea, in an isolated male society, the topic of female sexuality and its threat to male honor is quite common. The problem, it seems to me, is that men define their honor in terms of control over something they believe is essentially uncontrollable. The result is that shame rather than honor becomes the observable quality. The men believe that women have the power of shame (sexual receptivity and high orgastic capacity) and thus men's honor depends upon women *not* using that power.

On Kalymnos, as elsewhere in Greece and the Mediterranean, men are entitled to extramarital sexual liasons so long as they are not a threat to their wife's *philotimo*. "So long as you do not make a fool of your wife in her own house," one informant told me, "you do not have to worry that she will make a fool of you." While the sponge fishermen are at home on Kalymnos during the winter, they indulge themselves in food and drink and gambling to a large degree. Considering the general license of the sponge fisherman for antisocial behavior, an absence of adulterous unions (or perhaps great discretion in the conduct of such unions) indicates the sponge fishermen's fear of retaliation by their wives.

In fact, very little evidence was uncovered to support the claims made by the non-sponge fishermen Kalymnian males beyond their own accounts of their own sexual exploits with sponge fishermen's wives. Undoubtedly, these accounts were exaggerated although they were not without a grain of truth. Perhaps as many as six or more women were known to be habitual adulteresses, according to some informants. But in any case the number and percentage was very small.

A few cases of adultery are discovered each year. These are sufficient validation for the institutionalized rumoring and nervous joking that occurs as a result of a generalized male ambivalence towards women.

Some Kalymnian divers leave strict orders and rules for behavior during their absence. One diver left written instructions that his wife should not use cosmetics of any kind or sweet-smelling soaps or colognes during his absence. She should, he said, avoid making herself attractive to other men. She could not attend the cinema, even with other women. Beauty salons were off limits and she could not remain absent from the house after dark for any reason.

Residence patterns add to the difficulty of completing a secret tryst. While young couples usually establish their own homes, during the summer fishing season the sponger's wife goes to live with her own parents or her in-laws. Alternatively, her mother (or mother-in-law) may come to live with her if the latter is widowed. The high-density nature of residence and settlement patterns mitigates strongly against sexual indiscretions. Yet, the sponge fisherman's wife is always suspect, always liable to be maligned.

On the other hand, the sponge fisherman's wife is envied by many poor, lower-class women whose husbands are not absentees from the household. The reasons most commonly given for this are: (1) the sponge fisherman's wife is freed from the "burden" of catering to her husband's sexual demands; (2) she is freed from catering to her spouse's domestic needs for half the year; and (3) she is responsible to her husband only for the *welfare* of their children during his absence; she is not responsible to him for how she creates that welfare or how she raises the children.

In recent years much has happened to alter this situation. Many stresses have occurred within the system. The sponge industry has collapsed, and as a result the sponge fishermen have become a source of social discomfiture. Lacking capital for reinvestment, or education, and lacking skills to change jobs, sponge fishermen have become idle as the fleet has diminished. Lacking the mystique of their dangerous occupation, their "no-tomorrow" attitude and their antisocial behavior have become structural anachronisms.

As a result, women have taken the opportunity to capture a larger share of publicly recognized power. Several examples may illustrate. Each year during April, wandering musical bands, accompanied by female singers, used to arrive on Kalymnos to perform at the various taverns. The sponge fishermen collect large advance sums at this time against the impending trip. These sums are supposed to be used by the men's families until November. Several cases each year used to be recorded where a man would spend all his money at the taverns on music and drink. Sometimes the singers worked as prostitutes in their off-hours. Since 1960, as a result of a forceful lobby by Kalymnian women, captains are obliged to pay a man only one-half of his advance sum. The other half goes into the bank under the wife's name to insure her of at least minimal sustenance during her husband's absence. Men in the sponge industry, however, use subtle and technical countermaneuvers within the letter of the law to neutralize the gains of women.

In the early 1960s a sponge diver divorced his wife to marry an entertainer. In 1964 a group of women descended upon the mayor's office demanding that something be done to prevent such occurrences. In 1965 a law was passed forbidding female entertainment to accompany musicians on Kalymnos. Such displays of female power, it is supposed, would not have occurred in the days when sponge fishing was the mainstay of Kalymnian economy.

In the case of men leaving Kalymnos for foreign shores or the merchant marine in search of work, there has been much less social disorder. Traditional intra-familial roles have been more easily preserved. Men who leave the island in search of work are considered *timi i* (honorable). They are willing "to brave the trials of living in foreign environments" for long periods of time in order to support their families. Absent from Kalymnos, they present no threat to the plans of the economic aristocracy which is in the process of promoting tourism. Their very absence prevents the eruption of the battle between the sexes for power. The wives of these absent breadwinners are conspicuously careful to protect their husband's *philotimo*.

One of the prime motivations for Kalymnian men to opt for the merchant marine and for migration is the fact that they can thus maintain their traditional status most effectively. Women continue to occupy their position as "business widows" to be pitied and to be suspected. The burden thus continues to fall to them to preserve the masculine image of the father for the next generation of Kalymnian sons. The newly-won power of women, however, has placed part of the responsibility for the preservation of that image on the men themselves.

A Royal Sport

Jack London

While the ocean is a source of food and mineral resources for some, and a place from which to wrest a living for others, for countless numbers it is primarily a place of recreation—of sport, of challenge, and of escape from land-based pressures.

Jack (John Griffith) London, who spent much of his life in wandering and adventuring, went to sea when he was seventeen and worked in many parts of the Pacific, from the Bering Strait to the South Seas. His fascination with the sea was reflected in his book The Sea Wolf *(1904) and in* The Cruise of the Snark *(1911), an account of a voyage to the South Pacific in his fifty-foot ketch from which the following selection is excerpted. Here he captures the triumphs and dangers of wrestling with the sea in the "royal sport" of surfing that, sixty-five years later, has become one of the most popular of all ocean sports.*

*T*hat is what it is, a royal sport for the natural kings of earth. The grass grows right down to the water at Waikiki Beach, and within fifty feet of the everlasting sea. The trees also grow down to the salty edge of things, and one sits in their shade and looks seaward at a majestic surf thundering in on the beach to one's very feet. Half a mile out, where is the reef, the white-headed combers thrust suddenly skyward out of the placid turquoise-blue and come rolling in to shore. One after another they come, a mile long, with smoking crests, the white battalions of the infinite army of the sea. And one sits and listens to the perpetual roar, and watches the unending procession, and feels tiny and fragile before this tremendous force expressing itself in fury and foam and sound. Indeed, one feels microscopically small, and the thought that one may wrestle with this sea raises in one's imagination a thrill of apprehension, almost of fear. Why, they are a mile long, these bull-mouthed monsters, and they weigh a thousand tons, and they charge in to shore faster than a man can run. What chance? No chance at all, is the verdict of the shrinking ego; and one sits, and looks, and listens, and thinks the grass and the shade are a pretty good place in which to be.

And suddenly, out there where a big smoker lifts skyward, rising like a sea-god from out of the welter of spume and churning white, on the giddy, toppling, overhanging and downfalling, precarious crest appears the dark head of a man. Swiftly he rises through the rushing white. His black shoulders, his chest, his loins, his limbs—all is abruptly projected on one's vision. Where but the moment before was only the wide desolation and invincible roar, is now a man, erect, full-statured, not struggling frantically in that wild movement, not buried and crushed and buffeted by those mighty monsters, but standing above them all, calm and superb, poised on the giddy summit, his feet buried in the churning foam, the salt smoke rising to his knees, and all the rest of him in the free air and flashing sunlight, and he is flying through the air, flying forward, flying fast as the surge on which he stands. He is a Mercury—a brown Mercury. His heels are winged, and in them is the swiftness of the sea. In truth, from out of the sea he has leaped upon the back of the sea, and he is riding the sea that roars and bellows and cannot shake him from its back. But no frantic outreaching and balancing is his. He is impassive, motionless as a statue carved suddenly by some miracle out of the sea's depth from which he rose. And straight on toward shore he flies on his winged heels and the white crest of the breaker. There is a wild burst of foam, a long tumultuous

310

rushing sound as the breaker falls futile and spent on the beach at your feet; and there, at your feet steps calmly ashore a Kanaka, burnt golden and brown by the tropic sun. Several minutes ago he was a speck a quarter of a mile away. He has "bitted the bull-mouthed breaker" and ridden it in, and the pride in the feat shows in the carriage of his magnificent body as he glances for a moment carelessly at you who sit in the shade of the shore. He is a Kanaka—and more, he is a man, a member of the kingly species that has mastered matter and the brutes and lorded it over creation.

And one sits and thinks of Tristram's last wrestle with the sea on that fatal morning; and one thinks further, to the fact that that Kanaka has done what Tristram never did, and that he knows a joy of the sea that Tristram never knew. And still further one thinks. It is all very well, sitting here in cool shade of the beach, but you are a man, one of the kingly species, and what that Kanaka can do, you can do yourself. Go to. Strip off your clothes that are a nuisance in this mellow clime. Get in and wrestle with the sea; wing your heels with the skill and power that reside in you; bit the sea's breakers, master them, and ride upon their backs as a king should.

And that is how it came about that I tackled surf-riding. And now that I have tackled it, more than ever do I hold it to be a royal sport. But first let me explain the physics of it. A wave is a communicated agitation. The water that composes the body of a wave does not move. If it did, when a stone is thrown into a pond and the ripples spread away in an ever widening circle, there would appear at the centre an ever increasing hole. No, the water that composes the body of a wave is stationary. Thus, you may watch a particular portion of the ocean's surface and you will see the same water rise and fall a thousand times to the agitation communicated by a thousand successive waves. Now imagine this communicated agitation moving shoreward. As the bottom shoals, the lower portion of the wave strikes land first and is stopped. But water is fluid, and the upper portion has not struck any-

thing, wherefore it keeps on communicating its agitation, keeps on going. And when the top of the wave keeps on going, while the bottom of it lags behind, something is bound to happen. The bottom of the wave drops out from under and the top of the wave falls over, forward, and down, curling and cresting and roaring as it does so. It is the bottom of a wave striking against the top of the land that is the cause of all surfs.

But the transformation from a smooth undulation to a breaker is not abrupt except where the bottom shoals abruptly. Say the bottom shoals gradually for from quarter of a mile to a mile, then an equal distance will be occupied by the transformation. Such a bottom is that off the beach of Waikiki, and it produces a splendid surf-riding surf. One leaps upon the back of a breaker just as it begins to break, and stays on it as it continues to break all the way in to shore.

And now to the particular physics of surf-riding. Get out on a flat board, six feet long, two feet wide, and roughly oval in shape. Lie down upon it like a small boy on a coaster and paddle with your hands out to deep water, where the waves begin to crest. Lie out there quietly on the board. Sea after sea breaks before, behind, and under and over you, and rushes in to shore, leaving you behind. When a wave crests, it gets steeper. Imagine yourself, on your board, on the face of that steep slope. If it stood still, you would slide down just as a boy slides down a hill on his coaster. "But," you object, "the wave doesn't stand still." Very true, but the water composing the wave stands still, and there you have the secret. If ever you start sliding down the face of that wave, you'll keep on sliding and you'll never reach the bottom. Please don't laugh. The face of that wave may be only six feet, yet you can slide down it a quarter of a mile, or half a mile, and not reach the bottom. For, see, since a wave is only a communicated agitation or impetus, and since the water that composes a wave is changing every instant, new water is rising into the wave as fast as the wave travels. You slide down this new water, and yet remain in your old position on the wave, sliding

down the still newer water that is rising and
forming the wave. You slide precisely as fast
as the wave travels. If it travels fifteen miles an
hour, you slide fifteen miles an hour. Between
you and shore stretches a quarter of mile of
water. As the wave travels, this water obligingly
heaps itself into the wave, gravity does the rest,
and down you go, sliding the whole length of
it. If you still cherish the notion, while sliding,
that the water is moving with you, thrust your
arms into it and attempt to paddle; you will find
that you have to be remarkably quick to get a
stroke, for that water is dropping astern just as
fast as you are rushing ahead.

And now for another phase of the physics of
surf-riding. All rules have their exceptions. It
is true that the water in a wave does not travel
forward. But there is what may be called the
send of the sea. The water in the over-topping
crest does move forward, as you will speedily
realize if you are slapped in the face by it, or if
you are caught under it and are pounded by one
mighty blow down under the surface panting
and gasping for half a minute. The water in the
top of a wave rests upon the water in the bottom
of the wave. But when the bottom of the wave
strikes the land, it stops, while the top goes on.
It no longer has the bottom of the wave to hold
it up. Where was solid water beneath it, is now
air, and for the first time it feels the grip of gravity, and down it falls, at the same time being
torn asunder from the lagging bottom of the
wave and flung forward. And it is because of
this that riding a surf-board is something more
than a mere placid sliding down a hill. In truth,
one is caught up and hurled shoreward as by
some Titan's hand.

I deserted the cool shade, put on a swimming
suit, and got hold of a surfboard. It was too small
a board. But I didn't know, and nobody told
me. I joined some little Kanaka boys in shallow
water, where the breakers were well spent and

A ROYAL SPORT

small—a regular kindergarten school. I watched the little Kanaka boys. When a likely-looking breaker came along, they flopped upon their stomachs on their boards, kicked like mad with their feet, and rode the breakers in to the beach. I tried to emulate them. I watched them, tried to do everything that they did, and failed utterly. The breaker swept past, and I was not on it. I tried again and again. I kicked twice as madly as they did, and failed. Half a dozen would be around. We would all leap on our boards in front of a good breaker. Away our feet would churn like the stern-wheels of river steamboats, and away the little rascals would scoot while I remained in disgrace behind.

I tried for a solid hour, and not one wave could I persuade to boost me shoreward. And then arrived a friend, Alexander Hume Ford, a globe trotter by profession, bent ever on the pursuit of sensation. And he had found it at Waikiki. Heading for Australia, he had stopped off for a week to find out if there were any thrills in surf-riding, and he had become wedded to it. He had been at it every day for a month and could not yet see any symptoms of the fascination lessening on him. He spoke with authority.

"Get off that board," he said. "Chuck it away at once. Look at the way you're trying to ride it. If ever the nose of that board hits bottom, you'll be disembowelled. Here, take my board. It's a man's size."

I am always humble when confronted by knowledge. Ford knew. He showed me how to properly mount his board. Then he waited for a good breaker, gave me a shove at the right moment, and started me in. Ah, delicious moment when I felt that breaker grip and fling me. On I dashed, a hundred and fifty feet, and subsided with the breaker on the sand. From that moment I was lost. I waded back to Ford with his board. It was a large one, several inches thick, and weighed all of seventy-five pounds. He gave me advice, much of it. He had had no one to teach him, and all that he had laboriously learned in several weeks he communicated to me in half an hour. I really learned by proxy. And

inside of half an hour I was able to start myself and ride in. I did it time after time, and Ford applauded and advised. For instance, he told me to get just so far forward on the board and no farther. But I must have gone some farther, for as I came charging in to land, that miserable board poked its nose down to bottom, stopped abruptly, and turned a somersault, at the same time violently severing our relations. I was tossed through the air like a chip and buried ignominiously under the downfalling breaker. And I realized that if it hadn't been for Ford, I'd have been disembowelled. That particular risk is part of the sport, Ford says. Maybe he'll have it happen to him before he leaves Waikiki, and then, I feel confident, his yearning for sensation will be satisfied for a time.

When all is said and done, it is my steadfast belief that homicide is worse than suicide, especially if, in the former case, it is a woman. Ford saved me from being a homicide. "Imagine your legs are a rudder," he said. "Hold them close together, and steer with them." A few minutes later I came charging in on a comber. As I neared the beach, there, in the water, up to her waist, dead in front of me, appeared a woman. How was I to stop that comber on whose back I was? It looked like a dead woman. The board weighed seventy-five pounds, I weighed a hundred and sixty-five. The added weight had a velocity of fifteen miles per hour. The board and I constituted a projectile. I leave it to the physicists to figure out the force of the impact upon that poor, tender woman. And then I remembered my guardian angel, Ford. "Steer with your legs!" rang through my brain. I steered with my legs, I steered sharply, abruptly, with all my legs and with all my might. The board sheered around broadside on the crest. Many things happened simultaneously. The wave gave me a passing buffet, a light tap as the taps of waves go, but a tap sufficient to knock me off the board and smash me down through the rushing water to bottom, with which I came in violent collision and upon which I was rolled over and over. I got my head out for a breath of air and then gained my feet. There stood the woman before me. I felt like a hero. I had saved her life. And she laughed at me. It was not hysteria. She had never dreamed of her danger. Anyway, I solaced myself, it was not I but Ford that saved her, and I didn't have to feel like a hero. And besides, that leg-steering was great. In a few minutes more of practice I was able to thread my way in and out past several bathers and to remain on top my breaker instead of going under it.

"Tomorrow," Ford said, "I am going to take you out into the blue water."

I looked seaward where he pointed, and saw the great smoking combers that made the breakers I had been riding look like ripples. I don't know what I might have said had I not recollected just then that I was one of a kingly species. So all that I did say was, "All right, I'll tackle them tomorrow."

The water that rolls in on Waikiki Beach is just the same as the water that laves the shores of all the Hawaiian Islands; and in ways, especially from the swimmer's standpoint, it is wonderful water. It is cool enough to be comfortable, while it is warm enough to permit a swimmer to stay in all day without experiencing a chill. Under the sun or the stars, at high noon or at midnight, in midwinter or in midsummer, it does not matter when, it is always the same temperature—not too warm, not too cold, just right. It is wonderful water, salt as old ocean itself, pure and crystal-clear. When the nature of the water is considered, it is not so remarkable after all that the Kanakas are one of the most expert of swimming races.

So it was, next morning, when Ford came along, that I plunged into the wonderful water for a swim of indeterminate length. Astride of our surf-boards, or, rather, flat down upon them on our stomachs, we paddled out through the kindergarten where the little Kanaka boys were at play. Soon we were out in deep water where the big smokers came roaring in. The mere struggle with them, facing them and paddling seaward over them and through them,

was sport enough in itself. One had to have his wits about him, for it was a battle in which mighty blows were struck, on one side, and in which cunning was used on the other side—a struggle between insensate force and intelligence. I soon learned a bit. When a breaker curled over my head, for a swift instant I could see the light of day through its emerald body; then down would go my head, and I would clutch the board with all my strength. Then would come the blow, and to the onlooker on shore I would be blotted out. In reality the board and I have passed through the crest and emerged in the respite of the other side. I should not recommend those smashing blows to an invalid or delicate person. There is weight behind them, and the impact of the driven water is like a sandblast. Sometimes one passes through half a dozen combers in quick succession, and it is just about that time that he is liable to discover new merits in the stable land and new reasons for being on shore.

Out there in the midst of such a succession of big smoky ones, a third man was added to our party, one Freeth. Shaking the water from my eyes as I emerged from one wave and peered ahead to see what the next one looked like, I saw him tearing in on the back of it, standing upright on his board, carelessly poised, a young god bronzed with sunburn. We went through the wave on the back of which he rode. Ford called to him. He turned an airspring from his wave, rescued his board from its maw, paddled over to us and joined Ford in showing me things. One thing in particular I learned from Freeth, namely, how to encounter the occasional breaker of exceptional size that rolled in. Such breakers were really ferocious, and it was unsafe to meet them on top of the board. But Freeth showed me, so that whenever I saw one of that caliber rolling down on me, I slid off the rear end of the board and dropped down beneath the surface, my arms over my head and holding the board. Thus, if the wave ripped the board out of my hands and tried to strike me with it (a common trick of such waves), there would be a cushion of water a foot or more in depth, between my head and the blow. When the wave passed, I climbed upon the board and paddled on. Many men have been terribly injured, I learn, by being struck by their boards.

The whole method of surf-riding and surf-fighting, I learned, is one of non-resistance. Dodge the blow that is struck at you. Dive through the wave that is trying to slap you in the face. Sink down, feet first, deep under the surface, and let the big smoker that is trying to smash you go by far overhead. Never be rigid. Relax. Yield yourself to the waters that are ripping and tearing at you. When the undertow catches you and drags you seaward along the bottom, don't struggle against it. If you do, you are liable to be drowned, for it is stronger than you. Yield yourself to that undertow. Swim with it, not against it, and you will find the pressure removed. And, swimming with it, fooling it so that it does not hold you, swim upward at the same time. It will be no trouble at all to reach the surface.

The man who wants to learn surf-riding must be a strong swimmer, and he must be used to going under the water. After that, fair strength and common-sense are all that is required. The force of the big comber is rather unexpected. There are mix-ups in which board and rider are torn apart and separated by several hundred feet. The surf-rider must take care of himself. No matter how many riders swim out with him, he cannot depend upon any of them for aid. The fancied security I had in the presence of Ford and Freeth made me forget that it was my first swim out in deep water among the big ones. I recollected, however, and rather suddenly, for a big wave came in, and away went the two men on its back all the way to shore. I could have been drowned a dozen different ways before they got back to me.

One slides down the face of a breaker on his surf-board, but he has to get started to sliding. Board and rider must be moving shoreward at a good rate before the wave overtakes them. When you see the wave coming that you want to ride in, you turn tail to it and paddle shoreward with all your strength, using what is called the windmill stroke. This is a sort of spurt performed immediately in front of the wave. If the board is going fast enough, the wave accelerates it, and the board begins its quarter-of-a-mile slide.

I shall never forget the first big wave I caught out there in the deep water. I saw it coming, turned my back on it and paddled for dear life. Faster and faster my board went, till it seemed my arms would drop off. What was happening behind me I could not tell. One cannot look behind and paddle the windmill stroke. I heard the crest of the wave hissing and churning, and then my board was lifted and flung forward. I scarcely knew what happened the first half-minute. Though I kept my eyes open, I could not see anything, for I was buried in the rushing white of the crest. But I did not mind. I was chiefly conscious of ecstatic bliss at having caught the wave. At the end of the half-minute, however, I began to see things, and to breathe. I saw that three feet of the nose of my board was clear out of water and riding on the air. I shifted my weight forward, and made the nose come down. Then I lay, quite at rest in the midst of the wild movement, and watched the shore and the bathers on the beach grow distinct.

The Quest for the America's Cup: Lipton's Last Try

Henry R. Ilsley and Lincoln A. Werden

In tiny dinghies or luxurious yachts, thousands of sailors around the world thrill to the sport of sailing, some content with the challenge of wind and water to their skills as seamen, others seeking the excitement of competition. Most famous of all sailing races is the America's Cup in which a boat from another nation challenges the New York Yacht Club's entry for the cup first won in 1851 by the schooner America. *The race has been held an average of every five to six years since then, with the challenger—never successful—most frequently being British. The sportsmanship of yacht-racing was epitomized by Sir Thomas Lipton, the British tea merchant whose five attempts at the cup between 1899 and 1930 ended in failure. The following accounts from* The New York Times *of September 19, 1930 report on his final quest in what was then the sport of millionaires: Vincent Astor was the commodore of the New York Yacht Club, and J.P. Morgan Jr. watched as Harold S. Vanderbilt guided the winning* Enterprise *across the finish line.*

In 1974 the America's Cup was defended for the twenty-first time when Courageous *routed the Australian challenger,* Southern Cross. *Both sloops were aluminum. Interestingly,* Courageous *and the two other American yachts in the trials to choose a defender were owned by syndicates, one of which had raised three-quarters of a million dollars from 900 contributors. Nor was the winning captain, Frederick E. (Ted) Hood, a millionaire, but a sailmaker from Marblehead, Massachusetts.*

Charles Sheeler,
Pertaining to Yachts and Yachting, 1922

PHILADELPHIA MUSEUM OF ART

316

ENTERPRISE WINS SERIES AND AMERICA KEEPS CUP; HIS LAST TRY, SAYS LIPTON

SIR THOMAS MOURNS FAILURE AS FINAL

By Henry R. Ilsley,
Special to The New York Times.

NEWPORT, R.I. Sept. 18.—Sir Thomas Lipton has completed his quest of the America's Cup. He has issued his last challenge, built his last boat and sailed his last race for the trophy he rather would have gained than almost any other thing the world has to offer him. He fought a great fight and has gone down in honorable defeat, feeling that he has done his best.

"I will not challenge again. It's no use. We cannot win."

Sir Thomas made this positive and unqualified statement this afternoon standing on the after deck of his steam yacht Erin, shortly after Enterprise had rounded the second mark in the final contest of the 1930 series with a lead over Shamrock V that made her victory certain and decisive.

His declaration was made calmly and quietly after a full consideration of the subject, and the speaker was in a mood more serious than he allows himself to show to the casual acquaintance and rarely to his closer friends.

BRITISH PRESS PAYS TRIBUTE TO LIPTON

Holds He Is "King of Good Losers" and Has Helped Relations of Two Nations

LONDON, Sept. 18.—Sir Thomas Lipton may have failed to lift the America's Cup, but in the opinion of the British press he has earned by his demeanor in the face of a continuous but never ignominious succession of defeats the most enviable title of all—king of good losers.

He has done more. The Daily Herald holds he has triumphed where statesmen often have failed. "He has helped to weld together two nations in mutual respect and friendship." For no British commentator fails to note, The Herald observes, that America has seemed as anxious as Britain that Sir Thomas should gain his desire.

"New York," says The Herald, "was as despondent as London when it became clear that Enterprise was the better boat"; that fact is undisputed here.

Praises Victor's Afterguard

Enterprise, it is now accepted, as The Daily Telegraph again emphasizes editorially, owes her superiority partly to her duralumin mast, which weighs about a ton less than Shamrock's wooden spar, to that triangular boom and to those talented amateurs who sailed her—"a magnificent combination of talents," as The Telegraph says.

Where Sir Thomas and Lord Dunraven before have failed, who shall succeed? The Morning Post, which asks this question, can find no answer.

"The defender has the great advantage of racing in familiar waters," it says, "and the challenger has the disability of having to cross the Atlantic. Together these influences are so powerful that only marked superiority, either of design or seamanship or both, could suffice to countervail them.

Americans Able Sailors

"Such superiority has not been manifested and there appears to be no ground for supposing it will be developed. In the designing and sailing of yachts, the Americans have nothing to learn from anybody. We imagine the holders of the America's Cup need by no means keep it in a packing case ready for traveling."

On all sides regret is expressed over Lipton's withdrawing from future contests. "No man connected with the sea has done more for the art of sailing than Lipton," says The Daily Mail, "and if the cup ever were to come home it would have been most just that it should do so under his flag. It only remains to assure him that universal sympathy mingles with the congratulations we owe to the winners for their fine seamanship."

SPECTATORS AFLOAT SALUTE SHAMROCK

Fleet Stands by at Finish and Gives Ovation to Challenger as She Passes Buoy

By LINCOLN A. WERDEN,
Special to the New York Times.

NEWPORT, R.I., Sept. 18.—Newport Harbor took on the appearance of a normal harbor tonight as the last of the spectator fleet returned from the race and then began scattering to other ports.

The long procession of Coast Guard patrol boats steamed out this morning under a clear sky and into a strong wind that at times was blowing at a twenty-knot clip. But there were fewer boats in the spectator fleet to enjoy the best racing day of all and what proved to be the climax of the international series.

Secretary of the Navy Adams, skipper of the Boston cup entrant Yankee, that was defeated in the trials, was aboard Junius P. Morgan's Corsair and, after watching the entire race, had a fine view of the finish, since the handsome yacht came up on a line with the committee tug, Susan A, Moran.

A few of the larger yachts that were here during the first few days of the series remained to greet Harold S. Vanderbilt as he scored a clean sweep, but in true sportsman's fashion, not a single boat, large or small, of perhaps the 100 in all which waited about the finish, turned for home until they had seen Shamrock glide over the line.

Lipton Returns Fleet's Salute

When Shamrock had passed between the buoy and the committee boat she received an ovation by whistle and siren equal to that accorded the Enterprise. It was a compliment not only to the emerald-hulled sloop and her crew, who had their craft creeping up on the swift Enterprise over the last two legs of the course, but it was a tribute to Sir Thomas, aboard the Erin, which was on the outer ring of boats about the finish line.

And those of the returning spectator fleet that then speeded to Newport sounded their sirens three times going by the Erin, in salute to the challenger, who said tonight he would never toss his cap into the ring again.

The Big One

Zane Grey

Deep-sea sport fishing pits man's cunning and skill against the strength of some of the most majestic creatures of the ocean. The challenge of the chase and the struggle to land "the big one" are captured in the following account by Zane Grey of "The First Thousand-Pounder" (1930)—the first game fish of more than 1,000 pounds to be taken on rod and reel. Omitted here is the description of the attack on his hooked fish by sharks, those predators of the deep that fought him for his prize and robbed him of his record, which was subsequently disallowed because the sharks had mutilated the marlin's tail. Preceding the selection by Grey are some remarks by George Reiger, editor of Zane Grey: Outdoorsman, *a collection of hunting and fishing tales by the noted author of western stories, published in 1972 in commemoration of his centennial year.*

This is the story of the big one—the fish that Zane Grey spent a king's ransom pursuing across two oceans for more than two decades. But the triumph of this tale of perseverance is leavened with postmortem disappointment.

In 1939, the year Zane Grey died, Michael Lerner of the United States and Clive Firth of Australia founded an international association establishing guidelines for keeping records of salt-water game fish. Among the many restrictions insuring fair play are two that disqualified ZG's biggest catch: no one but the angler can touch any part of the tackle during the fight; and if a fish is shot, harpooned, or mutilated in any way, it is not eligible for a world record.

But the disappointment for Zane Grey fans runs deeper than the sacrifice of this first thousand-pounder. In recent years, the International Game Fish Association has created restrictions that eliminate nearly all the early records. Among other refinements, the organization insists on testing a sample of the line used to catch the submitted fish. The work of the pioneers is being jettisoned. The last of ZG's records, his 111-pound New Zealand yellowtail, has been replaced by another New Zealand fish of identical size caught in 1961. There's generally no memory of pre-1939 catches. While it's right and natural that larger fish be caught and honored, it's sad that the achievements of

the great pioneers are in no way commemorated. There is no recollection of Zane Grey's 582-pound broadbill swordfish, his 63-pound dolphin, his 758-pound bluefin tuna, or his 1,036-pound tiger shark. Only ichthyologists remember Zane Grey—and then rather modestly—for science has honored him by naming one of his favorite game fish, the Pacific sailfish, *Istiophorus greyi*.

There are other ironies associated with Zane Grey's success in Tahiti. Not long after publishing his discovery of giant marlin off Papeete, new and more convenient marlin grounds were discovered off the west coast of South America and the Kona coast of Hawaii. A fabulous fishing region was found in the vicinity of a little-known cape in Peru—Cabo Blanco. Several thousand-pounders were taken, and on August 4, 1953, Alfred C. Glassell, Jr., landed a 1,560-pound black marlin—still the heaviest billfish ever to be taken on rod and reel. More recently thousand-pound marlin have been caught off Cairns, Australia, while Zane Grey's achievements in Tahiti fade into the haze of big-game angling's past.

But ZG predicted that Tahiti would one day top the list of ocean angling hotspots, and in 1966 a huge marlin caught on a commercial handline by two local fishermen was brought into Papeete that vindicated his faith and staggered the angling world's imagination. When the fish was cut into sections and weighed for authentication, the pieces totaled *2,400 pounds!*

And even if the International Game Fish Association never recognized Zane Grey's catch as the forerunner to that ton-and-a-quarter marlin, ZG's story of its capture provides its own immortality.

Time is probably more generous to an angler than to any other individual. The wind, the sun, the open air, the colors and smells, the loneliness of the sea or the solitude of the stream, work some kind of magic.

Morning disclosed dark, massed, broken clouds, red-edged and purple-centered, with curtains of rain falling over the mountains.

I took down a couple of new feather jigs—silver-headed with blue eyes—just for good luck. They worked. We caught five fine bonito in the lagoon, right off the point where my cottage stands. Jimmy[1] held up five fingers: "Five bonito. Good!" he declared, which voiced all our sentiments. . . .

We headed out. A few black noddies skimmed the dark sea, and a few scattered bonito broke the surface. As usual—when we had them—we put out a big bonito on my big tackle and an ordinary one on the other. As my medium tackle holds one thousand yards of 39-thread line[2] it will seem interesting to anglers to speak of it as medium. The big outfit held fifteen hundred yards of line—one thousand of 39-thread and five hundred yards of 42[3] for backing; and this story will prove I needed it.

Off the east end there was a brightness of white and blue where the clouds broke, and in the west there were trade-wind clouds of gold and pearl, but for the most part a gray canopy overspread mountain and sea. All along the saw-toothed front of this range inshore the peaks were obscured and the canyons filled with down-dropping veils of rain.

What a relief from late days of sun and wind and wave! This was the kind of sea I loved to fish. The boat ran easily over a dark, low, lumpy swell. The air was cool, and as I did not have on any shirt, the fine mist felt pleasant to my skin. John[4] was at the wheel. Bob[5] sat up on top with Jimmy and Charley, learning to talk Tahitian. The teasers and heavy baits made a splashing, swishy sound that could be heard above the boil and gurgle of water from the propellers. We followed some low-skimming boobies for a while, and then headed for Captain M.'s boat, several miles farther out. A rain squall was obscuring the white tumbling reef and slowly moving toward us. Peter sat at my right, holding the line which had the larger bonito. He had both feet up on the gunwale. I noticed that the line on this reel was white and dry. I sat in the left chair, precisely as Peter, except that I had on two pairs of gloves with thumb-stalls in them. I have cut, burned, and skinned my hands too often on

---

[1] A six-foot, four-inch Tahitian who, according to Romer, was the most expert of any of ZG's crew at handling the heavy cane poles used in bonito fishing.

[2] 117-pound test.

[3] 126-pound test.

[4] John Loef was a California auto mechanic who asked Zane Grey what it was like to go fishing. One day ZG took him along and eventually trained him into one of his best boatmen.

[5] Bob Carney, a photographer and ZG's son-in-law.

a hard strike to go without gloves. They are a nuisance to wear all day, when the rest of you, almost, is getting pleasantly caressed by sun and wind, but they are absolutely necessary to an angler who knows what he is doing.

Peter and I were discussing plans for our New Zealand trip next winter—boats, camp equipment, and what not. And although our gaze seldom strayed from the baits, the idea of raising a fish was the farthest from our minds.

Suddenly I heard a sounding, vicious thump of water. Peter's feet went up in the air.

"*Ge-zus!*" he bawled.

His reel screeched. Quick as thought I leaned over to press my gloved hand on the whizzing spool of line. Just in time to save the reel from overrunning!

Out where Peter's bait had been showed a whirling, closing hole in the boiling white-green water. I saw a wide purple mass shooting away so close under the surface as to make the water look shallow. Peter fell out of the chair at the same instant I leaped up to straddle his rod. I had the situation in hand. My mind worked swiftly. It was an incredible wonderful strike. The other boys piled back to the cockpit to help Peter get my other bait and the teasers in.

Before this was even started, the fish ran out two hundred yards of line, then turning to the right he tore off another hundred. All in a very few seconds! Then a white splash, high as a tree, shot up, out of which leaped the most magnificent of all the leaping fish I had ever seen.

"GIANT MARLIN!" screamed Peter. What had happened to me I did not know, but I was cold, keen, hard, tingling, motivated to think and do the right thing. This glorious fish made a leap of thirty feet at least, low and swift, which gave me time to gauge his enormous size and his species. Here at last on the end of my line was the great Tahitian swordfish! He looked monstrous. He was pale, shiny gray in color, with broad stripes of purple. When he hit the water he sent up a splash like the flying surf on the reef.

By the time he was down I had the drag on and was winding the reel. Out he blazed again, faster, higher, longer, whirling the bonito around his head.

"Hook didn't catch!" yelled Peter, wildly. "It's on this side. He'll throw it."

"No, Peter! He's fast," I replied. Still I kept working like a windmill in a cyclone to get up the slack. The monster had circled in these two leaps. Again he burst out, a plunging leap which took him under a wall of rippling white spray. Next instant such a terrific jerk as I had never sustained nearly unseated me. He was away on his run.

"Take the wheel, Peter," I ordered, and released the drag. "Water! Somebody pour water on this reel! . . . *Quick!*"

The white line melted, smoked, burned off the reel. I smelled the scorching. It burned through my gloves. John was swift to plunge a bucket overboard and douse reel, rod, and me with water. That, too, saved us.

"After him, Pete!" I called, piercingly. The engines roared and the launch danced around to leap in the direction of the tight line.

"Full speed!" I added.

Then we had our race. It was thrilling in the extreme, and though brief it was far too long for me. Five hundred yards from us—over a third of a mile—he came up to pound and beat the water into a maelstrom.

"Slow up!" I sang out. We were bagging the line. Then I turned on the wheel-drag and began to pump and reel as never before in all my life. How precious that big spool—that big reel handle! They fairly ate up the line. We got back two hundred yards of the five hundred out before he was off again. This time, quick as I was, it took all my strength to release the drag, for when a weight is pulling hard it releases with extreme difficulty. No more risk like that! . . .

By the same tactics the swordfish sped off a hundred yards of line and by the same we recovered them and drew close to see him leap again, only two hundred feet off our starboard, a little ahead, and of all the magnificent fish I have ever seen, he excelled. His power to leap was beyond credence. Captain M.'s big fish, that broke off two years before, did not move like this one. True, he was larger. Nevertheless, this swordfish was so huge that when he came out in dazzling swift flight, my crew went simply mad. This was the first time my natives had been flabbergasted. They were as excited, as carried away, as Bob and John. Peter, however, stuck at the wheel as if he were after a wounded whale which might any instant turn upon him. I did not need to warn Peter not to let that fish hit us. If he had he would have made splinters out of our launch. Many an anxious glance did I cast toward Cappy's boat, two or three miles distant. Why did he not come? The peril was too great for us to be alone at the mercy of that beautiful brute if he charged us either by accident or by design. But Captain could not locate us, owing to the misty atmosphere, and missed seeing this grand fish in action. . . .

My swordfish, with short, swift runs took us five miles farther out, and then welcome to see, brought us back, all this while without leaping, though he broke water on the surface a number of times. He never sounded after that first dive. The bane of an angler is a sounding fish, and here in Tahitian waters, where there is no bottom, it spells catastrophe. The marlin slowed up and took to milling, a sure sign of a rattled fish. Then he rose again, and it happened to be when the rain had ceased. He made one high, frantic jump about two hundred yards ahead of us, and then threshed on the surface, sending the bloody spray high. All on board were quick to see that sign of weakening, of tragedy—blood. . . .

We ran for the nearest pass, necessarily fairly slow, with all that weight on our stern. The boat listed half a foot and tried to run in a circle. It was about one o'clock and the sky began to clear. Bob raved about what pictures he would take.

We were all wringing wet, and some of us as bloody as wet. I removed my soaked clothes and gave myself a brisk rub. I could not stand erect, and my hands hurt—pangs I endured gratefully.

We arrived at the dock about three o'clock, to find all our camp folk and a hundred natives assembled to greet us. Up and down had sped the news of the flags waving.

I went ashore and waited impatiently to see the marlin hauled out on the sand. It took a dozen men, all wading, to drag him in. And when they at last got him under the tripod, I approached, knowing I was to have a shock and prepared for it. . . .

[His] tail had a spread of five feet, two inches. His length was fourteen feet, two inches. His girth was six feet, nine inches. And his weight, as he was, 1,040 pounds.

Every drop of blood had been drained from his body, and this with at least 200 pounds of flesh the sharks took would have fetched his true and natural weight to 1,250 pounds. But I thought it best to have the record stand at the actual weight, without allowance for what he had lost. Nevertheless, despite my satisfaction and elation, as I looked up at his appalling shape, I could not help but remember the giant marlin Captain had lost in 1928, which we estimated at twenty-two or twenty-three feet, or the twenty-foot one I had raised at Tautira, or the twenty-eight foot one the natives had seen repeatedly alongside their canoes. And I thought of the prodigious leaps and astounding fleetness of this one I had caught. "My heaven!" I breathed. "What would a bigger one do?"

Solitude

Joshua Slocum

What makes individuals such as Sir Francis Chichester and Ann Davison turn their backs on human society and set out single-handedly to conquer the sea? The sea has always challenged the human spirit, and perhaps meeting that challenge alone is a supreme triumph of humanity over the brute and uncontrollable forces of nature. Whatever the reasons behind their voyages, all who have crossed the sea alone must have experienced an incomparable, utter solitude. In the following account from his book Sailing Alone Around the World, *Joshua Slocum himself never fully explained why he "resolved on a voyage around the world"; he commented only that he "was aware that no other vessel had sailed in this manner around the globe, but would have been loath to say another could not do it. . . . I was greatly amused, therefore, by the flat assertions of an expert that it could not be done." Forty-six thousand miles and three years later, Slocum had completed his epic voyage.*

The fog lifting before night, I was afforded a look at the sun just as it was touching the sea. I watched it go down and out of sight. Then I turned my face eastward, and there, apparently at the very end of the bowsprit, was the smiling full moon rising out of the sea. Neptune himself coming over the bows could not have startled me more. "Good evening, sir," I cried; "I'm glad to see you." Many a long talk since then I have had with the man in the moon; he had my confidence on the voyage.

About midnight the fog shut down again denser than ever before. One could almost "stand on it." It continued so for a number of days, the wind increasing to a gale. The waves rose high, but I had a good ship. Still, in the dismal fog I felt myself drifting into loneliness, an insect on a straw in the midst of the elements. I lashed the helm, and my vessel held her course, and while she sailed I slept.

During these days a feeling of awe crept over me. My memory worked with startling power. The ominous, the insignificant, the great, the small, the wonderful, the commonplace—all appeared before my mental vision in magical succession. Pages of my history were recalled which had been so long forgotten that they seemed to belong to a previous existence. I heard all the voices of the past laughing, crying, telling what I had heard them tell in many corners of the earth.

The loneliness of my state wore off when the gale was high and I found much work to do. When fine weather returned, then came the sense of solitude, which I could not shake off. I used my voice often, at first giving some order about the affairs of a ship, for I had been told that from disuse I should lose my speech. At the meridian altitude of the sun I called aloud, "Eight bells," after the custom on a ship at sea. Again from my cabin I cried to an imaginary man at the helm, "How does she head, there?" and again, "Is she on her course?" But getting no reply, I was reminded the more palpably of my condition. My voice sounded hollow on the empty air, and I dropped the practice. However, it was not long before the thought came to me that when I was a lad I used to sing; why not try that now, where it would disturb no one? . . . You should have seen the porpoises leap when I pitched my voice for the waves and the sea and all that was in it. Old turtles, with large eyes, poked their heads up out of the sea as I sang "Johnny Boker," and "We'll Pay Darby Doyl for his Boots," and the like. But the porpoises were, on the whole, vastly more appreciative than the turtles; they jumped a deal higher.

322

Across the Pacific by Raft

Thor Heyerdahl

Not all adventurers who risk their lives in seemingly frail and insignificant craft feel the need to voyage alone. But perhaps they all, at some point, find themselves wondering why they have accepted the challenge of the sea. One such voyager was Thor Heyerdahl, the Norwegian ethnologist who set out in 1947 on a balsa-wood raft with five companions to cross the Pacific Ocean from Peru to the Taumoto Islands in eastern Polynesia. The following excerpt is from his account of the trip, Kon-Tiki, *published in 1950.*

Once in a while you find yourself in an odd situation. You get into it by degrees and in the most natural way but, when you are right in the midst of it, you are suddenly astonished and ask yourself how in the world it all came about.

If, for example, you put to sea on a wooden raft with a parrot and five companions, it is inevitable that sooner or later you will wake up one morning out at sea, perhaps a little better rested than ordinarily, and begin to think about it.

On one such morning I sat writing in a dew-drenched logbook:

—*May 17. Norwegian Independence Day. Heavy sea. Fair wind. I am cook today and found seven flying fish on deck, one squid on the cabin roof, and one unknown fish in Torstein's sleeping bag. . . .*

Here the pencil stopped, and the same thought interjected itself: This is really a queer seventeenth of May—indeed, taken all round, a most peculiar existence. How did it all begin?

If I turned left, I had an unimpeded view of a vast blue sea with hissing waves, rolling by close at hand in an endless pursuit of an ever retreating horizon. If I turned right, I saw the inside of a shadowy cabin in which a bearded individual was lying on his back reading Goethe with his bare toes carefully dug into the latticework in the low bamboo roof of the crazy little cabin that was our common home.

"Bengt," I said, pushing away the green parrot that wanted to perch on the logbook, "can you tell me how the hell we came to be doing this?"

Goethe sank down under the red-gold beard. "The devil I do; you know best yourself. It was your damned idea, but I think it's grand."

He moved his toes three bars up and went on reading Goethe unperturbed. Outside the cabin three other fellows were working in the roasting sun on the bamboo deck. They were half-naked, brown-skinned, and bearded, with stripes of salt down their backs and looking as if they had never done anything else than float wooden rafts westward across the Pacific. Erik came crawling in through the opening with his sextant and a pile of papers.

"98° 46' west by 8° 2' south—a good day's run since yesterday, chaps!"

He took my pencil and drew a tiny circle on a chart which hung on the bamboo wall—a tiny circle at the end of a chain of nineteen circles that curved across from the port of Callao on the coast of Peru. Herman, Knut, and Torstein too came eagerly crowding in to see the new little circle that placed us a good 40 sea miles nearer the South Sea islands than the last in the chain.

"Do you see, boys?" said Herman proudly. "That means we're 850 sea miles from the coast of Peru."

"And we've got another 3,500 to go to get to the nearest islands," Knut added cautiously.

"And to be quite precise," said Torstein, "we're 15,000 feet above the bottom of the sea and a few fathoms below the moon."

So now we all knew exactly where we were, and I could go on speculating as to why. The parrot did not care; he only wanted to tug at the log. And the sea was just as round, just as sky-encircled, blue upon blue.

Perhaps the whole thing had begun the winter before, in the office of a New York museum. Or perhaps it had already begun ten years earlier, on a little island in the Marquesas group in the middle of the Pacific. . . .

I remembered very well one particular evening. The civilized world seemed incomprehensibly remote and unreal. We had lived on the island for nearly a year, the only white people there. . . .

We kept on sitting there and admiring the sea which, it seemed, was loath to give up demonstrating that here it came rolling in from eastward, eastward, eastward. . . . And we knew by ourselves, as we sat there, that far, far below that eastern horizon, where the clouds came up, lay the open coast of South America. There was nothing but 4,000 miles of open sea between.

We gazed at the driving clouds and the heaving moonlit sea, and we listened to an old man who squatted half-naked before us and stared down into the dying glow from a little smoldering fire.

"Tiki," the old man said quietly, "he was both god and chief. It was Tiki who brought my ancestors to these islands where we live now. Before that we lived in a big country beyond the sea."

When we crept to bed that night in our little pile hut, old Tei Tetua's stories of Tiki and the islanders' old home beyond the sea continued to haunt my brain, accompanied by the muffled roar of the surf in the distance. It sounded like a voice from far-off times, which, it seemed, had something it wanted to tell, out there in the night. I could not sleep. It was as though time no longer existed, and Tiki and his seafarers were just landing in the surf on the beach below. A thought suddenly struck me and I said to my wife: "Have you noticed that the huge stone figures of Tiki in the jungle are remarkably like the monoliths left by extinct civilizations in South America?"

I felt sure that a roar of agreement came from the breakers. And then they slowly subsided while I slept.

Swimming the Hellespont

André Maurois

Surfing, sailing, fishing—in all of these sports man meets the challenge of the sea assisted by a tool of his own invention, be it only a six-foot length of board. But in swimming, he plunges into the sea unaided, depending only on his own physical strength, and perhaps his knowledge of tides and currents, to sustain him. Perhaps that is why Lord Byron was so proud of his feat of swimming across the Dardanelles in 1810, valuing it more highly than all his literary accomplishments. The following excerpt is from André Maurois' biography Byron, *published in 1930.*

A pilgrim can make no halt. An English vessel, the *Pylades,* was setting out for Smyrna; they took passage on board. Amid the Isles they were rocked by the waves of the sea of Ulysses, waves the colour of wine-lees, with crests of opal. And at Smyrna Byron completed the second Canto of his poem. Hobhouse had no great opinion of it. Exaggerated sentiments, he kept saying, and rhetorical declamation; he preferred Pope. Byron himself, very fond of eighteenth-century poetry, was almost astounded by what had been the spontaneous expression of his emotions; but he stuffed the manuscript away at the bottom of his portmanteau. He would find some other path to fame.

The frigate which brought them from Smyrna to Constantinople put in at the Isle of Tenedos. From there he saw the entrance to the Dardanelles, the narrow cleft that separates two continents. A swift tide flowed like a river between two high banks, bare and dull-hued. And that was the Hellespont, where Leander had swum across the join his lover. Byron was eager to imitate him. He made two attempts. The first failed, but on May 3 he succeeded, swimming from Europe to Asia and remaining an hour and a half in the water. His companion, Mr. Ekenhead, beat him by five minutes. Neither of them was tired, only rather chilled, and Byron was prodigiously proud of himself. He wrote to his mother, to Hodgson, to the whole world, that he had swum from Sestos to Abydos, and the exploit, together with complaints about Fletcher, and the praise bestowed by Ali Pacha on the smallness of his ears, became one of the main themes of his letters. "I shall begin by telling you, having only told it you twice before, that I swam from Sestos to Abydos. I do this that you may be impressed with proper respect for me, the performer; for I plume myself over this achievement more than I could possibly do on any kind of glory, political, poetical, or rhetorical."

Winslow Homer, *Long Branch,
New Jersey,* 1869

THE LURE OF THE SEA
*While some seek to conquer the sea, others seek merely to
enjoy it in its ever-changing moods.*

327

Epilogue

New Interests, New Words
H. William Menard

Words are created, flourish, and fade away as the need for them vanishes. A hundred years ago the Nantucket whaler spent his idle hours doing scrimshaw work on busks for his lady's corset. Now the Nantucket whaler is gone, the old-fashioned corset is gone, and there is little need for a word meaning etching on ivory, or one meaning a reinforcement for a corset. Today there is a need for all the words in the following sentence: "We could not use the bathyscaph to explore the continental slope for manganese nodules because Chile extended its economic resource zone." The need arises because a large number of people are discovering, inventing, and doing new things related to the ocean. The surfer and the oceanographer need new words to describe waves; the geologist and the lawyer need new words to describe the sea floor and legal matters concerned with it; the oil company executive needs new words to order a platform capable of drilling for oil in deep water.

Confusion may be expected when new words are created. People use them in different ways, or they may use several new words to describe a single thing. Next, the meaning of the word may change. Then a word that is perfectly satisfactory for one purpose is used for another purpose for which it is not fitted. A book such as this inevitably contains a profusion of new words and concepts, some of which will be unfamiliar to the reader. This essay attempts to provide some guidance to the new words and to some of the basic concepts and new developments of oceanography. It should also alert the reader to the fact that one author may not be using a word in exactly the same way as another.

Nothing can be done about this. By the time the words are standardized and in the dictionaries, many of them will be as needed as "scrimshaw" and "busk."

Marine geology. The sea is level but the surface of the earth beneath the sea is as diversified as the more familiar surface of the land. There are rocky mountains and muddy plains, chasms, cliffs, and canyons. If we were in an orbiting spaceship some distance from the earth, we might have difficulty seeing the individual mountains and valleys of the land, but we would see that most of the land is at one "level"—that of the surrounding sea. This is because rivers erode the continents down to that level and transport mud to the sea. If the oceans magically evaporated, we would perceive another, more extensive "level" that is now under 12,000–20,000 feet of water. This is the level of the deep-sea floor. The continents stand higher than the sea floor because they are composed of different, less dense rocks.

Between the two levels is a slope, the **continental slope,** as steep and as variable as the face of a great mountain range. At the base of this continental slope is an almost smooth, gently sloping plain of sand and mud that has spread out from the continent. It is the **continental rise.**

From the shoreline to the top of the continental slope is another gently sloping plain called the **continental shelf.** Wherever it exists, it looks very like the coastal region next to it. Off the steep, rocky mountains of coastal California, the continental shelf is rocky, relatively steep, and very narrow. Off the broad, almost level coastal plain of Maryland and Virginia, the continental shelf is broad, gently sloping, and covered with sand and mud. The reason for the similarity is that sea level fluctuates during periods of thousands of years. Thus the continental shelf and the coastal plains are interchangeable depending on the temporary level of the sea. The shelf is the submerged edge of the continent.

Far from the edge of the continents, in the very center of the Atlantic Ocean is the Mid-Atlantic Ridge, part of an enormous submarine mountain range that winds around the earth. The range, or **Midocean Ridge,** exists because hot rock wells up from the interior of the earth along a central crack. The crack constantly widens and fills with more rock, thereby creating the rocky crust of the ocean floor. As the rock drifts away from the crack it gradually cools and contracts and sinks. Thus the oceanic crust is shallow where it is young and deepens as it ages. The deepest places are **trenches** that could easily contain Mount Everest and are thousands of miles long. They are the places where the aging oceanic crust plunges down and returns to the interior from whence it came.

At present, there is no doubt that large pieces of the earth's crust, **plates,** containing whole continents, drift about, but no one knows why. It is a mystery that probably will be solved with new ideas expressed in new words.

Marine resources. The term *resource* has a number of technical meanings that include the concepts of an available function or supply, whether discovered or not, that can be used or extracted, although not necessarily economically at present or with existing technology. In short, it has many rather vague meanings. The water of the ocean itself is a resource for many purposes, and it also contains both living and mineral resources of great value.

The main conventional value of the ocean derives from the fact that it is empty, broad, free, and easy to move over and thus provides a medium for cheap transportation. Were it not for these properties of the sea it would be unthinkable to move Middle East oil to the United States. However, the ocean water is valuable for many other purposes. It contains many elements in solution, and from it we extract table salt and magnesium. The ocean water is also a traditional place to dump sewage and other waste materials, including heat from power plants. It has been suggested that long-lived radioactive wastes be dumped in oceanic trenches whence they would be sucked back into the interior of the earth. The water is also a source of energy both because it moves in the tides and because it is at different temperatures at different depths. Even this short list makes it obvious that the various values conflict. Therein arises a need for new laws.

The living resources of the sea include plants and animals that are commonly classified for scientific, commercial, and legal purposes in many different ways. One of the most widespread systems of classification is based on how the organism moves. If it rests on the bottom, like a clam, it is part of the **benthos;** if it floats passively in the water it is **plankton;** and if it swims about it is **nekton. Benthomic fisheries** scoop organisms from the bottom, and the ownership of such a fishery is relatively easy to establish, and thus overfishing is relatively easy to control. **Pelagic fisheries** catch tuna and whales and the like on the high seas that belong, at the moment, to no one. Thus regulation of overfishing requires complex international treaties.

The importance of words in the lives of men and beasts can be seen in the story of the "lobster war." The lobster mainly stands on the bottom or walks a bit but swims just above the bottom with powerful thrusts of his tail when he is in a hurry. Brazil has a large lobster population on its continental shelf, and it claimed that the bottom-dwelling creature was under its control. France claimed that the free-swimming lobster was like any pelagic fish and thus not under national control at all. There is also a "cod war" between Britain and Iceland. Fishermen have traditionally ranged afar, so these are not new problems. One of the three items covered in the Treaty of Paris in which England agreed to American independence was the perpetual right of the United States to fish on the Grand Banks off Newfoundland.

The need for terms related to the mineral resources of the sea is relatively new simply because there was little cause to extract minerals from the sea floor until the cheaper ones on the land were nearing exhaustion. Interest in marine minerals began at the shoreline, went to shallow water nearby, and is only now extending to the continental rise and the deep-sea floor. Inasmuch as the continental shelf is merely a slightly submerged part of the continent, it contains exactly the same minerals as the dry land. However, it is more difficult to find submarine minerals and more expensive to extract them. Until recently only very valuable minerals such as gold, diamonds, and tin have been mined, and in general they have come from deposits that can be followed from land out to sea. Oil and gas in California and Louisiana can also be traced from land to sea and into deeper water all the way across the continental shelf. The continental slope offers little promise, but ultimately the thick piles of sediment in the continental rise may also yield oil and gas. If so, it will be even more expensive than the oil and gas from shallow water.

Except at the edges of the plates where the rocks are being created or destroyed, the deep-sea floor is a relatively quiet place compared to the land. Consequently, conditions are suitable for the very slow precipitation of layer upon layer of some of the elements dissolved in sea water. Among the most interesting of these for their commercial value are *manganese nodules,* which are strewn like a pavement of potatoes or oranges over enormous areas of the deep-sea floor, generally an older crust at some distance from the crest of midocean ridges. These nodules also intrigue scientists because no one really knows how they form.

The great opening crack in the sea floor emits not only liquid rock but also solutions containing valuable metals. These solutions tend to be dissipated in the enormous volume of the open ocean, but in confined places, such as the Red Sea, they are concentrated in hot, salty brine and underlying mud in small chasms in the center of the sea. By current estimates these brines contain billions of dollars worth of metallic ores, and they are only half as deep as most manganese nodules.

Ships, boats, rafts, and buoys. Surface vessels have gradually evolved over the centuries as capabilities have increased and needs have changed. This evolution has accelerated during the last few decades because we are doing so many new things or doing them on a scale that is vastly greater than in the past. The oil tanker, for example, has grown and changed to the point where it has a new name, *very large crude carrier* (VLCC). Most sailors are probably happy that these enormous, barely maneuverable oil cans are not called ships. They are too big to enter most harbors, so they commonly load and discharge their cargoes in deep water terminals or simply connect to pipelines that are attached to enormous buoys.

The Coast Guard has for many years anchored buoys near harbor approaches to collect information about currents automatically. In the same way the Weather Bureau, now part of the National Oceanic and Atmospheric Administration (NOAA), has collected data on land and from ships to help predict the weather. Now we are trying to make longer range and more accurate predictions, and we need devices cheaper than ships that will monitor the properties of the water and the air in the open ocean. For this purpose we have *monster buoys* and similar devices. These buoys, although anchored to stay in place, move up and down with waves and tides. For scientific purposes we also need buoys that are relatively motionless as waves go by them. The only such device is *FLIP* (floating instrument platform), which resembles a submarine when it is towed but sinks its tail and floats upright in the water when it is working.

The large and costly machines that drill and produce offshore oil have rapidly evolved and include almost every conceivable variation of ship, raft, buoy, and island. The only consistent characteristic has been that the equipment stands high enough to avoid damage by storm waves. Off populous California there are artificial islands with drilling rigs disguised as high-rise buildings. In very shallow water off Louisiana the drilling platform may rest on pilings like a short section of a pier. For slightly deeper water, it is more economical to build a platform, on a framework like the Eiffel Tower, in a shipyard, float it to the drilling site, and sink the feet of the framework to the bottom. Alternatively, a *jack-up rig* may be used. It resembles a raft with large poles at the corners. These are lowered to the bottom and then the raft elevates itself above the water surface. In still deeper water it becomes too expensive to build a framework from the bottom to the platform. One solution is to drill from a surface ship or barge called a *semi-submersible rig,* which is like a platform rising above two submarines. The alternative solution to the depth problem is to drill from a ship but to mount a permanent, remote-controlled *submerged production system* on the bottom. This system has not yet been used, but it is in the process of development.

The American drilling ship *Glomar Challenger* is not used for oil drilling but for scientific research. It can drill down at least a hundred feet into the sea floor anywhere in the oceans except in trenches and has reached a maximum penetration of over 4,000 feet.

Diving. We have greatly increased our ability to do things under water during the last few decades. Prior to 1940 a diver could plunge down with a lungful of air and work for a few minutes at shallow depth. To stay down longer or go a little deeper he could use a diving suit and air hoses. Below that, he took an air supply in tanks and was encased in ever more massive containers. Naval submarines cruised about at a few hundred feet but could do nothing else. The ultimate for its time was the bathysphere, a very strong steel sphere with viewing ports, which could be lowered half a mile on a steel cable. The cable set the limit. Two Americans, Otis Barton and William Beebe, used it to make observations but they could do nothing but look.

The development of the *aqualung* by Jacques Cousteau and Emile Gagnan during World War II freed divers from cumbersome diving suits and air hoses, revolutionizing diving as a sport and opening up the shallow sea to whole new fields of science and engineering.

A breakthrough in deep diving was made by Auguste Piccard, who in 1948 developed the *bathyscaph* that enabled humans to descend to the deepest parts of the sea.

Marvelous though it was, the bathyscaph was not very useful because it could move only slowly and for small distances, and the observer could not handle things outside the sphere. Cousteau, among others, developed little submersibles that have become diving work boats. His famous solution was the *diving saucer* (soucoup sousmarine), which is like a flying saucer, goes up and down like a normal military submarine rather than a balloon, and has the kinds of outside handling devices that are used to manipulate radioactive materials. There are now many types of *deep submersibles* that can work almost anywhere except in trenches.

Some underwater tasks require prolonged diving, involving physiological changes that necessitate very slow ascent to the surface. Depending on the depth and time, a point is reached where it becomes reasonable to put a prefabricated house on the sea floor so the diver does not have to spend so much time going up and down. Such pressurized, self-contained living and working spaces are the devices called *Sealab* and *Conshelf* and *Tektite.* Scientists have lived in and worked out of them for as long as two months at a depth of fifty feet and for weeks at greater depths. There appears to be no reason why larger underwater habitats should not be put to use should the need arise.

Location of the shoreline. In order to establish who owns or controls or is responsible for something, it is usually necessary to determine where the thing is. I may think I own a garage, but if a survey shows that it is on my neighbor's lot, it belongs to him. Surveys determine two things: where a point is horizontally relative to some accepted point or line, and where it is vertically relative to some accepted level surface. For a wide range of human activities, including those at sea, the definition of the accepted reference points, lines, and surfaces can be very important.

It probably indicates a vital characteristic of the human mind that we determine the elevation of a thing such as solid rock by referring it to an idea called *mean sea level.* The sea is widespread, and if it would just stop moving it would be level and therefore an ideal surface from which to measure elevations. The trouble is it moves constantly because of waves generated by the wind, the tides caused by the gravitational attraction of the moon, and millennia-long heaving caused by changes in volume as glaciers freeze or melt.

Since we cannot measure just single sea level that means very much, in the United States we measure sea level every hour for nineteen years and then take the average value and call it mean sea level. For some purposes, however, it is not the average level that is important but the average of the high tides or of the low tides, so we have *mean high water* or *mean low water.* A ship is more apt to run aground at low tide than at high tide, for example, so nautical charts give depths relative to mean low water.

With regard to coastal property there are further complications because within historical times the land itself may rise or fall enough to shoal a sandbank or submerge a beach. Furthermore, in some regions such as southern California, waves of sand drift along the beaches over periods of years. A house may be built at the seaward boundary of a sandy lot, namely at the mean high water line, only to have the sand drift away and expose the foundation.

All these variations in sea level pose many problems in the management of the *coastal zone,* but at least none of the horizontal boundaries are affected except those on the seaward side. How much more complex are the boundaries at sea, which are determined relative to the shoreline—which we know is shifting? In very flat regions, such as the Mississippi delta in Louisiana, the shoreline is augmented by the mud in every river flood and shifted by every storm. On the average, augmentation wins and the shoreline builds outward over historical times. What was the continental shelf when the United States purchased Louisiana may now be soggy land.

There is another type of complication if you want to measure a certain agreed-upon distance from shore. The shoreline curves, thus it is necessary to agree from which point the offshore distance is measured. The resulting problems concern the legal definitions of bays and the extensions of natural boundaries out to sea.

Marine law. The Congress has the task of producing or ratifying fair and reasonable laws concerning the ocean that will enable us to get along with each other and with other nations. It will be evident from what has gone before that this will be difficult because of conflicting human interests and the perversity of nature in not lending itself to classification. It is easy to describe the sea floor in a way that satisfies a marine geologist and is useful for thinking about developing natural resources. Unfortunately it is not easy to describe the sea floor in a way that is universally satisfying to lawyers and diplomats.

Everybody on the crowded shore wants to use the sea because of its virtue, namely that it is empty. In the United States we are trying to solve the problems of a congestion of multiple uses by means of the National Environmental Protection Act (NEPA). It requires that an organization contemplating an action that will significantly affect the environment must prepare an Environmental Impact Statement (EIS) to spell out the consequences before seeking governmental approval to proceed. NEPA applies to all parts of the country. In addition, for the seacoast we have the Coastal Zone Management Act, which requires that states and local governments be given the opportunity to prepare management plans to integrate different overlapping interests before significant changes occur.

Until recently the international law of the sea was fairly straight-forward. Everyone agreed that a nation had sovereignty over *territorial waters* that extended three miles from shore. Sovereignty was not quite as absolute as on land, chiefly because peaceful ships of any nation could sail through the waters. It may seem that there would be difficulties in defining the three-mile limit because the shoreline moves about and is embayed, but no one cared very much. Fish were plentiful and so were minerals on land, and there was no way of removing minerals from the sea floor even if they existed.

Circumstances are now entirely different. The world is running out of protein, oil, gas, and metals; the peaceful ships are capable of spilling enough oil to pollute an entire coast; and some developing nations claim some joint title to large areas of the sea floor. Even in the United States problems are complex because we have both federal and state governments. States, in general, have the rights to offshore minerals within territorial waters. The nation, however, claims jurisdiction over minerals on the *outer continental shelf* (OCS) to a distance of 200 miles. That sounds clear, but oilfields are concentrated off Louisiana, and where in the brown muddy water and brown watery mud is the shore? If identifiable, when is the appropriate date to identify it? When the nation was founded? When we made the Louisiana Purchase? Now?

It is more difficult to obtain international agreements than internal ones because of the diversity of interests. Thus as the resources of the continental shelf became more important in the 1950s, lawyers sought to define national jurisdiction of this new region in neutral, scientific terms. Perhaps a nation could claim title to the minerals of a region that scientists called the "continental shelf." The shelf clearly was not definable by a width but perhaps it was by a depth. Nautical charts usually show a line where the depth is 100 fathoms (600 feet) or 200 meters (660 feet) because that is where a ship returning from the deep sea ought to be concerned about running aground on some stray rock. That was usually defined as the edge of the continental shelf until just about those same 1950s when modern sounding instruments came

into common use. Then it developed that the actual change from gentle continental shelf to steep continental slope generally occurred at a depth of roughly 400 feet but that it ranged from 200 feet to 2,000 feet. Moreover, in some places the sea floor really did not lend itself to a useful description by the terms "continental shelf" and "slope."

How then does a nation, or a group of nations, define the limits of the adjacent sea over which it claims some sort of power? There was an abortive attempt to define jurisdiction over minerals as extending to the depth to which a nation's technology permitted the extraction of minerals. This hoped-for limitation is already meaningless because the United States, and probably several other nations, can already mine or drill at any depth.

Thus a nation's extended boundaries are what it says they are and other nations will agree to, and this is also true of the degree of control it exerts within the boundary. Some nations propose to extend the almost complete control of territorial waters for large distances from shore. Some nations wish to control an *economic resource zone* but not extend territorial waters and so on. It appears that only power and politics can carve up the liquid sea.

The Authors

MATTHEW ARNOLD (1822–1888), English poet and critic, served as inspector of elementary schools for thirty-five years, advocating a state-regulated system of primary and secondary education. His numerous books of poetry expressed the nineteenth century romantic pessimism found in "Dover Beach." An important literary and social critic, he published *On Translating Homer, Essays in Criticism,* and *On the Study of Celtic Literature* while he was professor of poetry at Oxford from 1857 to 1867. Several later books, such as *St. Paul and Protestantism,* expressed his interest in religion. He wrote *Discourses in America* after a lecture tour in the United States in the 1880s.

W(YSTAN) H(UGH) AUDEN (1907–1973), a major Anglo-American poet and dramatist, was born and educated in England. During the 1930s he was leader of a left-wing literary group that included Christopher Isherwood, with whom he wrote three verse plays and also *Journey to a War* (1939), based on their experiences in China. His many books of poetry, including the 1948 Pulitzer-Prize winner, *The Age of Anxiety,* range in subject matter from politics to modern psychology and Christianity. In 1939 he moved to the United States, where he taught in several colleges and universities and became an American citizen in 1946. He returned to England in 1956 to assume a poetry professorship at Oxford but maintained his United States residence until 1972.

H. DAVID BALDRIDGE is a senior research associate at the Mote Marine Laboratory in Sarasota, Florida. As a naval officer attached to the Naval Aerospace Medical Center in Pensacola, he undertook a study for the Office of Naval Research on factors associated with known instances of shark attacks on man. His report was published in 1973. He was also a member of the Shark Research Panel of the American Institute of Biological Sciences.

WILLARD N. BASCOM is director of the Southern California Coastal Water Research Project in El Segundo, which is studying the effects of waste disposal in the ocean. A mining engineer, he was associated with the University of California, Berkeley, studying waves and beaches, and with Scripps Institution of Oceanography, participating in several Pacific expeditions. He joined the staff of the National Academy of Sciences in 1954 and served as executive secretary of the Committee on Meteorology and the Maritime Research Advisory Committee and eventually became director of the Mohole project to drill through the earth's crust. His account of that project appears in *Hole in the Bottom of the Sea.* Other books include *Deep Water, Ancient Ships, Great Sea Poetry,* and *Waves and Beaches.* He also successfully prospected for diamonds under the sea and recovered Spanish treasure from an old galleon.

ROBERT M. BENDINER (1909–), on the editorial board of the *New York Times* since 1969, is former managing editor of *The Nation.* From 1957 to 1961 he was American correspondent for *The New Statesman,* London. The recipient of numerous honors, including the Benjamin Franklin Magazine Award and a Guggenheim fellowship, he has been a free-lance contributor to many outstanding periodicals. His books include *The Riddle of the State Department, White House Fever, Obstacle Course on Capitol Hill, Just Around the Corner, The Politics of Schools,* and *The Strenuous Decade,* which he coedited with Daniel Aaron.

H(ARVEY) RUSSELL BERNARD (1940–) has been teaching anthropology at West Virginia University since 1972, having previously been a member of the faculty at Washington State University and a research associate in marine affairs at Scripps Institution of Oceanography. He was also a lecturer at the University of Athens and a scientific collaborator at the Democritos Nuclear Research Center in Athens. A specialist in technology and cultural change and in marine cultures, he is coeditor of *Technology and Social Change* and *Introduction to Chicano Studies* and author of several articles in professional journals.

MAX BLUMER (1923–), an organic geochemist, has been on the staff of Woods Hole Oceanographic Institution in Massachusetts since 1959, serving as senior scientist since 1964. Born in Basel, Switzerland, he was a faculty member at the University of Basel before coming to the United States in 1950. His work and publications have been in the field of the alteration of organic products in the sea and the origin of petroleum.

PHILIP BOOTH (1925–), a poet, is an English professor at Syracuse University, where he joined the faculty in 1961. He previously taught at Bowdoin and Wellesley College. He has spent much of his time on the Maine coast, where he is past commodore of the Castine Yacht Club. His many awards include the Hokin Prize of *Poetry* magazine, the Lamont Prize of the Academy of American Poets, the Saturday Review Poetry Award, the National Institute of Arts and Letters Award, and a Guggenheim fellowship. Among his collections of poetry are *Letter from a Distant Land, The Islanders, Weathers and Edges, Beyond Our Fears,* and *Margins.*

ELISABETH MANN BORGESE is a founder of the International Ocean Institute (IOI). Headquartered in Malta, the IOI organized the annual *Pacem in Maribus* conferences, aimed at producing a peaceful international regime of the oceans. In addition to serving as the official observer for IOI at the United Nations Conferences on the Law of the Sea, Ms. Borgese is executive secretary of the board of editors of the *Encyclopedia Britannica* and a senior fellow of the Center for the Study of Democratic Institutions. She is the editor of *Pacem in Maribus* and the author of *The Drama of the Oceans,* published in the spring of 1976.

SEYOM BROWN, a political scientist, joined the Brookings Institution in 1969 as a senior fellow in foreign policy studies, having previously worked with the Rand Corporation. He has also been a consultant to the State Department and has taught at UCLA and at the Johns Hopkins School of Advanced International Studies.

RACHEL CARSON (1907–1964), a science writer and biologist, wrote *The Sea Around Us* (1951), which won the National Book Award. She was also the author of *Under the Sea Wind* and *The Edge of the Sea,* and her *Silent Spring* (1962) brought international attention to the problems of environmental pollution. She was on the staff of the University of Maryland and Johns Hopkins University prior to 1936, when she joined the Bureau of Fisheries (now the Fish and Wildlife Service) as a biologist; she served as its editor-in-chief from 1949 to 1952. She was a Guggenheim and Royal Society limited fellow, and her honors include awards from the New York Zoological Society, the American Association for the Advancement of Science, and the National Wildlife Federation.

LUTHER J. CARTER is currently a staff writer for *Science* magazine, specializing in public policy issues. A former congressional fellow of the American Political Science Association, he was city hall reporter and Washington correspondent for the *Virginia Pilot* of Norfolk from 1956 to 1965. He is the author of *The Florida Experience: Land and Water Policy in a Growth State,* written while he was a research associate at Resources for the Future.

FREDERICK CHAMIER (1796–1870) was an English novelist. He served in the Royal Navy from 1809 to 1856, and many of his stories, including *Tom Bowling,* were about the sea.

LINDA CHARLTON is a reporter with the Washington bureau of the *New York Times*. Before joining the *Times* in 1969, she was a reporter for *Newsday* and associate editor of *Listening* magazine. She is the recipient of the Best Feature Award of the New York Newspaper Women's Association.

GEOFFREY CHAUCER (c. 1340–1400) was the leading English poet of the Middle Ages. A civil servant and diplomat, he wrote for the courts of Edward III and Richard II. Best known for *The Canterbury Tales,* a collection of verse stories told by a group of pilgrims journeying to the shrine of Thomas à Becket, he was the first English poet to use "heroic verse"— rhymed couplets in iambic pentameter.

SIR FRANCIS CHICHESTER (1901–1972) was a British adventurer. In 1929 he made one of the first solo flights from England to Australia, and he made the first east-west flight across the Tasman Sea between New Zealand and Australia in 1931. His plan to fly around the world was cut short by a plane accident. Taking up ocean sailing in 1953, he won the first solo transatlantic race in 1960, and in 1966/1967 he sailed around the world alone in his fifty-foot yacht, the *Gipsy Moth IV*. His books include his autobiography, *The Lonely Sea and the Sky,* and *The Gipsy Moth Circles the World.*

EUGENIE CLARK (1922–) is a professor of zoology at the University of Maryland, where she joined the faculty in 1968. She was a research assistant at Scripps Institution of Oceanography, at the New York Zoological Society, and at the American Museum of Natural History in New York before serving as executive director of the Cape Haze Marine Laboratory in Sarasota, Florida from 1955 to 1967. A specialist in icthyology, she has received awards from the Underwater Society of America, the American Littoral Society, and the Gold Medal Award of the Society of Women Geographers. She has been named a fellow of the Atomic Energy Commission and the American Association for the Advancement of Science, among others. She is author of *Lady with a Spear, The Lady and the Sharks,* and numerous articles, including two cover stories for *National Geographic* in 1975.

SAMUEL TAYLOR COLERIDGE (1772–1834) was an English poet, critic, and philosopher. In 1798 he and William

Wordsworth jointly published *Lyrical Ballads,* which contained his most widely read work, "The Rime of the Ancient Mariner." An influential figure in the romantic movement, his other poems include "Kubla Khan" and "Christabel." His analysis of Wordsworth's poetry is contained in his best-known work of criticism, *Biographia Literaria.*

EDWARD CORINO, a scientist with the Esso Research and Engineering Company, has written on problems of oil pollution in the ocean.

JACQUES-IVES COUSTEAU (1910–), French naval officer and ocean explorer, attended the naval school at Brest, France and served in the French underground during World War II. Since 1950 he has been commander of the *Calypso,* which has been familiar to millions as base ship for television filming of underseas investigations. He has been director of the Oceanographic Museum of Monaco since 1957, and he is a foreign associate of the United States National Academy of Sciences. Inventor of the aqualung and a process for using television under water, he heads the Conshelf Saturation Dive Program. He has written and produced films about the oceans, and his books include *The Silent World, The Living Sea,* and the multi-volume *Undersea World of Jacques Cousteau.*

STEPHEN CRANE (1871–1900) was a novelist, poet, and author of short stories. While a starving free-lance writer in New York's Bowery, he wrote *Maggie: A Girl of the Streets* (1893), but it was *The Red Badge of Courage* (1895) that first brought him critical acclaim and later a position as a war correspondent in Cuba. His mastery of the short-story form is particularly notable in the title stories of *The Open Boat and Other Tales* (1898) and *The Monster and Other Stories* (1899). His books of free verse, *The Black Rider* and *War Is Kind,* were influential on twentieth-century poets. He died at twenty-eight in Germany, where he had gone for treatment of tuberculosis.

RICHARD HENRY DANA (1815–1882) interrupted his studies at Harvard to ship as a common sailor around Cape Horn to California. His account of the voyage, *Two Years Before the Mast* (1840), expressed his desire to better the lot of the sailor, a concern he carried into his subsequent practice of law. His book *The Seaman's Friend* (1841) became a standard manual of maritime law. An early supporter of the Free Soil party, he gave free legal aid to victims of the Fugitive Slave Law.

ROBERT S. DIETZ (1914–) has been with the Atlantic Oceanographic and Meteorological Labs of the National Oceanic and Atmospheric Administration in Miami, Florida since 1970. A geophysicist and oceanographer, he previously served with the Scripps Institution of Oceanography, with the Navy Electronics Laboratory, and with the United States Coast and Geodetic Survey. In 1961 he developed the theory of sea-floor spreading. He also published in the field of selenography (the study of the moon's physical features) and meteoritics. He is coauthor of *Seven Miles Down,* with Jacques Piccard, and author of many technical reports for the United States Office of Naval Research.

MARNE A. DUBS is director of the Ocean Resources Department of the Kennecott Copper Corporation. He also serves as chairman of the Committee on Undersea Mineral Resources of the American Mining Congress.

LARRY L. FABIAN is a research associate in the Foreign Policy Studies Program of the Brookings Institution, studying the impact of science and technology on American foreign policy. A former member of the research staff of the Carnegie Endowment for International Peace, he currently serves on the State Department's Advisory Committee on International Organization Affairs.

ZANE GREY (1875–1939) was the author of more than sixty books, most of which were adventure stories of the American West. In such books as *Riders of the Purple Sage, The Last of the Plainsmen,* and *The Lone Star Ranger,* he established the lone gunfighter as a hero. By the 1960s, his books had sold more than 30 million copies. Grey was also an ardent outdoorsman and conservationist, and he wrote a number of stories of outdoor adventure, drawing on his expert knowledge of fishing.

ALLEN L. HAMMOND, a geophysicist, is the research news editor for *Science* magazine. He has written more than 100 articles for *Science* and other journals, and he is the coauthor, with W. Metz and T. Maugh, of *Energy and the Future,* which has been translated into six languages.

THOR HEYERDAHL (1914–), Norwegian explorer and ethnologist, theorized that there had been contact across oceans between ancient civilizations. In 1947 he sailed in a balsa-wood raft, *Kon-Tiki,* from Peru to Polynesia to prove that Polynesians may have originated in South America. In 1970 he and an international crew of seven sailed a facsimile of an ancient Egyptian reed boat, *Ra-II,* from Morocco to Barbados to prove that the ancient Egyptians could have influenced the pre-Columbian cultures of the New World. In addition to *Kon-Tiki* and *The Ra*

Expeditions, Heyerdahl wrote *Aku-Aku,* a study of the stone monuments of Easter Island.

C(LARENCE) P(URVIS) IDYLL (1916–) is currently study director of the National Ocean Policy Study with the National Oceanic and Atmospheric Administration in Rockville, Maryland. A native of Canada, he is a specialist in fish populations, marine ecology, and the development of international fisheries. He was a biologist with the International Pacific Salmon Fisheries Commission from 1941 to 1948 before going to the University of Miami, where he served as professor and chairman of the Division of Fisheries and Estuarine Ecology in the School of Marine and Atmospheric Sciences, and as executive secretary and then chairman of the Gulf and Caribbean Fisheries Institute. He also served as senior research advisor to the United Nations Food and Agricultural Organization. He is the author of *The Sea Against Hunger* and *Abyss: The Deep Sea and the Creatures That Live in It,* and is coauthor and editor of *Exploring the Ocean World: A History of Oceanography.*

HENRY R. ILSLEY contributed articles on sports to the *New York Times* in the 1930s.

HIROSHI KASAHARA is with the Department of Fisheries of the United Nations Food and Agricultural Administration in Rome, Italy. He was formerly a professor in the College of Fisheries at the University of Washington. He is coauthor with William T. Burke of *North Pacific Fisheries Management* and of several articles on fisheries and the problem of world food supply.

BOSTWICK H. KETCHUM (1912–) is senior scientist and associate director of the Woods Hole Oceanographic Institution in Massachusetts, where he first joined the staff as a marine biologist in 1940. He has also been lecturer in biological oceanography and an associate member of the department of biology at Harvard University. A specialist in the physiology of algae and in pollution of the sea, he has participated in many studies of the relationship of science to social problems associated with the oceans. He is the author of more than seventy scientific papers, coauthor of *Marine Fouling and its Prevention* with A.C. Redfield, and editor of *The Water's Edge: Critical Problems of the Coastal Zone.*

HENRY A. KISSINGER (1923–), a political scientist, has been secretary of state since 1973. Born in Germany, Kissinger came to the United States in 1938. He was a member of the government department at Harvard University from 1958 to 1971, serving also with Harvard's Center for International Affairs and directing the Harvard Defense Studies Program. A consultant to various government agencies, he was named assistant to the President for National Security Affairs when Nixon took office in 1969. His books include the prize-winning *Nuclear Weapons and Foreign Policy, The Necessity for Choice: Prospects of American Foreign Policy, The Troubled Partnership,* and *American Foreign Policy.* He was awarded the Nobel Peace Prize in 1973.

FRANK W. LANE (1908–), a British civil servant, worked for twenty years in a technical library. Since the forties he has been a photographic and literary agent, specializing in natural science. He has done much to popularize science, appearing on more than 100 British Broadcasting Corporation programs. His books include *The Elements Rage, Animal Wonder World, Nature Parade,* and *Kingdom of the Octopus.*

FRANCIS (FRANK) L. LAQUE (1904–), a Canadian-born metallurgist, joined the development and research division of the International Nickel Company in 1927, retiring in 1969 as vice-president. He continues to serve as a consultant and since 1970 has been a senior lecturer at Scripps Institution of Oceanography. He also served as president of the International Organization for Standardization. His many awards include the F.N. Speller Award in corrosion engineering, the Howard Coonley Medal of the American Standards Association, and the Acheson Medal of the Electrochemical Society.

JACK (JOHN GRIFFITH) LONDON (1876–1916) was a novelist, short-story writer, playwright, and essayist. At the age of seventeen he became a seaman on a sealing vessel to Japan and the Bering Sea. At various times in his life he was an oyster pirate, a prospector in the Klondike gold rush, a newspaper correspondent during the Russo-Japanese War, and a war correspondent in Mexico. He drew on these personal experiences for his stories, which included *The Call of the Wild, White Fang,* and *The Sea Wolf.* In 1907 he sailed to the South Pacific, a voyage hs recounted in *The Cruise of the Snark.* A socialist, he considered his social tracts, *The People of the Abyss* and *The Iron Heel,* to be his most important works.

J(OHN) V(ICTOR) LUCE (1920–) is professor of classics at Trinity College, Dublin. His books include *Lost Atlantis, Homer and the Heroic Age,* and *The End of Atlantis.*

ALFRED THAYER MAHAN (1840–1914), an early advocate of sea power, spent nearly forty years on active duty in the United States Navy. His various positions at

337

sea and on shore included the presidency of the Naval War College in Newport, Rhode Island. His two major works, *The Influence of Sea Power Upon History, 1660–1783* (1890) and *The Influence of Sea Power Upon the French Revolution and Empire, 1793–1812* (1892) had a great impact on strategic thinking and the build-up of naval forces in the period prior to World War I. He also wrote *The Interest of America in Sea Power, Present and Future,* a biography of Nelson, and a naval history of the American Revolution.

LUIS MARDEN (1913–) has been a member of the staff of *National Geographic* magazine since 1934, serving as chief of the foreign editorial staff for the past ten years. A specialist in underwater research and photography and deep-sea diving journalism, he has produced several documentary films. He is the author of numerous articles and of *Color Photography With the Miniature Camera.*

JOHN MASEFIELD (1878–1967), English poet, playwright, and fiction writer, ran away to sea at the age of thirteen. He began his literary career in 1897, working for some time for the *Manchester Guardian.* His first books of poetry, *Salt-Water Ballads* (1902) and *Ballads,* earned him the title "Poet of the Sea," but his long narrative poems, including "Dauber," won greater critical acclaim. He was named poet laureate in 1930. Among his other works are the novels *Multitude and Solitude* and *The Bird of Dawning* and an autobiography, *So Long to Learn.*

ANDRÉ MAUROIS (1885–1967) was the pen name of Emile Herzog, a French biographer, novelist, and essayist. Among his best-known works were lives of Shelley, Byron, Disraeli, George Sand, Victor Hugo, and three generations of the Alexandre Dumas family (*The Titans*). He was also the author of *Climats,* a novel, and popular histories of France, England, and the United States.

HERMAN MELVILLE (1819–1891) is today regarded as one of the greatest of all American authors, although he met with mixed critical success in his own time. His novels, a blend of fact, fiction, adventure, and symbolism, drew on his personal experiences. *Redburn* was based on his voyage as a cabin boy aboard a ship to England; his trip aboard the whaler *Acushnet* from 1841 to 1842 provided the background for his greatest novel, *Moby-Dick;* and his capture by a tribe of cannibals in the Marquesas Islands, after he jumped ship, was the basis for *Typee.* Other books included *Omoo, White Jacket, Mardi,* and *Billy Budd,* the story of the clash between innocence and evil, published posthumously.

H. WILLIAM MENARD (1920–) joined the faculty of the Scripps Institution of Oceanography as a professor of geology in 1956 after serving as a naval officer in World War II and as an oceanographer with the Naval Electronics Laboratory in San Diego. He has participated in a score of deep-sea oceanographic expeditions and has made more than a thousand scientific dives with an aqualung. He served on the White House science staff in 1965/1966 and is a consultant to Congress, various federal agencies, learned societies, and industry. A member of both the National Academy of Sciences and the American Academy of Arts and Sciences, he discovered the first fracture zones in the floor of the ocean. He is the author of more than 100 scientific papers and of *Marine Geology of the Pacific, Anatomy of an Expedition, Science: Growth and Change,* and *Geology, Resources and Society.*

J(OHN) WILLIAM MIDDENDORF II (1924–) became secretary of the navy in 1974 after serving as undersecretary for a year. Active in Republican politics, he was treasurer of the Republican National Committee from 1964 to 1969, when he was named United States ambassador to the Netherlands. Prior to entering public service, he was a financial analyst, and he is the author of *Investment Policies of Fire and Casualty Insurance Companies.*

SAMUEL ELIOT MORISON (1887–) has been a member of the history faculty at Harvard University since 1915. Although officially retired in 1955, he continues to be an active writer. During World War II he served as historian of naval operations, rising to the rank of rear admiral, and he wrote the fifteen-volume *History of United States Naval Operations in World War II.* Twice awarded the Pulitzer Prize, for *Admiral of the Ocean Sea,* a biography of Columbus, and for *John Paul Jones,* he is the author of more than twenty-five books, including *The Oxford History of the American People* and the widely used text *The Growth of the American Republic* (with Commager and Leuchtenburg). He received the first Balzan Foundation Award in History (1963) and the Presidential Medal of Freedom (1964).

NOËL MOSTERT, now a Canadian citizen, was born and educated in South Africa, where he was shipping correspondent for the *Cape Times.* In 1947 he became parliamentary correspondent for the United Press in Ottawa, and later he was foreign correspondent and columnist for the *Montreal Star.* He was a frequent contributor to *The Reporter* magazine and numerous other American periodicals. He has lived in the Mediterranean area since 1964. *Supership* is his first book.

338

SIR JOHN MURRAY (1841–1914) was a marine zoologist. Born in Canada, he studied there and in Edinburgh. After a voyage on a whaler, he was appointed one of the naturalists for the *Challenger* Expedition. He was assistant editor and then editor-in-chief of the *Challenger Reports*, and he wrote the *Narrative* of the expedition as well as the report on deep-sea deposits. He contributed innumerable papers on biology and oceanography to scientific journals.

ARVID PARDO was Malta's ambassador to the United Nations from 1964 to 1971. In a speech before the General Assembly in 1967, he called attention to the resources of the ocean and the need for a new law of the sea to regulate their use. He is currently an associate of the Center for the Study of Democratic Institutions and a member of the faculty at the University of Southern California.

JOHN H. PARRY (1914–) is Gardiner Professor of Oceanic History and Affairs at Harvard University, a position he has held since he came to the United States in 1965. Born in England, he was a fellow of Clare College and a university lecturer in history at Cambridge University. He served in the Royal Navy from 1939 to 1946. He was a professor of history at the University of the West Indies and principal of University College, Ibadan, Nigeria, and University College, Swansea, Wales, prior to becoming vice-chancellor of the University of Wales in 1963. His many books include *The Spanish Theory of Empire, Europe and a Wider World, A Short History of the West Indies, The Age of Reconnaissance, The Spanish Seaborne Empire, Trade and Dominion,* and *The Discovery of the Sea.*

JACQUES PICCARD (1922–) is a Swiss oceanographic engineer. With his father, Auguste, he designed and operated the deep-diving bathyscaph. Hired as a scientific consultant to the United States Navy when it bought the bathyscaph *Trieste*, Piccard lived for a few years in San Diego. With Don Walsh of the navy, he made a record dive of 35,800 feet in 1960. He returned to Lausanne, where he designed the mesoscaph *Auguste Piccard*, a sightseeing submarine, and several other submarines for scientific observations. His many honors include the Distinguished Public Service Award, given by President Eisenhower, and the Theodore Roosevelt Distinguished Service Award. He is the author, with Robert S. Dietz, of *Seven Miles Down* and numerous articles.

GEORGE H. QUESTER (1936–) is a professor of government at Cornell University, where he joined the faculty in 1970. He had previously taught at Harvard University. A specialist in defense policy and arms control, he is the author of *Deterrence Before Hiroshima, Nuclear Diplomacy,* and *Politics of Nuclear Proliferation.*

ROGER REVELLE (1909–) holds a joint appointment as Richard Saltonstall Professor of Population Policy at Harvard, where he was a director of Population Studies from 1964 until 1975, and as professor of science and public policy at the University of California, San Diego. He became interested in problems of human population and their resources after leading a study panel on agricultural development in West Pakistan. Earlier he had a distinguished career as an oceanographer, rising from research assistant to director of the Scripps Institution of Oceanography in La Jolla, California, where he is professor emeritus. His geophysical explorations of the deep Pacific contributed significantly to theories of sea-floor spreading. In recognition of his help in establishing the San Diego campus of the University of California, the first college was named after him. Recent books include *America's Changing Environment, The Survival Equation,* and *Population and Social Change.*

EMMA ROBERTS (mid-19th century) journeyed to India on one of the merchant ships that regularly carried passengers. Her account of the voyage, *The East India Voyage, or The Outward Bound,* was published in London in 1845.

EUGENIO DE SALAZAR (16th century) was a Spanish official who served as governor of Tenerife in the Canaries, as judge in Santo Domingo and Mexico, and as councillor of the Indies. He wrote a considerable number of allegorical poems and private letters.

WILLIAM SHAKESPEARE (1564–1616) is generally acknowledged as the greatest literary genius of the English language, and probably of the world. Actor, playwright, and poet, he was the leader of the most prosperous theatrical troupe in Elizabethan London. His plays may be divided into the histories—*Henry IV, Henry V, Henry VI, Richard II,* and *Richard III;* the comedies, which include *As You Like It, Twelfth Night, A Midsummer Night's Dream,* and *The Taming of the Shrew;* the tragicomedies or romances, such as *The Tempest* and *The Winter's Tale;* and the tragedies, including *Hamlet, Othello, King Lear, Macbeth,* and *Antony and Cleopatra.*

JOSHUA SLOCUM (1844–1909), sea captain and adventurer, was born in Nova Scotia and ran away to sea as a youth, spending several years sailing on British coal and grain ships. He became a United States citizen and commanded several ships in the waters of

Alaska and the South Seas, fishing for salmon and carrying cargo. Stranded in Brazil after a shipwreck, he built a thirty-five foot craft that he sailed back to the United States in 1888. In 1895 he set sail from Boston in the thirty-four foot *Spray* on the first solo voyage around the world. His account, *Sailing Alone Around the World,* was published in 1900.

ROGER B. STEIN is an associate professor of English at the State University of New York at Binghamton. He assembled an exhibit of American marine painting for the Whitney Museum of American Art and published the accompanying book, *Seascape and the American Imagination,* in 1975. He is also the author of *John Ruskin and Aesthetic Thought in America.*

SIR CHARLES WYVILLE THOMSON (1830–1882), a Scottish naturalist, taught natural history at several colleges in Scotland and Ireland before becoming a professor at the University of Edinburgh in 1870. From 1872 to 1876 he directed the scientific staff of the *Challenger* Expedition, writing several volumes of the reports of the expedition upon his return. His work in sounding and dredging was instrumental in proving that animal life is abundant down to at least 650 fathoms and that deep-sea temperatures are not constant.

GERHARD TIMMERMANN is head of the technical committee on "The History of Shipbuilding" of the Schiffbautechnische Gesellschaft in Hamburg, Germany. He was formerly head of the maritime section of the Altonaer Museum.

BILL TRUCK (mid-19th century) was senior boatswain of the Royal Navy College of Greenwich, England. He was the author of *The Man-o'-War's Man,* published in London in 1843.

C. JEREMY TUNSTALL (1934–) has been professor of sociology at City University, London, since 1974. He was previously a fellow of the University of Essex and senior lecturer in sociology at the Open University. His books include *The Fishermen, The Advertising Man, Old and Alone,* and *Journalista at Work.*

PAUL C. TYCHSEN (1916–), a geologist, served with the United States Geological Survey from 1946 to 1949. He has been chairman of the department of geology at the University of Wisconsin, Superior, since 1952. His special fields of interest are stratigraphy and geomorphology.

JULES VERNE (1828–1905), a French novelist, is considered by many to be the father of modern science fiction. He wrote several operetta librettos before turning to the tales of adventure and fantasy that brought him enormous popularity. Carrying his readers beneath the earth as well as above it and across it, he forecast the inventions of the airplane, the submarine, television, guided missiles, and space satellites. Among his best-known works are *Twenty Thousand Leagues Under the Sea, Around the World in Eighty Days, A Journey to the Center of the Earth,* and *From the Earth to the Moon.*

ALAN J. VILLIERS (1903–), seaman and author, knows both the sea and ships intimately. Born in Australia, he has sailed since boyhood on every manner of ship— whalers in the Antarctic, a four-masted barque, a Danish schoolship in which he sailed 58,000 miles around the world, Arabian dhows in the Persian Gulf, Portuguese cod-fishing schooners, World War II landing craft, and the nuclear ship *Savannah.* He was master of the replica of the *Mayflower,* which he sailed to the United States in 1957, and he has commanded many of the four-masted ships used in movies. He has written more than twenty-five books, including *The Set of the Sails, The Way of a Ship, Men, Ships and the Sea,* and *Captain Cook.* He is a frequent contributor to *National Geographic.*

EDWARD WENK JR. (1920–) has been a professor of engineering and public affairs at the University of Washington since 1970. He previously held a variety of positions with the federal government, including chief of the Science Policy Research Division of the Library of Congress, technical assistant to the President's science advisor, and executive secretary of the National Council on Marine Resources and Engineering Development. A frequent consultant in ocean engineering, he is currently serving on the United States Congress' Technology Assessment Council. He is the author of *The Politics of the Ocean.*

LINCOLN A. WERDEN first joined the *New York Times* in 1928. For more than forty years he was a sports writer, specializing in golf in his later years.

WILLIAM WERTENBAKER, a staff writer for *The New Yorker* magazine, has written extensively on scientific subjects. He won recognition from the American Association for the Advancement of Science in 1974 as an outstanding science writer for his three-part series on Maurice Ewing. These articles later became the basis for his book, *The Floor of the Sea.*

HELEN L. WINSLOW has written on scrimshaw for the Historical Society of Nantucket, Massachusetts.